AF770783

Essentials of Biochemistry

Essentials of Biochemistry

Gustavo Cobb

SYRAWOOD
PUBLISHING HOUSE
New York

Published by Syrawood Publishing House,
750 Third Avenue, 9th Floor,
New York, NY 10017, USA
www.syrawoodpublishinghouse.com

Essentials of Biochemistry
Gustavo Cobb

© 2019 Syrawood Publishing House

International Standard Book Number: 978-1-68286-829-4 (Hardback)

This book contains information obtained from authentic and highly regarded sources. All chapters are published with permission under the Creative Commons Attribution Share Alike License or equivalent. A wide variety of references are listed. Permissions and sources are indicated; for detailed attributions, please refer to the permissions page. Reasonable efforts have been made to publish reliable data and information, but the authors, editors and publisher cannot assume any responsibility for the validity of all materials or the consequences of their use.

The publisher's policy is to use permanent paper from mills that operate a sustainable forestry policy. Furthermore, the publisher ensures that the text paper and cover boards used have met acceptable environmental accreditation standards.

Trademark Notice: Registered trademark of products or corporate names are used only for explanation and identification without intent to infringe.

Cataloging-in-Publication Data

Essentials of biochemistry / Gustavo Cobb.
p. cm.
Includes bibliographical references and index.
ISBN 978-1-68286-829-4
1. Biochemistry. 2. Chemistry. 3. Biology. I. Cobb, Gustavo.
QD415 .E87 2019
572--dc23

Printed in China.

TABLE OF CONTENTS

PREFACE

Biochemistry is a vast field of study. It is primarily concerned with all the varied chemical processes related to living organisms. The in-depth study of biomolecules like proteins, carbohydrates, lipids, etc. is a major focus of this field. The understanding of these biomolecules facilitates the study of tissues, organs as well as the processes occurring within living cells. Biochemistry also delves into the processes of metabolism. It is the mechanism by which energy is harnessed by the cells via chemical reactions. This textbook provides comprehensive insights into the diverse areas of biochemistry. Different approaches, evaluations and methodologies have also been included herein. As this field is emerging at a rapid pace, the concepts of this book will help the readers understand the fundamental concepts and applications of this subject.

A detailed account of the significant topics covered in this book is provided below:

Chapter 1- The in-depth study of processes occurring at a chemical and molecular level within an organism is referred to as biochemistry. This is an introductory chapter which aims to provide an overview of the subject matter.

Chapter 2- Carbohydrate is a widely studied biomolecule. The primary role of carbohydrates is energy storage. They are present in foods like cereals, sugar, potatoes, etc. This chapter elaborates the structure, functions and major types of carbohydrates.

Chapter 3- Lipids are biological molecules, which help in forming cell membranes. They also perform other processes such as energy storage and signaling. This all-inclusive chapter delves into the types of lipids, their metabolism and other biological functions facilitated by them.

Chapter 4- The scientific field of the study of proteins is known as proteomics. Proteins are macromolecules which perform an array of functions in living organisms. Some of their primary functions include biosynthesis and chemical synthesis, cell signaling, DNA replication, etc. The following chapter discusses the structure, design and metabolism of proteins in detail.

Chapter 5- Nucleic acids are the most significant biopolymers present in all living organisms. They store information in the nucleus. The three primary types of nucleic acids are deoxyribonucleic acid, artificial nucleic acid and ribonucleic acid. The following chapter elucidates the diverse aspects related to the structure and metabolism of nucleic acids.

I would like to make a special mention of my publisher who considered me worthy of this opportunity and also supported me throughout the process. I would also like to thank the editing team at the back-end who extended their help whenever required.

Gustavo Cobb

Chapter 1

Biochemistry: An Overview

The in-depth study of processes occurring at a chemical and molecular level within an organism is referred to as biochemistry. This is an introductory chapter which aims to provide an overview of the subject matter.

Biochemistry is the study of chemical reactions in living beings, and of biological molecules in general. It is important to cell biology and physiology. The study of biochemistry involves enzymes, nucleic acids, carbohydrates, sugars, proteins, and lipids. In the body, most of the molecules are polymers built of long chains of carbon atoms with hydrogen, oxygen and other atoms added.

Macromolecules

The four types of large biological molecules (called macromolecules) are listed here.

Nucleic Acids

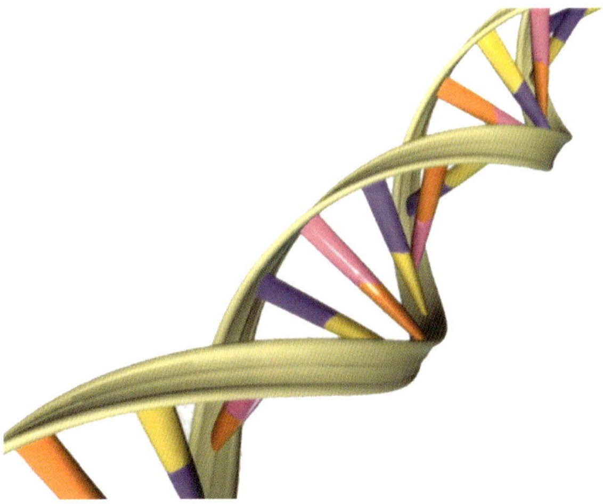

DNA, a nucleic acid, is made up of a double helix.

Nucleic acids are long-chain carbon molecules which make DNA and RNA. Their building blocks are called nucleotides. DNA is a common type of nucleic acid, which is usually joined up in a *double helix*. It is the substance of heredity and holds the information for life which passes from generation to generation. RNA is the other main type of nucleic acid. It acts to make the information from DNA work inside the cells of the body. There are many different types of RNA, each of which has a function inside cells.

Proteins

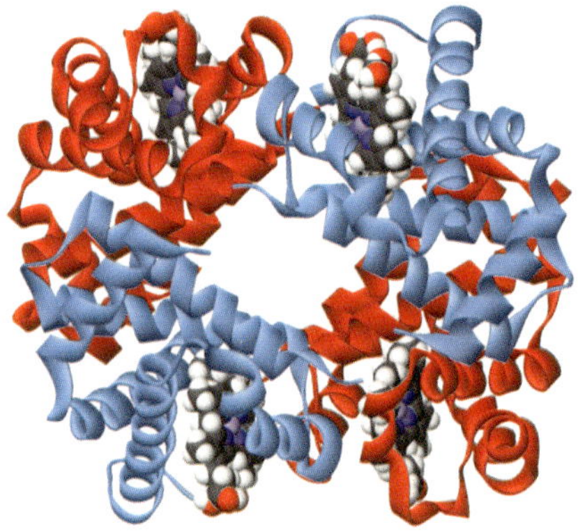

A ribbon diagram is one way biochemists describe the shape of proteins. This ribbon diagram is of the protein Haemoglobin which is in your blood, and is responsible for carrying oxygen.

Proteins are polymers of amino acids. There are twenty different common types of amino acid. Broadly speaking, they have two kinds of function. The first is structural: they make up many of the body's key structures in cells and tissues. Muscle, for example, is mainly made of protein. The second is as enzymes, which are long-chain proteins which may include an inorganic group as a co-enzyme.

Enzymes

Enzymes are special types of protein which greatly speed up the chemical reactions in a living thing. Their function is to speed up and help chemical reactions, by lowering the enzyme's activation energy. There are a few enzymes that are not proteins but instead made of RNA, which are called ribozymes, and are in fact nucleic acids.

Carbohydrates

Carbohydrates include sugars and starches. The simplest carbohydrates are the monosaccharides, meaning "single sugar". Examples of monosaccharides are glucose and fructose. Polysaccharides are long molecules made from many units joined together. Examples are starch, glycogen, and cellulose. Carbohydrates have a number of functions, but the most important is to act as a ready source of energy for the body's metabolism. By breaking the chemical bonds in carbohydrates, energy is released and can be used by the body.

Lipids

Lipids are fats, and waxes. Saturated lipids contain single bonds, and are found in butter and lard. Unsaturated lipids have one or more double bonds, and are often found in oils. The human body stores lipids as an energy source. When the body needs a large amount of energy, lipid molecules are broken down to release that energy.

Chapter 2

Carbohydrates: Structure and Functions

Carbohydrate is a widely studied biomolecule. The primary role of carbohydrates is energy storage. They are present in foods like cereals, sugar, potatoes, etc. This chapter elaborates the structure, functions and major types of carbohydrates.

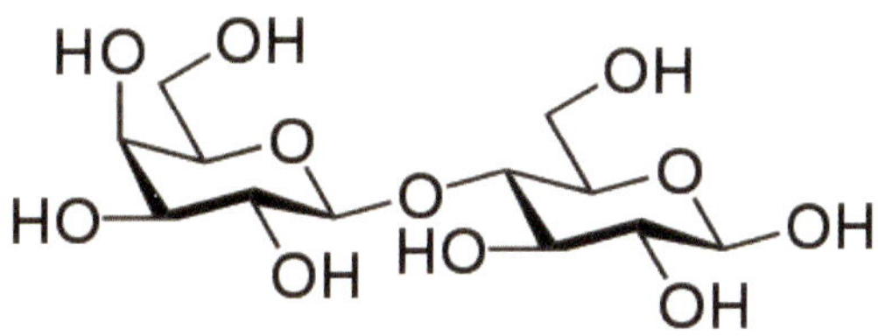

Lactose is a disaccharide found in animal milk. It consists of a molecule of D-galactose and a molecule of D-glucose bonded by beta-1-4 glycosidic linkage. It has a formula of $C_{12}H_{22}O_{11}$.

A carbohydrate is a biomolecule consisting of carbon (C), hydrogen (H) and oxygen (O) atoms, usually with a hydrogen–oxygen atom ratio of 2:1 (as in water); in other words, with the empirical formula $C_m(H_2O)_n$ (where m could be different from n). This formula holds true for monosaccharides. Some exceptions exist; for example, deoxyribose, a sugar component of DNA, has the empirical formula $C_5H_{10}O_4$. The carbohydrates are technically hydrates of carbon; structurally it is more accurate to view them as aldoses and ketones.

The term is most common in biochemistry, where it is a synonym of 'saccharide', a group that includes sugars, starch, and cellulose. The saccharides are divided into four chemical groups: monosaccharides, disaccharides, oligosaccharides, and polysaccharides. Monosaccharides and disaccharides, the smallest (lower molecular weight) carbohydrates, are commonly referred to as sugars. The word *saccharide* comes from the meaning "sugar". While the scientific nomenclature of carbohydrates is complex, the names of the monosaccharides and disaccharides very often end in the suffix -ose. For example, grape sugar is the monosaccharide glucose, cane sugar is the disaccharide sucrose, and milk sugar is the disaccharide lactose.

Carbohydrates perform numerous roles in living organisms. Polysaccharides serve for the storage of energy (e.g. starch and glycogen) and as structural components (e.g. cellulose in plants and chitin in arthropods). The 5-carbon monosaccharide ribose is an important component of coenzymes (e.g. ATP, FAD and NAD) and the backbone of the genetic molecule known as RNA. The related deoxyribose is a component of DNA. Saccharides and their derivatives include many other important biomolecules that play key roles in the immune system, fertilization, preventing pathogenesis, blood clotting, and development.

Carbohydrates are found in a wide variety of foods. The important sources are cereals (wheat, maize, rice), potatoes, sugarcane, fruits, table sugar (sucrose), bread, milk, etc. Starch and sugar are the important carbohydrates in our diet. Starch is abundant in potatoes, maize, rice and other cereals. Sugar appears in our diet mainly as sucrose (table sugar), which is added to drinks and many prepared foods such as jam, biscuits and cakes, and glucose and fructose which occur naturally in many fruits and some vegetables.

Glycogen is a carbohydrate found in the liver and muscles (as animal source). Cellulose in the cell wall of all plant tissue is a carbohydrate. It is important in our diet as fibre which helps to maintain a healthy digestive system.

Division

Carbohydrates are polyhydroxy aldehydes, ketones, alcohols, acids, their simple derivatives and their polymers having linkages of the acetal type. They may be classified according to their degree of polymerization and may be divided initially into three principal groups, namely sugars, oligosaccharides and polysaccharides.

The major dietary carbohydrates

Class (DP*)	Subgroup	Components
Sugars (1–2)	Monosaccharides	Glucose, galactose, fructose, xylose
	Disaccharides	Sucrose, lactose, maltose, trehalose
	Polyols	Sorbitol, mannitol
Oligosaccharides (3–9)	Malto-oligosaccharides	Maltodextrins
	Other oligosaccharides	Raffinose, stachyose, fructo-oligosaccharides
Polysaccharides (>9)	Starch	Amylose, amylopectin, modified starches
	Non-starch polysaccharides	Cellulose, hemicellulose, pectins, hydrocolloids

DP * = Degree of polymerization.

Carbohydrate Synthesis

Carbohydrate synthesis is a sub-field of organic chemistry concerned specifically with the generation of natural and unnatural carbohydrate structures. This can include the synthesis of monosaccharide residues or structures containing more than one monosaccharide, known as oligosaccharides.

Background

Generally speaking, carbohydrates can be classified into two groups, simple sugars and complex carbohydrates. Simple sugars, also called monosaccharides, are carbohydrates which can not be converted into smaller sugars by hydrolysis. When two or more monosaccharide units are connected to one another via a glycoside linkage, complex carbohydrates are formed. Complex carbohydrates, according to the different number of monosaccharide units, can be classed into three groups, disaccharides, oligosaccharides, and polysaccharides. A disaccharide is formed from two monosaccharides. Oligosaccharides can be formed by a small number of monosaccharides linked together. Higher oligosaccharides are called polysaccharides. It is now well known that glycoconjugates play an indispensable role in many biological processes. These biological processes in which carbohydrates are involved are typically associated not to monosaccharides, but to oligosaccharides structures of glycoconjugates. Therefore, the oligosaccharide synthesis becomes more and more important in studying the biological activities.

Oligosaccharide Synthesis

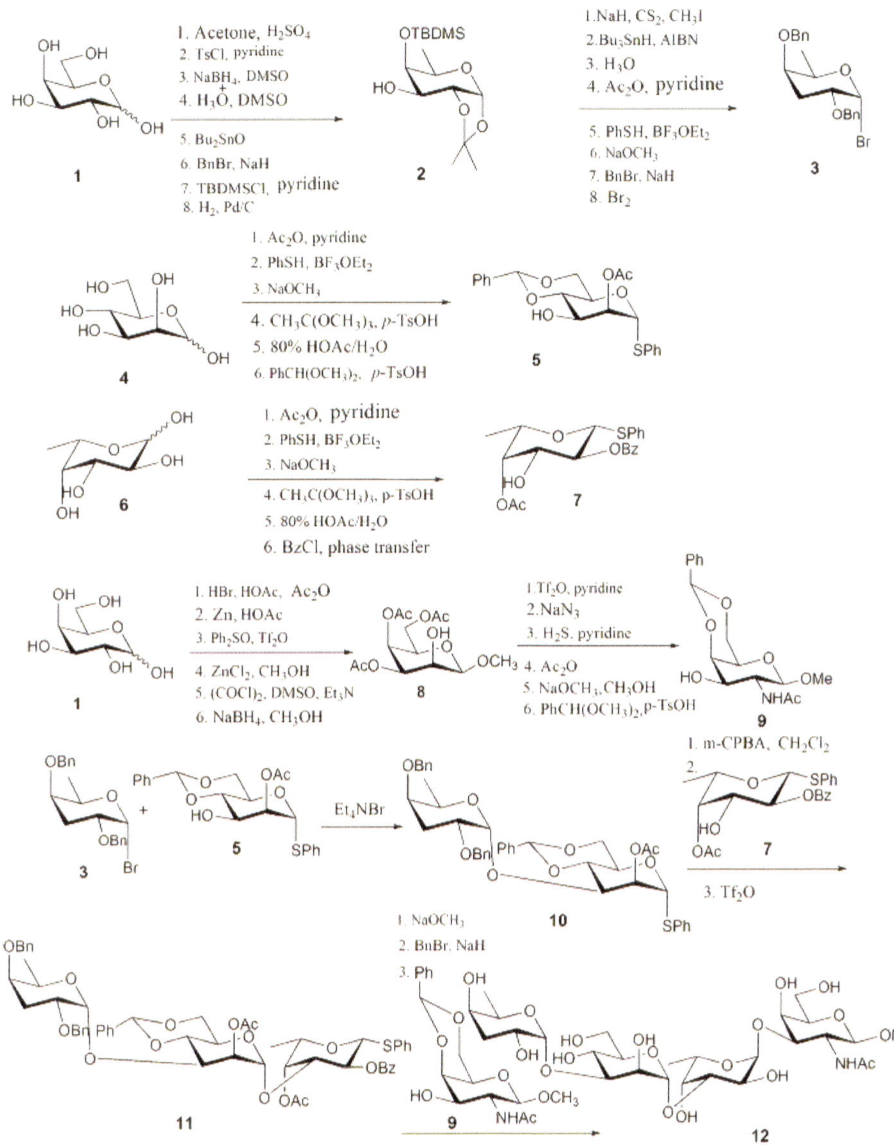

Scheme

Oligosaccharides have diverse structures. The number of monosaccharides, ring size, the different anomeric stereochemistry, and the existence of the branched-chain sugars all contribute to the amazing complexity of the oligosaccharide structures. The essence of the reducing oligosaccharide synthesis is connecting the anomeric hydroxyl of the glycosyl donors to the alcoholic hydroxyl groups of the glycosyl acceptors. Protection of the hydroxyl groups of the acceptor with the target alcoholic hydroxyl group unprotected can assure the regiochemical control. Additionally, factors such as the different protecting groups, the solvent, and the glycosylation methods can influence the anomeric configurations. This concept is illustrated by an oligosaccharide synthesis in Scheme. Oligosaccharide synthesis normally consists of four parts: preparation of the glycosyl donors, preparation of the glycosyl acceptors with a single unprotected hydroxyl group, the coupling of them, and the deprotection process.

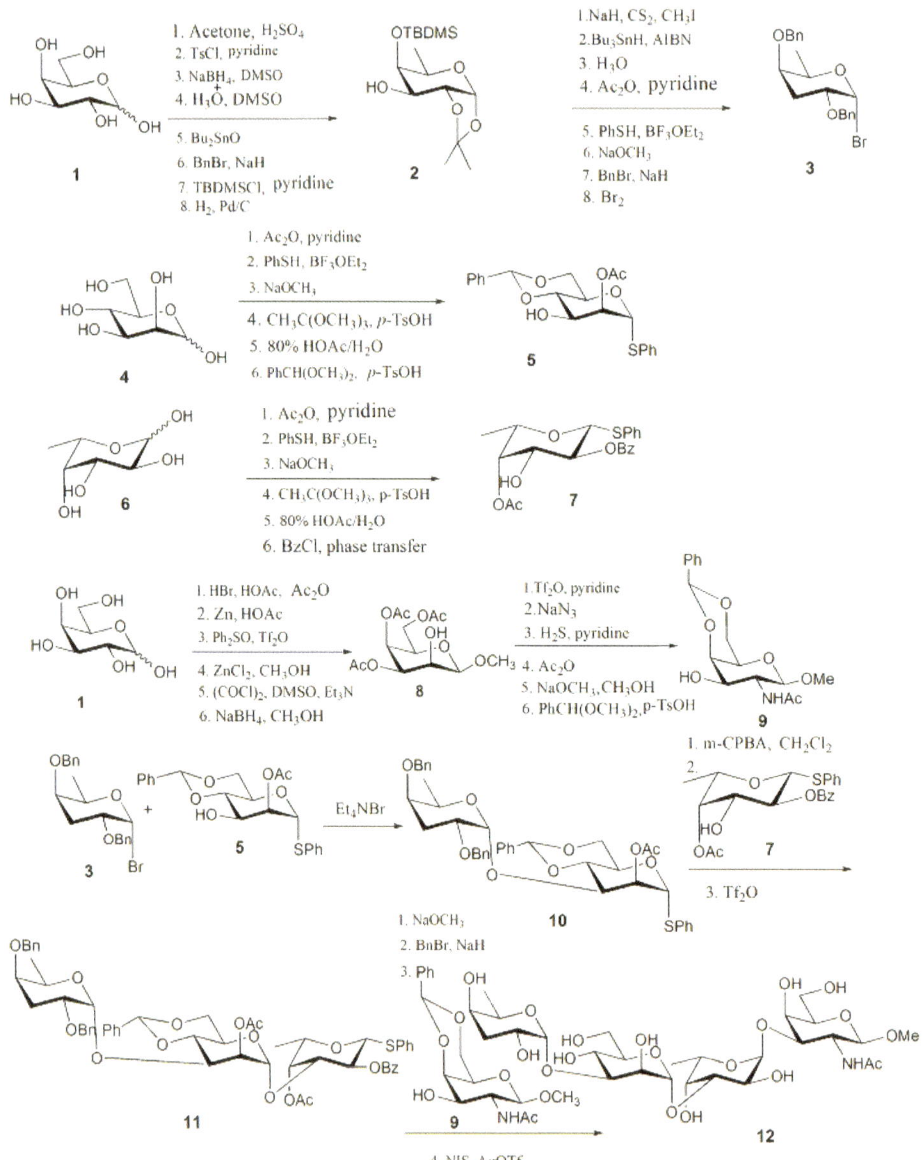

Scheme

Building Blocks

Common donors in oligosaccharide synthesis are glycosyl halides, glycosyl acetates, thioglycosides, trichloroacetimidates, pentenyl glycosides, and glycals. Of all these donors, glycosyl halides are classic donors, which played a historical role in the development of glycosylation reactions. Thioglycoside and trichloroacetimidate donors are used more than others in contemporary glycosylation methods. When it comes to the trichloroacetimidate method, one of the advantages is that there is no need to introduce heavy metal reagents in the activation process. Moreover, using different bases can selectively lead to different anomeric configurations. (Scheme) As to the thioglycosides, the greatest strength is that they can offer a temporary protection to the anomeric centre because they can survive after most of the activation processes. Additionally, a variety of activation methods can be employed, such as NIS/ AgOTf, NIS/ TfOH, IDCP (Iodine dicollidine perchlorate), iodine, and Ph_2SO/ Tf_2O. Furthermore, in the preparation of 1, 2-trans glycosidic linkage, using thioglycosides and imidates can promote the rearrangement of the orthoester by-products, since the reaction mixtures are acidic enough.

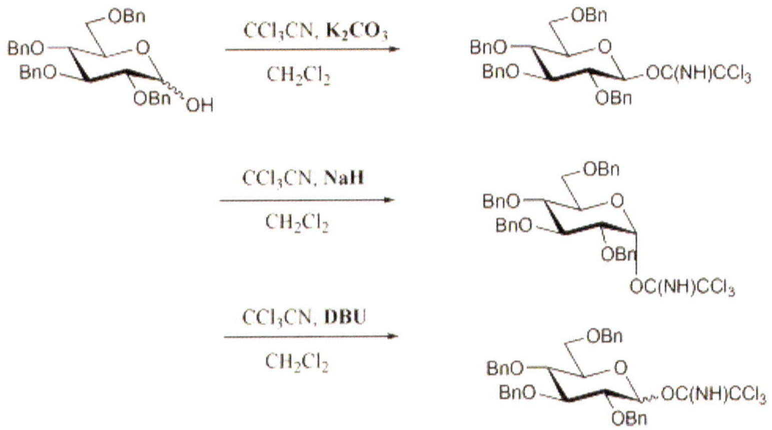

Scheme

Stereoselectivity

The structures of acceptors play a critical role in the rate and stereoselectivity of glycosylations. Generally, the unprotected hydroxyl groups are less reactive when they are between bulky protecting groups. That is the reason why the hydroxyl group at OH-4 in pyranosides is unreactive. Hyperconjugation is involved when OH-4 is anti-periplanar to the ring oxygen, which can also reduce its reactivity. (Scheme) Furthermore, acyl protecting groups can reduce the reactivity of the acceptors compared with alkyl protecting groups because of their electron-withdrawing ability. Hydroxyl group at OH-4 of N-acetylglucosamine derivatives is particularly unreactive.

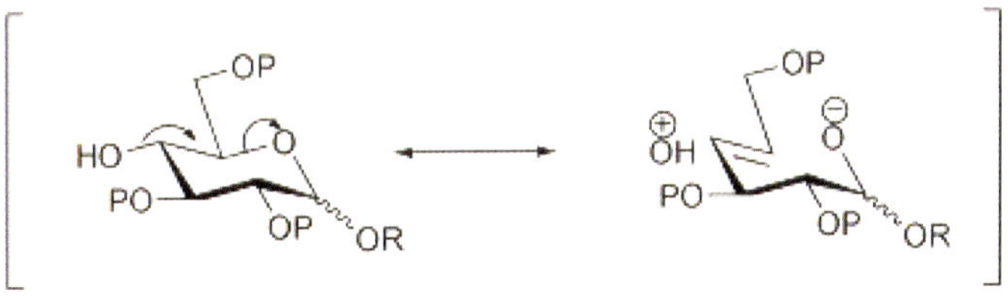

Scheme

The glycosidic bond is formed from a glycosyl donor and a glycosyl acceptor. There are four types of glycosidic linkages: 1, 2-trans-α, 1, 2-trans-beta, 1, 2-cis-α, and 1, 2-cis-beta linkages. 1, 2-trans glycosidic linkages can be easily achieved by using 2-O-acylated glycosyl donors (neighboring group participation). To prevent the accumulation of the orthoester intermediates, the glycosylation condition should be slightly acidic.

Difficult Linkages

It is somewhat more difficult to prepare 1, 2-cis-β-glycosidic linkages stereoselectively. Typically, when non-participating groups on O-2 position, 1, 2-cis-β-linkage can be achieved either by using the historically important halide ion methods, or by using 2-O-alkylated glycosyl donors, commonly thioglycosides or trichloroacetimidates, in nonpolar solvents.

In the early 1990s, it was still the case that the beta mannoside linkage was too challenging to be attempted by amateurs. However, the method introduced by Crich (Scheme), with 4,6-benzylidene protection a prerequisite and anomeric alpha triflate a key intermediate leaves this problem essentially solved. The concurrently developed but rather more protracted intramolecular aglycon delivery (IAD) approach is a little used but nevertheless stereospecific alternative.

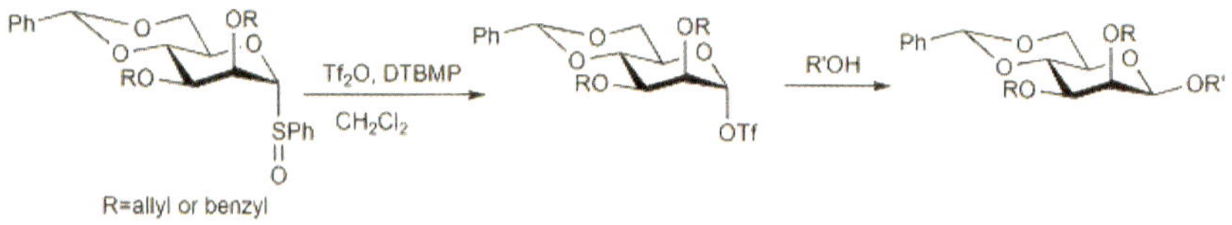

Carbohydrate Conformation

Carbohydrate conformation refers to the overall three-dimensional structure adopted by a carbohydrate (saccharide) molecule as a result of the through-bond and through-space physical forces it experiences arising from its molecular structure. The physical forces that dictate the three-dimensional shapes of all molecules—here, of all monosaccharide, oligosaccharide, and polysaccharide molecules—are sometimes summarily captured by such terms as "steric interactions" and "stereoelectronic effects".

Saccharide and other chemical conformations can be reasonably shown using two-dimensional structure representations that follow set conventions; these capture for a trained viewer an understanding of the three-dimensional structure via structure drawings they are also represented by stereograms on the two dimensional page, and increasingly using 3D display technologies on computer monitors.

Formally and quantitatively, conformation is captured by description of a molecule's angles—e.g., sets of three sequential atoms (bond angles) and four sequential atoms (torsion angles, dihedral angles), where the locations and angular directions of nonbonding electrons ("lone pair electrons") must sometimes also be taken into account.

Conformations adopted by saccharide molecules in response to the physical forces arising from their bonding and nonbonding electrons, modified by the molecule's interactions with its aqueous or other solvent environment, strongly influence their reactivity with and recognition by other molecules (processes which in turn can alter conformation). Chemical transformations and biological signalling mediated by conformation-dependent molecular recognition between molecules underlie all essential processes in living organisms.

Conformations of Carbohydrates

Monosaccharide Conformation

Pyranose and furanose forms can exist in different conformers and one can interconvert between the different conformations if an energy requirement is met. For the furanose system there are two possible conformers: Twist (T) and Envelope (E). In the pyranose system five conformers are possible: Chair (C), Boat (B), Skew (S), Half-Chair (H) or Envelope (E). In all cases there are four or more atoms that make up a plane. In order to define which atoms are above and below the plane one must orient the molecule so that the atoms are numbered clockwise when looking from the top. Atoms above the plane are prefixed as a superscript and atoms below the plane are suffixed as a subscript. If the ring oxygen is above or below the plane it must be prefixed or suffixed appropriately.

Conformational Analysis

The chair conformation of six membered rings have a dihedral angle of 60° between adjacent substituents thus usually making it the most stable conformer. Since there are two possible chair conformation steric and stereoelectronic effects such as the anomeric effect, 1,3 diaxial interactions, dipoles and intramolecular hydrogen bonding must be taken into consideration when looking at relative energies. Conformations with 1,3 diaxial interactions are usually disfavored due to steric congestion and can shift equilibrium to the other chair form (example: ${}^{1}C_{4}$ to ${}^{4}C_{1}$). The size of the substituents greatly affects this equilibrium. However, intramolecular hydrogen bonding can be an example of a stabilizing 1,3 diaxial interaction. Dipoles also play a role in conformer stability, aligned dipoles lead to an increase in energy while opposed dipoles lead to a lowering of energy hence a stabilizing effect, this can be complicated by solvent effects. Polar solvents tend to stabilize aligned dipoles. All interaction must be taken into account when determining a preferred conformation.

Conformations of five membered rings are limited to two, envelope and twist. The envelope conformation has four atoms in a plane while the twist form only has three. In the envelope form two different scenarios can be envisioned; one where the ring oxygen is in the four atom plane and one where it is puckered above or below the plane. When the ring oxygen is not in the plane the substituents eclipse and when it is in the plane torsional strain is relieved. Conformational analysis for the twist form is similar thus leading to the two forms being very close in energy.

Anomers and Related Effects

Anomers are diastereoisomers of glycosides, hemiacetals or related cyclic forms of sugars, or related molecules differing in configuration only at C-1. When the stereochemistry of the first carbon matches the stereochemistry of the last stereogenic center the sugar is the α-anomer when they are opposite the sugar is the β-anomer.

Anomeric Effect

Anomers can be interconverted through a process known as mutarotation. The anomeric effect more accurately called the endo-anomeric effect is the propensity for heteroatoms at C-1 to be oriented axially. This is counter intuitive as one would expect the equatorially anomer to be the thermodynamic product. This effect has been rationalized through dipole-dipole repulsion and n-σ* arguments.

Reverse Anomeric Effect

The reverse anomeric effect, proposed in 1965 by R. U. Lemieux, is the tendency for electropositive groups at the anomeric position to be oriented equatorially. Original publication reported this phenomenon with N-(2,3,4,6-tetra-O-acetyl-α-D-glucopyranosyl)-4-methylpyridinium bromide. However, further studies have shown the effect to be a solvation and steric issue. It is accepted that there is no generalized reverse anomeric effect.

Hydroxymethyl Conformation

Rotation around the C-5/C-6 bond is described by the angle ω. Three possible staggered conformations possible; gauche-trans (gt), gauche-gauche (gg), trans-gauche(tg). The name indicates the interaction between O-5 and OH-6 first followed by the interaction between OH-6 and C-4.

Oligosaccaharide Conformation

In addition to the factors affecting monosaccharide residues, conformational analysis of oligosaccharides and polysaccharides requires consideration additional factors.

Exo-anomeric Effect

Exo-anomeric effect is similar to the endo-anomeric effect. The difference being that the lone pair being donated is coming from the substituent at C-1. However, since the substituent can be either axial or equatorial there are two types of exo-anomeric effects, one from axial glycosides and one from equatorial glycosides as long as the donating orbital is anti-periplanar to the accepting orbital.

Glycosidic Torsion Angles

Three angles are described by φ, ψ and ω (in the case of glycosidic linkages via O-6). Steric considerations and anomeric effects need to be taken into consideration when looking at preferred angles.

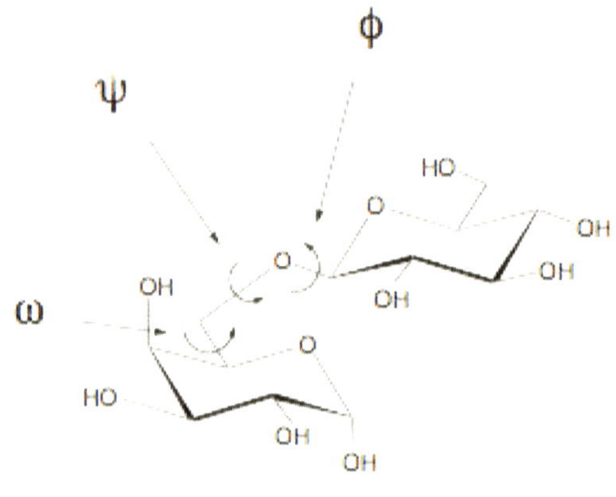

Conformations in Solution

In solution, reducing monosaccharides exist in equilibrium between their acyclic and cyclic forms with less than 1% in the acyclic form. The open chain form can close to give the pyranose and furanose with both the α and β anomers present for each. The equilibrium population of conformers depends on their relative energies which can be determined to a rough approximation using steric and stereoelectronic arguments. It has been shown that cations in solution can shift the equilibrium.

Sugar

Closeup of raw crystals of unrefined, unbleached sugar

Sugar is the generic name for sweet-tasting, soluble carbohydrates, many of which are used in food. There are various types of sugar derived from different sources. Simple sugars are called monosaccharides and include glucose (also known as dextrose), fructose, and galactose. The "table sugar" or "granulated sugar" most customarily used as food is sucrose, a disaccharide of glucose and fructose. Sugar is used in prepared foods (e.g., cookies and cakes) and it is added to some foods and beverages (e.g., coffee and tea). In the body, sucrose is hydrolysed into the simple sugars fructose and glucose. Other disaccharides include maltose from malted grain, and lactose from milk. Longer chains of sugars are called oligosaccharides or polysaccharides. Some other chemical substances, such as glycerol and sugar alcohols may also have a sweet taste, but are not classified as sugars. Diet food substitutes for sugar include aspartame and sucralose, a chlorinated derivative of sucrose.

Sugars are found in the tissues of most plants and are present in sugarcane and sugar beet in sufficient concentrations for efficient commercial extraction. The world production of sugar in 2011 was about 168 million tonnes. The average person consumes about 24 kilograms (53 lb) of sugar each year (33.1 kg in developed countries), equivalent to over 260 food calories per person, per day. Since the latter part of the twentieth century, it has been questioned whether a diet high in sugars, especially refined sugars, is good for human health. Over-consumption of sugar has been implicated in the occurrence of obesity, diabetes, cardiovascular disease, dementia, and tooth decay. Numerous studies have been undertaken to try to clarify the position, but with varying results, mainly because of the difficulty of finding populations for use as controls that do not consume or are largely free of any sugar consumption.

Chemistry

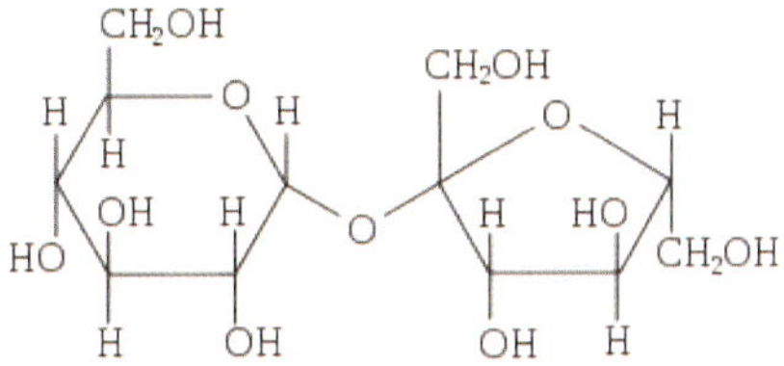

Sucrose: a disaccharide of glucose (left) and fructose (right), important molecules in the body.

Scientifically, *sugar* loosely refers to a number of carbohydrates, such as monosaccharides, disaccharides, or oligosaccharides. Monosaccharides are also called "simple sugars," the most important being glucose. Most monosaccharides have a formula that conforms to $C_nH_{2n}O_n$ with n between 3 and 7 (deoxyribose being an exception). Glucose has the molecular formula $C_6H_{12}O_6$. The names of typical sugars end with *-ose*, as in "glucose" and "fructose". Sometimes such words may also refer to any types of carbohydrates soluble in water. The acyclic mono- and disaccharides contain either aldehyde groups or ketone groups. These carbon-oxygen double bonds (C=O) are the reactive cen-ters. All saccharides with more than one ring in their structure result from two or more monosaccharides joined by glycosidic bonds with the resultant loss of a molecule of water (H_2O) per bond.

Monosaccharides in a closed-chain form can form glycosidic bonds with other monosaccharides, creating disaccharides (such as sucrose) and polysaccharides (such as starch). Enzymes must hydrolyze or otherwise break these glycosidic bonds before such compounds become metabolized. After digestion and absorption the principal monosaccharides present in the blood and internal tissues include glucose, fructose, and galactose. Many pentoses and hexoses can form ring structures. In these closed-chain forms, the aldehyde or ketone group remains non-free, so many of the reactions typical of these groups cannot occur. Glucose in solution exists mostly in the ring form at equilibrium, with less than 0.1% of the molecules in the open-chain form.

Natural Polymers

Biopolymers of sugars are common in nature. Through photosynthesis, plants produce glyceraldehyde-3-phosphate (G3P), a phosphated 3-carbon sugar that is used by the cell to make monosaccharides such as glucose ($C_6H_{12}O_6$) or (as in cane and beet) sucrose ($C_{12}H_{22}O_{11}$). Monosaccharides may be further converted into structural polysaccharides such as cellulose and pectin for cell wall construction or into energy reserves in the form of storage polysaccharides such as starch or inulin. Starch, consisting of two different polymers of glucose, is a readily degradable form of chemical energy stored by cells, and can be converted to other types of energy. Another polymer of glucose is cellulose, which is a linear chain composed of several hundred or thousand glucose units. It is used by plants as a structural component in their cell walls. Humans can digest cellulose only to a very limited extent, though ruminants can do so with the help of symbiotic bacteria in their gut. DNA and RNA are built up of the monosaccharides deoxyribose and ribose, respectively. Deoxyribose has the formula $C_5H_{10}O_4$ and ribose the formula $C_5H_{10}O_5$.

Flammability and Response to Heating

Magnification of grains of refined sucrose, the most common free sugar.

Because sugars burn easily when exposed to flame, the handling of sugars risks dust explosion. The

2008 Georgia sugar refinery explosion, which killed 14 people and injured 40, and destroyed most of the refinery, was caused by the ignition of sugar dust.

In its culinary use, sugar that is exposed to a heat source causes caramelization to take place. As the process occurs, volatile chemicals such as diacetyl are released, producing the characteristic caramel flavor.

Monosaccharide

Monosaccharides, also called simple sugars, are the most basic units of carbohydrates. They are fundamental units of carbohydrates and cannot be further hydrolyzed to simpler compounds. The general formula is $C_nH_{2n}O_n$. They are the simplest form of sugar and are usually colorless, water-soluble, and crystalline solids. Some monosaccharides have a sweet taste. Examples of monosaccharides include glucose (dextrose), fructose (levulose) and galactose. Monosaccharides are the building blocks of disaccharides (such as sucrose and lactose) and polysaccharides (such as cellulose and starch). Further, each carbon atom that supports a hydroxyl group (so, all of the carbons except for the primary and terminal carbon) is chiral, giving rise to a number of isomeric forms, all with the same chemical formula. For instance, galactose and glucose are both aldohexoses, but have different physical structures and chemical properties.

Structure and Nomenclature

With few exceptions (e.g., deoxyribose), monosaccharide have this chemical formula: $C_x(H_2O)_y$, where conventionally $x \geq 3$. Monosaccharides can be classified by the number x of carbon atoms they contain: triose (3) tetrose (4), pentose (5), hexose (6), heptose (7), and so on.

The most important monosaccharide, glucose, is a hexose. Examples of heptoses include the ketoses mannoheptulose and sedoheptulose. Monosaccharides with eight or more carbons are rarely observed as they are quite unstable. In aqueous solutions monosaccharides exist as rings if they have more than four carbons.

Linear-chain Monosaccharides

Simple monosaccharides have a linear and unbranched carbon skeleton with one carbonyl (C=O) functional group, and one hydroxyl (OH) group on each of the remaining carbon atoms. Therefore, the molecular structure of a simple monosaccharide can be written as $H(CHOH)_n(C{=}O)(CHOH)_mH$, where $n + 1 + m = x$; so that its elemental formula is $C_xH_{2x}O_x$.

By convention, the carbon atoms are numbered from 1 to x along the backbone, starting from the end that is closest to the C=O group.

If the carbonyl is at position 1 (that is, n or m is zero), the molecule begins with a formyl group H(C=O)– and is technically an aldehyde. In that case, the compound is termed an aldose. Otherwise, the molecule has a keto group, a carbonyl –(C=O)– between two carbons; then it is formally a ketone, and is termed a ketose. Ketoses of biological interest usually have the carbonyl at position 2.

The various classifications above can be combined, resulting in names such as "aldohexose" and "ketotriose".

A more general nomenclature for open-chain monosaccharides combines a prefix to indicate the number of carbons (tri-, tetr-, pent-, hex-, etc.) with the suffixes "-ose" for aldoses and "-ulose" for ketoses. In the latter case, if the carbonyl is not at position 2, its position is then indicated by a numeric infix. So, for example, $H(C{=}O)(CHOH)_4H$ is pentose, $H(CHOH)(C{=}O)(CHOH)_3H$ is pentulose, and $H(CHOH)_2(C{=}O)(CHOH)_2H$ is pent-3-ulose.

Open-chain Stereoisomers

Two monosaccharides with equivalent molecular graphs (same chain length and same carbonyl position) may still be distinct stereoisomers, whose molecules differ in the three-dimensional arrangement of the bonds of certain atoms. This happens only if the molecule contains a stereogenic center, specifically a carbon atom that is chiral (connected to four distinct molecular sub-structures). Those four bonds can have any of two configurations in space distinguished by their handedness. In a simple open-chain monosaccharide, every carbon is chiral except the first and the last atoms of the chain, and (in ketoses) the carbon with the keto group.

For example, the triketose H(CHOH)(C=O)(CHOH)H (glycerone, dihydroxyacetone) has no stereogenic center, and therefore exists as a single stereoisomer. The other triose, the aldose $H(C{=}O)(CHOH)_2H$ (glyceraldehyde), has one chiral carbon — the central one, number 2 — which is bonded to groups –H, –OH, $-C(OH)H_2$, and –(C=O)H. Therefore, it exists as two stereoisomers whose molecules are mirror images of each other (like a left and a right glove). Monosaccharides with four or more carbons may contain multiple chiral carbons, so they typically have more than two stereoisomers. The number of distinct stereoisomers with the same diagram is bounded by 2^c, where c is the total number of chiral carbons.

The Fischer projection is a systematic way of drawing the skeletal formula of an acyclic monosaccharide so that the handedness of each chiral carbon is well specified. Each stereoisomer of a simple open-chain monosaccharide can be identified by the positions (right or left) in the Fischer diagram of the chiral hydroxyls (the hydroxyls attached to the chiral carbons).

Most stereoisomers are themselves chiral (distinct from their mirror images). In the Fischer projection, two mirror-image isomers differ by having the positions of all chiral hydroxyls reversed right-to-left. Mirror-image isomers are chemically identical in non-chiral environments, but usually have very different biochemical properties and occurrences in nature.

While most stereoisomers can be arranged in pairs of mirror-image forms, there are some non-chiral stereoisomers that are identical to their mirror images, in spite of having chiral centers. This happens whenever the molecular graph is symmetrical, as in the 3-ketopentoses $H(CHOH)_2(CO)(CHOH)_2H$, and the two halves are mirror images of each other. In that case, mirroring is equivalent to a half-turn rotation. For this reason, there are only three distinct 3-ketopentose stereoisomers, even though the molecule has two chiral carbons.

Distinct stereoisomers that are not mirror-images of each other usually have different chemical properties, even in non-chiral environments. Therefore, each mirror pair and each non-chiral stereoisomer may be given a specific monosaccharide name. For example, there are 16 distinct aldohexose stereoisomers, but the name "glucose" means a specific pair of mirror-image aldohexoses. In the Fischer projection, one of the two glucose isomers has the hydroxyl at left on C3, and at right on C4 and C5; while the other isomer has the reversed pattern. These specific monosaccharide names have conventional three-letter abbreviations, like "Glu" for glucose and "Thr" for threose.

Generally, a monosaccharide with n asymmetrical carbons has 2^n stereoisomers. The number of open chain stereoisomers for an aldose monosaccharide is larger by one than that of a ketose monosaccharide of the same length. Every ketose will have $2^{(n-3)}$ stereoisomers where $n > 2$ is the number of carbons. Every aldose will have $2^{(n-2)}$ stereoisomers where $n > 2$ is the number of carbons. These are also referred to as epimers which have the different arrangement of –OH and –H groups at the asymmetric or chiral carbon atoms (this does not apply to those carbons having the carbonyl functional group).

Configuration of Monosaccharides

Like many chiral molecules, the two stereoisomers of glyceraldehyde will gradually rotate the polarization direction of linearly polarized light as it passes through it, even in solution. The two stereoisomers are identified with the prefixes D- and L-, according to the sense of rotation: D-glyceraldehyde is dextrorotatory (rotates the polarization axis clockwise), while L-glyceraldehyde is levorotatory (rotates it counterclockwise).

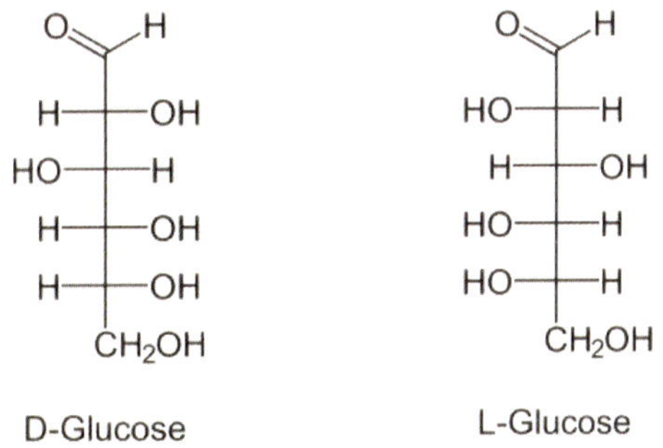

The D- and L- prefixes are also used with other monosaccharides, to distinguish two particular stereoisomers that are mirror-images of each other. For this purpose, one considers the chiral carbon that is furthest removed from the C=O group. Its four bonds must connect to –H, –OH, –C(OH)H, and the rest of the molecule. If the molecule can be rotated in space so that the directions of those four groups match those of the analog groups in D-glyceraldehyde's C2, then the isomer receives the D- prefix. Otherwise, it receives the L- prefix.

In the Fischer projection, the D- and L- prefixes specifies the configuration at the carbon atom that is second from bottom: D- if the hydroxyl is on the right side, and L- if it is on the left side.

Note that the D- and L- prefixes do not indicate the direction of rotation of polarized light, which is a combined effect of the arrangement at all chiral centers. However, the two enantiomers will always rotate the light in opposite directions, by the same amount.

Cyclisation of Monosaccharides

A monosaccharide often switches from the acyclic (open-chain) form to a cyclic form, through a nucleophilic addition reaction between the carbonyl group and one of the hydroxyls of the same molecule. The reaction creates a ring of carbon atoms closed by one bridging oxygen atom. The resulting molecule has an hemiacetal or hemiketal group, depending on whether the linear form was an aldose or a ketose. The reaction is easily reversed, yielding the original open-chain form.

In these cyclic forms, the ring usually has 5 or 6 atoms. These forms are called furanoses and pyranoses, respectively — by analogy with furan and pyran, the simplest compounds with the same carbon-oxygen ring (although they lack the double bonds of these two molecules). For example, the aldohexose glucose may form a hemiacetal linkage between the hydroxyl on carbon 1 and the oxygen on carbon 4, yielding a molecule with a 5-membered ring, called glucofuranose. The same reaction can take place between carbons 1 and 5 to form a molecule with a 6-membered ring, called glucopyranose. Cyclic forms with a 7-atom ring (the same of oxepane), rarely encountered, are called heptoses.

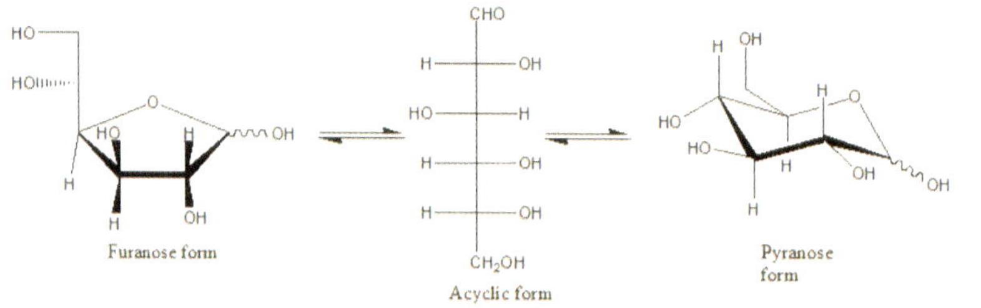

Conversion between the furanose, acyclic, and pyranose forms of D-glucose.

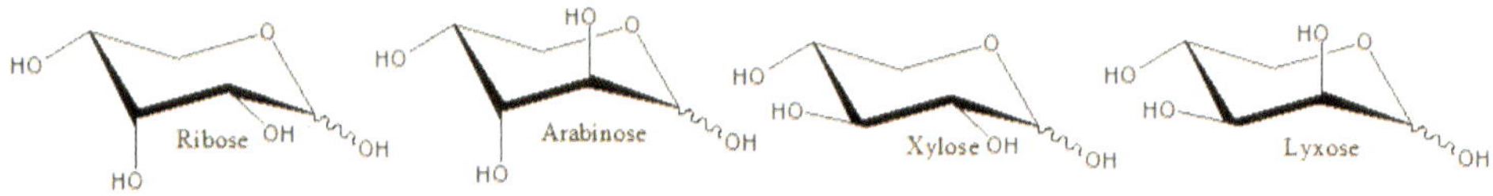

Pyranose forms of some pentose sugars.

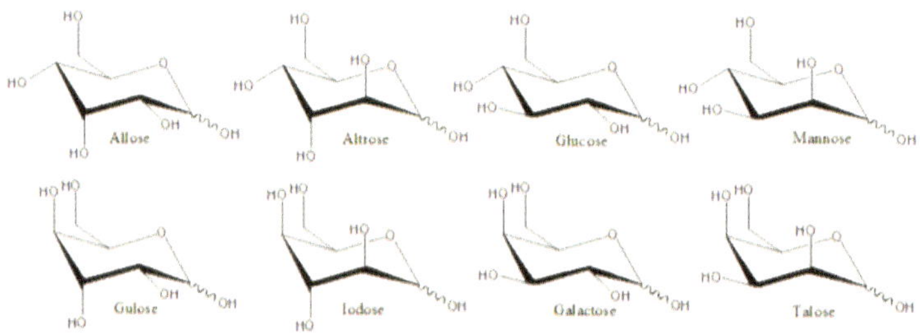

Pyranose forms of some hexose sugars.

For many monosaccharides (including glucose), the cyclic forms predominate, in the solid state and in solutions, and therefore the same name commonly is used for the open- and closed-chain isomers. Thus, for example, the term "glucose" may signify glucofuranose, glucopyranose, the open-chain form, or a mixture of the three.

Cyclization creates a new stereogenic center at the carbonyl-bearing carbon. The –OH group that replaces the carbonyl's oxygen may end up in two distinct positions relative to the ring's midplane. Thus each open-chain monosaccharide yields two cyclic isomers (anomers), denoted by the prefixes α- and β-. The molecule can change between these two forms by a process called mutarotation, that consists in a reversal of the ring-forming reaction followed by another ring formation.

Haworth Projection

The stereochemical structure of a cyclic monosaccharide can be represented in a Haworth projection. In this diagram, the α-isomer for the pyranose form of a D-aldohexose has the -OH of the anomeric carbon below the plane of the carbon atoms, while the β-isomer has the -OH of the anomeric carbon above the plane. Pyranoses typically adopt a chair conformation, similar to that of cyclohexane. In this conformation, the α-isomer has the -OH of the anomeric carbon in an axial position, whereas the β-isomer has the OH- of the anomeric carbon in equatorial position (considering D-aldohexose sugars).

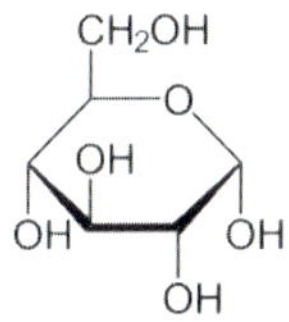

α-D-Glucopyranose

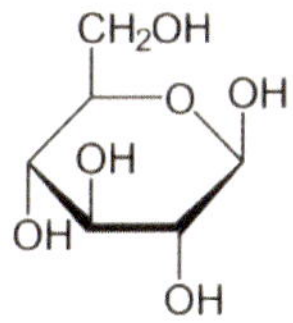

β-D-Glucopyranose

Glucose

Glucose is a simple sugar with the molecular formula $C_6H_{12}O_6$, which means that it is a molecule that is made of six carbon atoms, twelve hydrogen atoms, and six oxygen atoms. Glucose circulates in the blood of animals as blood sugar. It is made during photosynthesis from water and carbon dioxide, using energy from sunlight. It is the most important source of energy for cellular respiration. Glucose is stored as a polymer, in plants as starch and in animals as glycogen.

With six carbon atoms, it is classed as a hexose, a subcategory of the monosaccharides. D-Glucose is one of the sixteen aldohexose stereoisomers. The D-isomer, D-glucose, also known as dextrose, occurs widely in nature, but the L-isomer, L-glucose, does not. Glucose can be obtained by hydrolysis of carbohydrates such as milk sugar (lactose), cane sugar (sucrose), maltose, cellulose, glycogen, etc. It is commonly commercially manufactured from cornstarch by hydrolysis via pressurized steaming at controlled pH in a jet followed by further enzymatic depolymerization.

In 1747, Andreas Marggraf was the first to isolate glucose. Glucose is on the World Health Organization's List of Essential Medicines, the most important medications needed in a basic health system. The name glucose derives through the French from the which means "sweet," in reference to must, the sweet, first press of grapes in the making of wine. The suffix "-ose" is a chemical classifier, denoting a carbohydrate.

Structure and Nomenclature

Glucose is a monosaccharide with formula $C_6H_{12}O_6$ or H-(C=O)-$(CHOH)_5$-H, whose five hydroxyl (OH) groups are arranged in a specific way along its six-carbon back.

Open-chain Form

In its fleeting open-chain form, the glucose molecule has an open (as opposed to cyclic) and unbranched backbone of six carbon atoms, C-1 through C-6; where C-1 is part of an aldehyde group H(C=O)-, and each of the other five carbons bears one hydroxyl group -OH. The remaining bonds of the backbone carbons are satisfied by hydrogen atoms -H. Therefore, glucose is both a hexose and an aldose, or an aldohexose. The aldehyde group makes glucose a reducing sugar giving a positive reaction with the Fehling test.

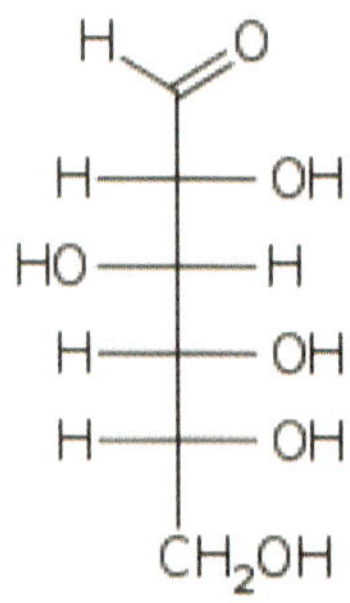

D-Glucose in Fischer projection

Each of the four carbons C-2 through C-5 is a stereocenter, meaning that its four bonds connect to four different substituents. (Carbon C-2, for example, connects to -(C=O)H, -OH, -H, and -$(CHOH)_4H$.) In D-glucose, these four parts must be in a specific three-dimensional arrangement. Namely, when the molecule is drawn in the Fischer projection, the hydroxyls on C-2, C-4, and C-5 must be on the right side, while that on C-3 must be on the left side.

The positions of those four hydroxyls are exactly reversed in the Fischer diagram of L-glucose. D- and L-glucose are two of the 16 possible aldohexoses; the other 14 are allose, altrose, galactose, gulose, idose, mannose, and talose, each with two enantiomers, "D-" and "L-".

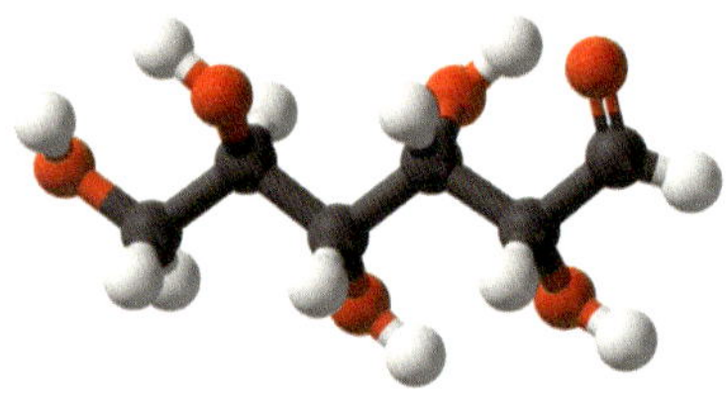
The aldehyde form of glucose

It is important to note that the linear form of glucose makes up less than 3% of the glucose molecules in a water solution. The rest is one of two cyclic forms of glucose that are formed when the hydroxyl group on carbon 5 (C5) bonds to the aldehyde carbon 1 (C1).

Cyclic Forms

In solutions, the open-chain form of glucose (either "D-" or "L-") exists in equilibrium with several cyclic isomers, each containing a ring of carbons closed by one oxygen atom. In aqueous solution however, more than 99% of glucose molecules, at any given time, exist as pyranose forms. The open-chain form is limited to about 0.25% and furanose forms exists in negligible amounts. The terms "glucose" and "D-glucose" are generally used for these cyclic forms as well. The ring arises from the open-chain form by an intramolecular nucleophilic addition reaction between the aldehyde group (at C-1) and either the C-4 or C-5 hydroxyl group, forming a hemiacetal linkage, -C(OH)H-O-.

The reaction between C-1 and C-5 yields a six-membered heterocyclic system called a pyranose, which is a monosaccharide sugar (hence "–ose") containing a derivatised pyran skeleton. The (much rarer) reaction between C-1 and C-4 yields a five-membered furanose ring, named after the cyclic ether furan. In either case, each carbon in the ring has one hydrogen and one hydroxyl attached, except for the last carbon (C-4 or C-5) where the hydroxyl is replaced by the remainder of the open molecule (which is -$(C(CH_2OH)HOH)$-H or -(CHOH)-H, respectively).

Cyclic forms of glucose

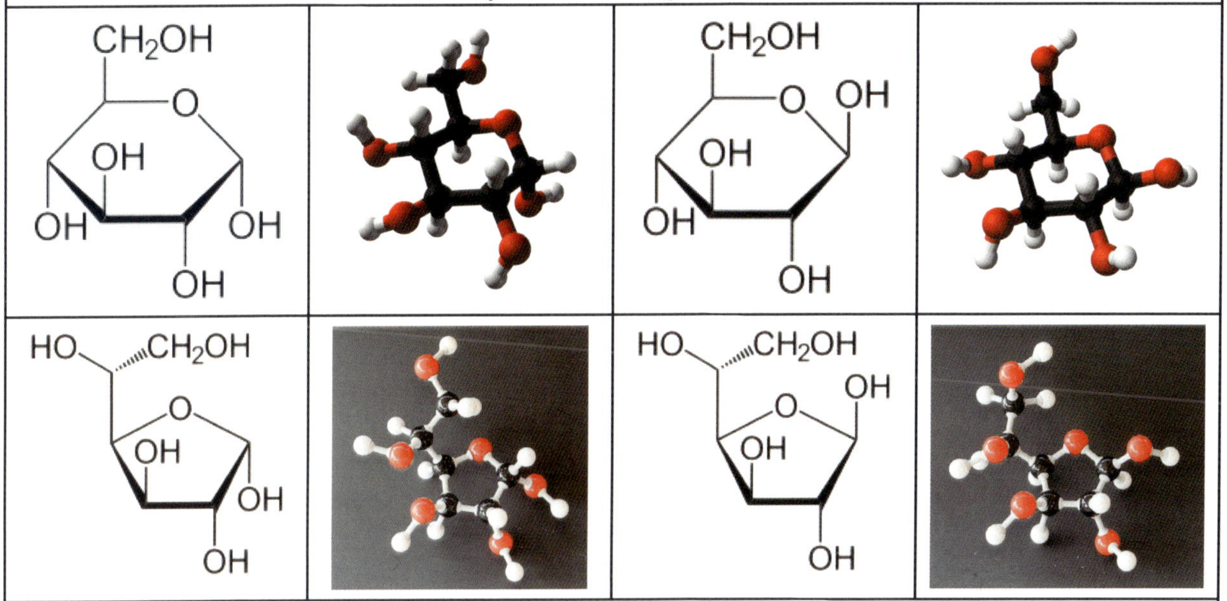

From left to right: Haworth projections and ball-and-stick structures of the α- and β- anomers of D-glucopyranose (top row) and D-glucofuranose (bottom row)

The ring-closing reaction makes carbon C-1 chiral, too, since its four bonds lead to -H, to -OH, to carbon C-2, and to the ring oxygen. These four parts of the molecule may be arranged around C-1 (the anomeric carbon) in two distinct ways, designated by the prefixes "α-" and "β-". When a glucopyranose molecule is drawn in the Haworth projection, the designation "α-" means that the hydroxyl group attached to C-1 and the $-CH_2OH$ group at C-5 lies on opposite sides of the ring's plane (a *trans* arrangement), while "β-" means that they are on the same side of the plane (a *cis* arrangement). Therefore, the open-chain isomer D-glucose gives rise to four distinct cyclic isomers: α-D-glucopyranose, β-D-glucopyranose, α-D-glucofuranose, and β-D-glucofuranose. These five structures exist in equilibrium and interconvert, and the interconversion is much more rapid with acid catalysis.

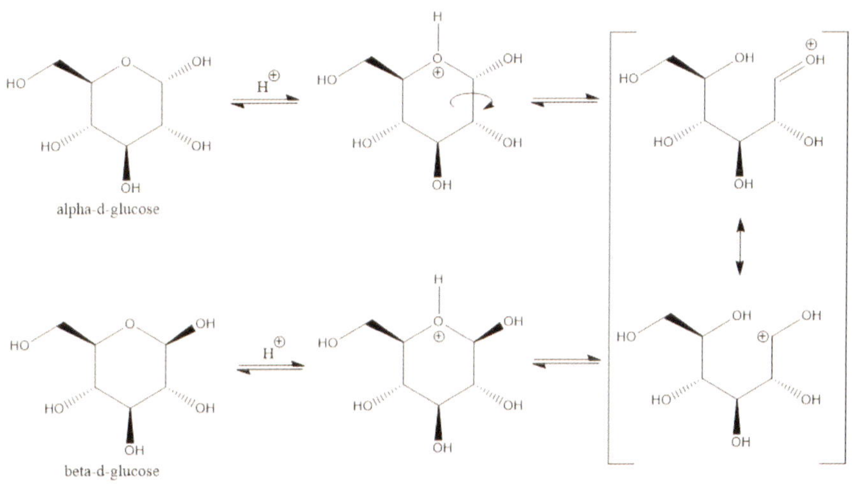

The other open-chain isomer L-glucose similarly gives rise to four distinct cyclic forms of L-glucose, each the mirror image of the corresponding D-glucose.

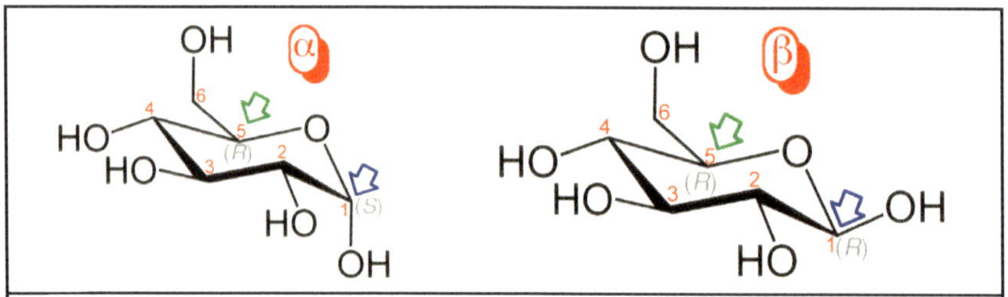

Chair conformations of α- (left) and β- (right) D-glucopyranose

The rings are not planar, but are twisted in three dimensions. The glucopyranose ring (α or β) can assume several non-planar shapes, analogous to the "chair" and "boat" conformations of cyclohexane. Similarly, the glucofuranose ring may assume several shapes, analogous to the "envelope" conformations of cyclopentane.

In the solid state, only the glucopyranose forms are observed, forming colorless crystalline solids that are highly soluble in water and acetic acid but poorly soluble in methanol and ethanol. They melt at 146 °C (295 °F) (α) and 150 °C (302 °F) (β), and decompose at higher temperatures into carbon and water.

Rotational Isomers

Each glucose isomer is subject to rotational isomerism. Within the cyclic form of glucose, rotation may occur around the O6-C6-C5-O5 torsion angle, termed the ω-angle, to form three staggered rotamer conformations called *gauche-gauche* (gg), *gauche-trans* (gt) and *trans-gauche* (tg). There is a tendency for the ω-angle to adopt a *gauche* conformation, a tendency that is attributed to the gauche effect.

Sources and Absorption

Most dietary carbohydrates contain glucose, either as their only building block (as in the polysaccharides starch and glycogen), or together with another monosaccharide (as in the hetero-polysaccharides sucrose and lactose). Unbounded glucose is one of the main ingredients of honey.

In the lumen of the duodenum and small intestine, the glucose oligo- and polysaccharides are broken down to monosaccharides by the pancreatic and intestinal glycosidases. Other polysaccharides cannot be processed by the human intestine and require assistance by intestinal flora if they are to be broken down; the most notable exceptions are sucrose (fructose-glucose) and lactose (galactose-glucose). Glucose is then transported across the apical membrane of the enterocytes by SLC5A1 (SGLT1), and later across their basal membrane by SLC2A2 (GLUT2).

Some glucose is converted to lactic acid by astrocytes, which is then utilized as an energy source by brain cells; some glucose is used by intestinal cells and red blood cells, while the rest reaches the liver, adipose tissue and muscle cells, where it is absorbed and stored as glycogen (under the influence of insulin). Liver cell glycogen can be converted to glucose and returned to the blood when insulin is low or absent; muscle cell glycogen is not returned to the blood because of a lack of enzymes. In fat cells, glucose is used to power reactions that synthesize some fat types and have other purposes. Glycogen is the body's "glucose energy storage" mechanism, because it is much more "space efficient" and less reactive than glucose itself.

Galactose

Galactose, sometimes abbreviated Gal, is a monosaccharide sugar that is about as sweet as glucose, and about 30% as sweet as sucrose. It is a C-4 epimer of glucose.

Galactan is a polymeric form of galactose found in hemicellulose, and forming the core of the galactans, a class of natural polymeric carbohydrates.

Structure and Isomerism

Galactose exists in both open-chain and cyclic form. The open-chain form has a carbonyl at the end of the chain.

Four isomers are cyclic, two of them with a pyranose (six-membered) ring, two with a furanose (five-membered) ring. Galactofuranose occurs in bacteria, fungi and protozoa , and is recognized by a putative chordate immune lectin intelectin through its exocyclic 1,2-diol. In the cyclic form there are two anomers, named alpha and beta, since the transition from the open-chain form to the cyclic form involves the creation of a new stereocenter at the site of the open-chain carbonyl. In the beta form, the alcohol group is in the equatorial position, whereas in the alpha form, the alcohol group is in the axial position.

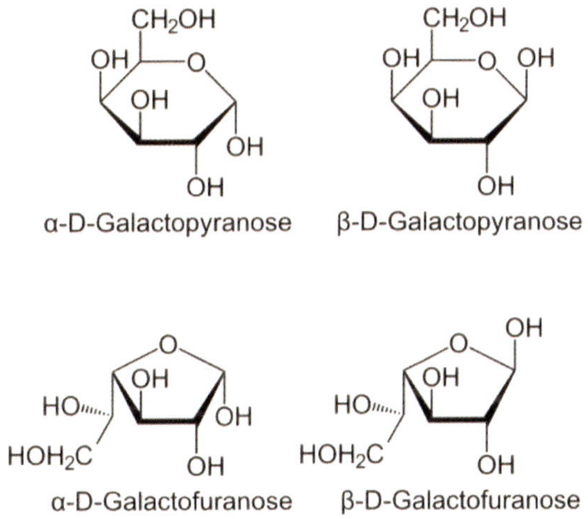

Cyclic forms of galactose

Relationship to Lactose

Galactose is a monosaccharide. When combined with glucose (monosaccharide), through a condensation reaction, the result is the disaccharide lactose. The hydrolysis of lactose to glucose and galactose is catalyzed by the enzymes lactase and β-galactosidase. The latter is produced by the *lac* operon in *Escherichia coli*.

In nature, lactose is found primarily in milk and milk products. Consequently, various food products made with dairy-derived ingredients can contain lactose. Galactose metabolism, which converts galactose into glucose, is carried out by the three principal enzymes in a mechanism known as the Leloir pathway. The enzymes are listed in the order of the metabolic pathway: galactokinase (GALK), galactose-1-phosphate uridyltransferase (GALT), and UDP-galactose-4'-epimerase (GALE).

In human lactation, glucose is changed into galactose via hexoneogenesis to enable the mammary glands to secrete lactose. However, most lactose in breast milk is synthesized from galactose taken up from the blood, and only 35±6% is made from galactose from *de novo* synthesis. Glycerol also contributes some to the mammary galactose production.

Metabolism

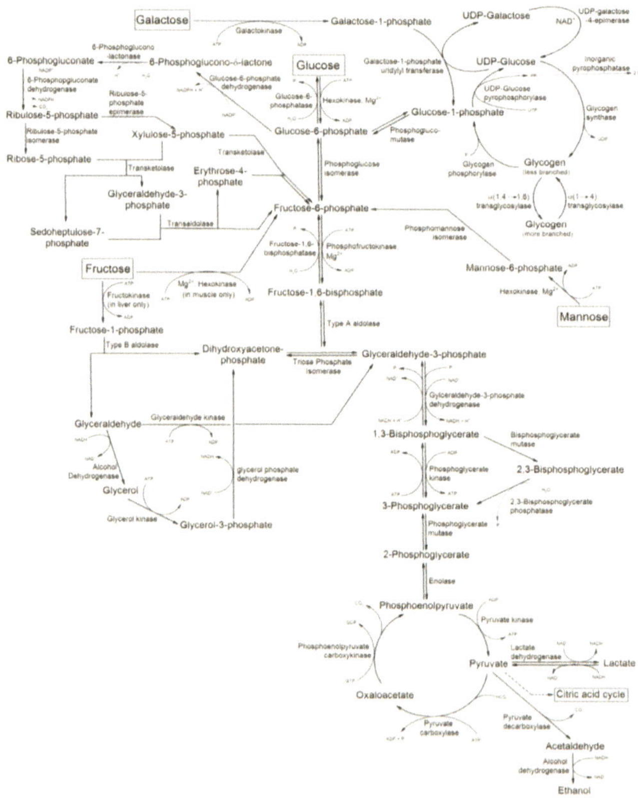

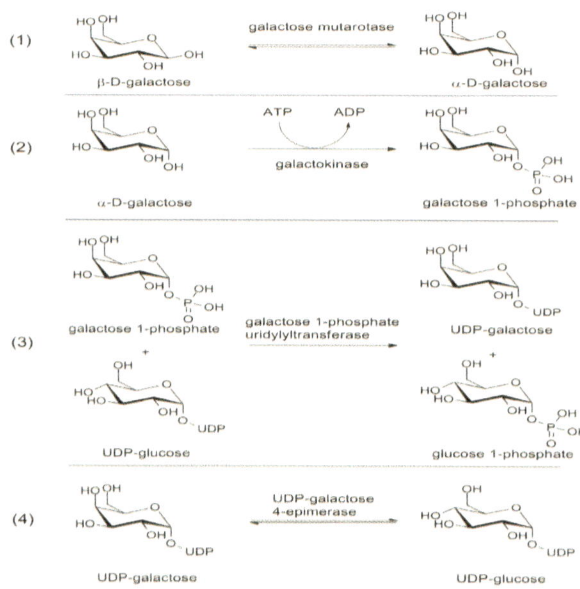

Galactose metabolism

Glucose is the primary metabolic fuel for humans. It is more stable than galactose and is less susceptible to the formation of nonspecific glycoconjugates, molecules with at least one sugar attached to a protein or lipid. Many speculate that it is for this reason that a pathway for rapid conversion from galactose to glucose has been highly conserved among many species.

The main pathway of galactose metabolism is the Leloir pathway; humans and other species, however, have been noted to contain several alternate pathways, such as the De Ley Doudoroff pathway. The Leloir pathway consists of the latter stage of a two-part process that converts β-D-galactose to UDP-glucose. The initial stage is the conversion of β-D-galactose to α-D-galactose by the enzyme, mutarotase (GALM). The Leloir pathway then carries out the conversion of α-D-galactose to UDP-glucose via three principal enzymes: Galactokinase (GALK) phosphorylates α-D-galactose to galactose-1-phosphate, or Gal-1-P; Galactose-1-phosphate uridyltransferase (GALT) transfers a UMP group from UDP-glucose to Gal-1-P to form UDP-galactose; and finally, UDP galactose-4'-epimerase (GALE) interconverts UDP-galactose and UDP-glucose, thereby completing the pathway.

Galactosemia is an inability to properly break down galactose due to a genetically inherited mutation in one of the enzymes in the Leloir pathway. As a result, the consumption of even small quantities is harmful to galactosemics.

Sources

Galactose is found in dairy products, avocados, sugar beets, other gums and mucilages. It is also synthesized by the body, where it forms part of glycolipids and glycoproteins in several tissues; and is a by-product from the third-generation ethanol production process (from macroalgae).

Fructose

Fructose, or fruit sugar, is a simple ketonic monosaccharide found in many plants, where it is often bonded to glucose to form the disaccharide, sucrose. It is one of the three dietary monosaccharides, along with glucose and galactose, that are absorbed directly into blood during digestion. Fructose was discovered by French chemist Augustin-Pierre Dubrunfaut in 1847. The name "fructose" was coined in 1857 by the English chemist, William Allen Miller. Pure, dry fructose is a sweet, white, odorless, crystalline solid, and is the most water-soluble of all the sugars. Fructose is found in honey, tree and vine fruits, flowers, berries, and most root vegetables.

Commercially, fructose is derived from sugar cane, sugar beets, and maize. Crystalline fructose is the monosaccharide, dried, ground, and of high purity. High-fructose corn syrup is a mixture of glucose and fructose as monosaccharides. Sucrose is a compound with one molecule of glucose covalently linked to one molecule of fructose. All forms of fructose, including fruits and juices, are commonly added to foods and drinks for palatability and taste enhancement, and for browning of some foods, such as baked goods. About 240,000 tonnes of crystalline fructose are produced annually.

As for any sugar, excessive consumption of fructose may contribute to insulin resistance, obesity, elevated LDL cholesterol and triglycerides, leading to metabolic syndrome, type 2 diabetes and cardiovascular disease. The European Food Safety Authority stated that fructose is preferable over sucrose and glucose in sugar-sweetened foods and beverages because of its lower effect on postprandial blood sugar levels, and also noted that "high intakes of fructose may lead to metabolic complications such as dyslipidaemia, insulin resistance, and increased visceral adiposity". Further, the UK's Scientific Advisory Committee on Nutrition in 2015 disputed the claims of fructose causing metabolic disorders, stating that "there is insufficient evidence to demonstrate that fructose intake leads to adverse health outcomes independent of any effects related to its presence as a component of total and free sugars.

Chemical Properties

Fructose is a 6-carbon polyhydroxyketone. Crystalline fructose adopts a cyclic six-membered structure owing to the stability of its hemiketal and internal hydrogen-bonding. This form is formally called D-fructopyranose. In solution, fructose exists as an equilibrium mixture of 70% fructopyranose and about 22% fructofuranose, as well as small amounts of three other forms, including the acyclic structure.

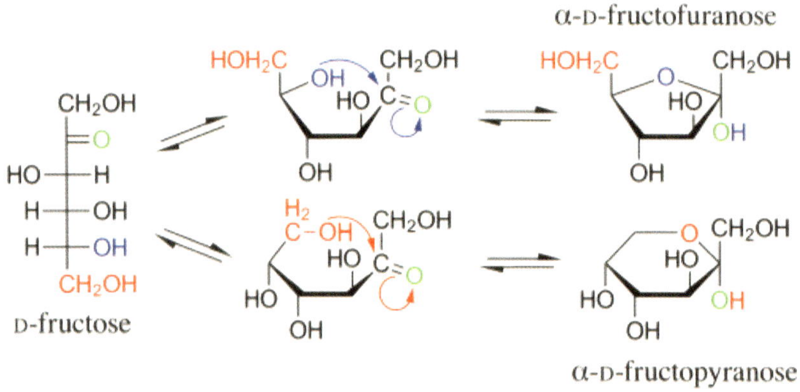

Relationship between the acyclic and the cyclic (hemiketal) isomers of fructose

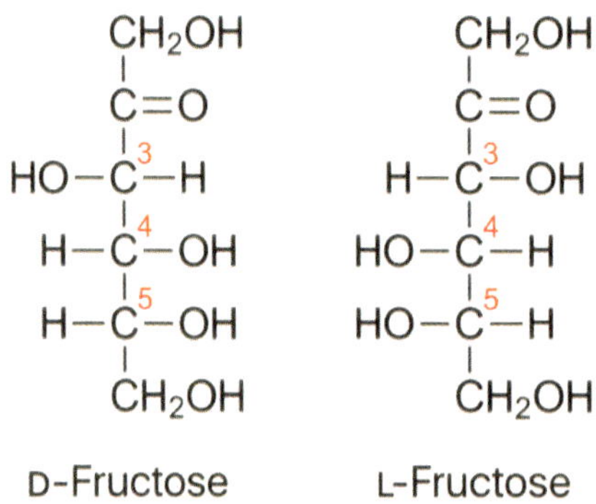

D- and L-isomers of fructose (open-chain form)

Reactions

Fructose and Fermentation

Fructose may be anaerobically fermented by yeast or bacteria. Yeast enzymes convert sugar (glucose, or fructose) to ethanol and carbon dioxide. The carbon dioxide released during fermentation will remain dissolved in water, where it will reach equilibrium with carbonic acid, unless the fermentation chamber is left open to the air. The dissolved carbon dioxide and carbonic acid produce the carbonation in bottled fermented beverages.

Fructose and Maillard Reaction

Fructose undergoes the Maillard reaction, non-enzymatic browning, with amino acids. Because fructose exists to a greater extent in the open-chain form than does glucose, the initial stages of the Maillard reaction occur more rapidly than with glucose. Therefore, fructose has potential to contribute to changes in food palatability, as well as other nutritional effects, such as excessive browning, volume and tenderness reduction during cake preparation, and formation of mutagenic compounds.

Dehydration

Fructose readily dehydrates to give hydroxymethylfurfural ("HMF"). This process, in the future, may become part of a low-cost, carbon-neutral system to produce replacements for petrol and diesel from plants.

Fructose Metabolism

All three dietary monosaccharides are transported into the liver by the GLUT2 transporter. Fructose and galactose are phosphorylated in the liver by fructokinase (K_m= 0.5 mM) and galactokinase (K_m = 0.8 mM). By contrast, glucose tends to pass through the liver (K_m of hepatic glucokinase = 10 mM) and can be metabolised anywhere in the body. Uptake of fructose by the liver is not regulated by insulin. However, insulin is capable of increasing the abundance and functional activity of GLUT5 in skeletal muscle cells.

Fructolysis

The initial catabolism of fructose is sometimes referred to as fructolysis, in analogy with glycolysis, the catabolism of glucose. In fructolysis, the enzyme fructokinase initially produces fructose 1-phosphate, which is split by aldolase B to produce the trioses dihydroxyacetone phosphate (DHAP) and glyceraldehyde . Unlike glycolysis, in fructolysis the triose glyceraldehyde lacks a phosphate group. A third enzyme, triokinase, is therefore required to phosphorylate glyceraldehyde, producing glyceraldehyde 3-phosphate. The resulting trioses are identical to those obtained in glycolysis and can enter the gluconeogenic pathway for glucose or glycogen synthesis, or be further catabolized through the lower glycolytic pathway to pyruvate.

Metabolism of Fructose to DHAP and Glyceraldehyde

The first step in the metabolism of fructose is the phosphorylation of fructose to fructose 1-phosphate by fructokinase, thus trapping fructose for metabolism in the liver. Fructose 1-phosphate then undergoes hydrolysis by aldolase B to form DHAP and glyceraldehydes; DHAP can either be isomerized to glyceraldehyde 3-phosphate by triosephosphate isomerase or undergo reduction to glycerol 3-phosphate by glycerol 3-phosphate dehydrogenase. The glyceraldehyde produced may also be converted to glyceraldehyde 3-phosphate by glyceraldehyde kinase or further converted to glycerol 3-phosphate by glycerol 3-phosphate dehydrogenase. The metabolism of fructose at this point yields intermediates in the gluconeogenic pathway leading to glycogen synthesis as well as fatty acid and triglyceride synthesis.

Synthesis of Glycogen from DHAP and Glyceraldehyde 3-phosphate

The resultant glyceraldehyde formed by aldolase B then undergoes phosphorylation to glyceraldehyde 3-phosphate. Increased concentrations of DHAP and glyceraldehyde 3-phosphate in the liver drive the gluconeogenic pathway toward glucose and subsequent glycogen synthesis. It appears that fructose is a better substrate for glycogen synthesis than glucose and that glycogen replenishment takes precedence over triglyceride formation. Once liver glycogen is replenished, the intermediates of fructose metabolism are primarily directed toward triglyceride synthesis.

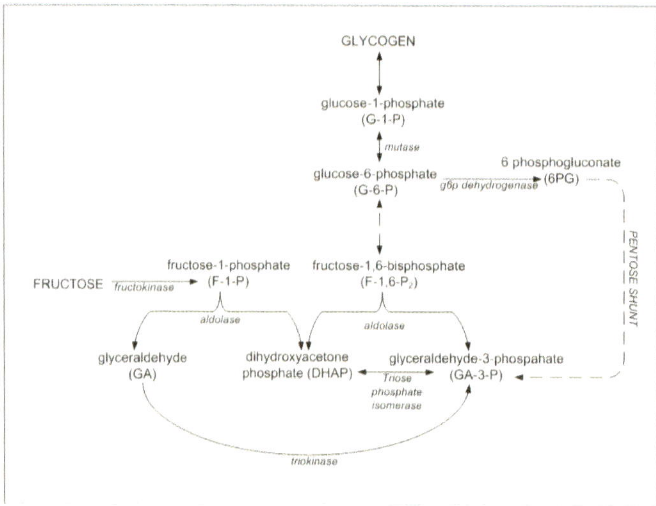

Metabolic conversion of fructose to glycogen in the liver

Synthesis of Triglyceride from DHAP and Glyceraldehyde 3-phosphate

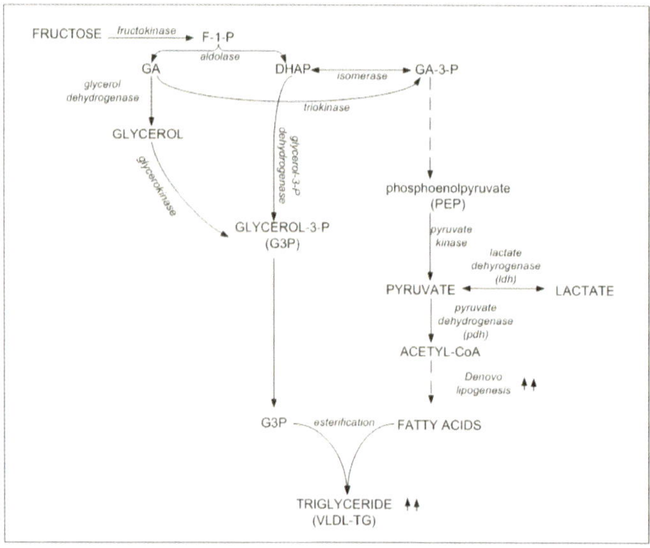

Metabolic conversion of fructose to triglyceride in the liver

Carbons from dietary fructose are found in both the free fatty acid and glycerol moieties of plasma triglycerides. High fructose consumption can lead to excess pyruvate production, causing a buildup of Krebs cycle intermediates. Accumulated citrate can be transported from the mitochondria into the cytosol of hepatocytes, converted to acetyl CoA by citrate lyase and directed toward fatty acid synthesis.

In addition, DHAP can be converted to glycerol 3-phosphate, providing the glycerol backbone for the triglyceride molecule. Triglycerides are incorporated into very-low-density lipoproteins (VLDL), which are released from the liver destined toward peripheral tissues for storage in both fat and muscle cells.

Disaccharide

A disaccharide (also called a double sugar or biose) is the sugar formed when two monosaccharides (simple sugars) are joined by glycosidic linkage. Like monosaccharides, disaccharides are soluble in water. Three common examples are sucrose, lactose, and maltose.

Disaccharides are one of the four chemical groupings of carbohydrates (monosaccharides, disaccharides, oligosaccharides, and polysaccharides). The most common types of disaccharides—sucrose, lactose, and maltose—have twelve carbon atoms, with the general formula $C_{12}H_{22}O_{11}$. The differences in these disaccharides are due to atomic arrangements within the molecule.

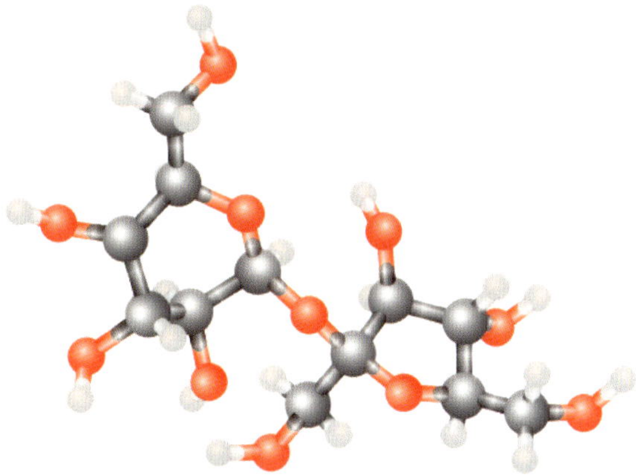

Sucrose, a disaccharide formed from condensation of a molecule of glucose and a molecule of fructose

The joining of simple sugars into a double sugar happens by a condensation reaction, which involves the elimination of a water molecule from the functional groups only. Breaking apart a double sugar into its two simple sugars is accomplished by hydrolysis with the help of a type of enzyme called a disaccharidase. As building the larger sugar ejects a water molecule, breaking it down consumes a water molecule. These reactions are vital in metabolism. Each disaccharide is broken down with the help of a corresponding disaccharidase (sucrase, lactase, and maltase).

Classification

There are two functionally different classes of disaccharides:

- Reducing disaccharides, in which one monosaccharide, the reducing sugar of the pair, still has a free hemiacetal unit that can perform as a reducing aldehyde group; cellobiose and maltose are examples of reducing disaccharides, each with one hemiacetal unit, the other occupied by the glycosidic bond, which prevents it from acting as a reducing agent.

- Non-reducing disaccharides, in which the component monosaccharides bond through an acetal linkage between their anomeric centers. This results in neither monosaccharide being left with a hemiacetal unit that is free to act as a reducing agent. Sucrose and trehalose are examples of non-reducing disaccharides because their glycosidic bond is between their respective hemiacetal carbon atoms. The reduced chemical reactivity of the non-reducing sugars in comparison to reducing sugars, may be an advantage where stability in storage is important.

Formation

The formation of a disaccharide molecule from two monosaccharide molecules proceeds by displacing a hydroxyl radical from one molecule and a hydrogen nucleus (a proton) from the other, so that the now vacant bonds on the monosaccharides join the two monomers together. The vacant bonds on the hydroxyl radical and the proton unite in their turn, forming a molecule of water, that then goes free. Because of the removal of the water molecule from the product, the term of convenience for such a process is "dehydration reaction" (also "condensation reaction" or "dehydration synthesis"). For example, milk sugar (lactose) is a disaccharide made by condensation of one molecule of each of the monosaccharides glucose and galactose, whereas the disaccharide sucrose in sugar cane and sugar beet, is a condensation product of glucose and fructose. Maltose, another common disaccharide, is condensed from two glucose molecules.

The dehydration reaction that bonds monosaccharides into disaccharides (and also bonds monosaccharides into more complex polysaccharides) forms what are called glycosidic bonds.

Properties

The glycosidic bond can be formed between any hydroxyl group on the component monosaccharide. So, even if both component sugars are the same (e.g., glucose), different bond combinations (regiochemistry) and stereochemistry (*alpha*- or *beta*-) result in disaccharides that are diastereoisomers with different chemical and physical properties.

Depending on the monosaccharide constituents, disaccharides are sometimes crystalline, sometimes water-soluble, and sometimes sweet-tasting and sticky-feeling.

Assimilation

Digestion involves breakdown to the monosaccharides.

Sucrose

Sucrose is common table sugar. It is a disaccharide, a molecule composed of the two monosaccharides, glucose and fructose. Sucrose is produced naturally in plants, from which table sugar is refined. It has the formula $C_{12}H_{22}O_{11}$.

For human consumption, sucrose is extracted, and refined, from either sugar cane or sugar beet. Sugar mills are located where sugar cane is grown to crush the cane and produce raw sugar which is shipped around the world for refining into pure sucrose. Some sugar mills also process the raw sugar into pure sucrose. Sugar beet factories are in colder climates where the beet is grown and process the beets directly into refined sugar. The sugar refining process involves washing the raw sugar crystals before dissolving them into a sugar syrup which is filtered and then passed over carbon to remove any residual colour. The by now clear sugar syrup is then concentrated by boiling under vacuum and crystallised as the final purification process to produce crystals of pure sucrose. These crystals are clear, odourless, and have a sweet taste. En masse, the crystals appear white.

Sugar is often an added ingredient in food production and food recipes. About 175 million tonnes of sugar were produced worldwide in 2013.

Physical and Chemical Properties

Structural O-α-D-glucopyranosyl-(1→2)-β-D-fructofuranoside

In sucrose, the components glucose and fructose are linked via an ether bond between C1 on the glucosyl subunit and C2 on the fructosyl unit. The bond is called a glycosidic linkage. Glucose exists predominantly as two isomeric "pyranoses" (α and β), but only one of these forms links to the fructose. Fructose itself exists as a mixture of "furanoses", each of which having α and β isomers, but only one particular isomer links to the glucosyl unit. What is notable about sucrose is that, unlike most disaccharides, the glycosidic bond is formed between the reducing ends of both glucose and fructose, and not between the reducing end of one and the nonreducing end of the other. This linkage inhibits further bonding to other saccharide units. Since it contains no anomeric hydroxyl groups, it is classified as a non-reducing sugar.

Sucrose crystallizes in the monoclinic space group $P2_1$ with room-temperature lattice parameters a = 1.08631 nm, b = 0.87044 nm, c = 0.77624 nm, β = 102.938°.

The purity of sucrose is measured by polarimetry, through the rotation of plane-polarized light by a solution of sugar. The specific rotation at 20 °C using yellow "sodium-D" light (589 nm) is +66.47°. Commercial samples of sugar are assayed using this parameter. Sucrose does not deteriorate at ambient conditions.

Thermal and Oxidative Degradation

Solubility of sucrose in water vs. temperature	
T (°C)	S (g/dL)
50	259
55	273
60	289
65	306
70	325
75	346
80	369
85	394
90	420

Sucrose does not melt at high temperatures. Instead, it decomposes—at 186 °C (367 °F)—to form caramel. Like other carbohydrates, it combusts to carbon dioxide and water. Mixing sucrose with the oxidizer potassium nitrate produces the fuel known as rocket candy that is used to propel amateur rocket motors.

$$C_{12}H_{22}O_{11} + 6\ KNO_3 \rightarrow 9\ CO + 3\ N_2 + 11\ H_2O + 3\ K_2CO_3$$

This reaction is somewhat simplified though. Some of the carbon does get fully oxidized to carbon dioxide, and other reactions, such as the water-gas shift reaction also take place. A more accurate theoretical equation is:

$$C_{12}H_{22}O_{11} + 6.288\ KNO_3 \rightarrow 3.796\ CO_2 + 5.205\ CO + 7.794\ H_2O + 3.065\ H_2 + 3.143\ N_2 + 2.998\ K_2CO_3 + 0.274\ KOH$$

Sucrose burns with chloric acid, formed by the reaction of hydrochloric acid and potassium chlorate:

$$8\ HClO_3 + C_{12}H_{22}O_{11} \rightarrow 11\ H_2O + 12\ CO_2 + 8\ HCl$$

Sucrose can be dehydrated with sulfuric acid to form a black, carbon-rich solid, as indicated in the following idealized equation:

H_2SO_4(catalyst) + $C_{12}H_{22}O_{11} \rightarrow 12\ C + 11\ H_2O$ + Heat (and some $H_2O + SO_3$ as a result of the heat).

The formula for sucrose's decomposition can be represented as 2 step reaction, first simplified reaction is dehydration of sucrose to pure Carbon and water then carbon oxidises to CO_2 with O_2 from air.

$$C_{12}H_{22}O_{11} + \text{heat} \rightarrow 12\ C + 11\ H_2O$$

$$12\ C + 12\ O_2 \rightarrow 12\ CO_2$$

Hydrolysis

Hydrolysis breaks the glycosidic bond converting sucrose into glucose and fructose. Hydrolysis is, however, so slow that solutions of sucrose can sit for years with negligible change. If the enzyme sucrase is added, however, the reaction will proceed rapidly. Hydrolysis can also be accelerated with acids, such as cream of tartar or lemon juice, both weak acids. Likewise, gastric acidity converts sucrose to glucose and fructose during digestion, the bond between them being an acetal bond which can be broken by an acid.

Synthesis and Biosynthesis of Sucrose

The biosynthesis of sucrose proceeds via the precursors UDP-glucose and fructose 6-phosphate, catalyzed by the enzyme sucrose-6-phosphate synthase. The energy for the reaction is gained by the cleavage of Uridine diphosphate (UDP). Sucrose is formed by plants and cyanobacteria but not by other organisms. Sucrose is found naturally in many food plants along with the monosaccharide fructose. In many fruits, such as pineapple and apricot, sucrose is the main sugar. In others, such as grapes and pears, fructose is the main sugar.

Chemical Synthesis

Although sucrose is almost invariably isolated from natural sources, its chemical synthesis was first achieved in 1953 by Raymond Lemieux.

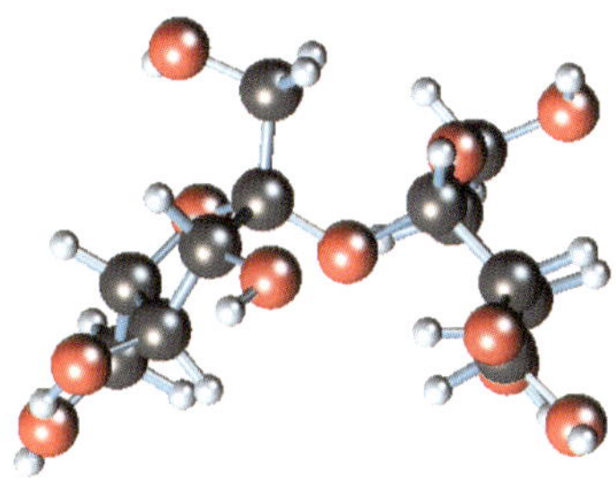

Model of sucrose molecule

Lactose

Lactose is a disaccharide molecule; a sugar composed of galactose and glucose that is found in milk. Lactose makes up around 2–8% of milk (by weight), although the amount varies among species and individuals, and milk with a reduced amount of lactose also exists. It can be extracted from sweet or sour whey. It has a formula of $C_{12}H_{22}O_{11}$ and the hydrate formula $C_{12}H_{22}O_{11}·H_2O$, making it an isomer of sucrose.

Structure and Reactions

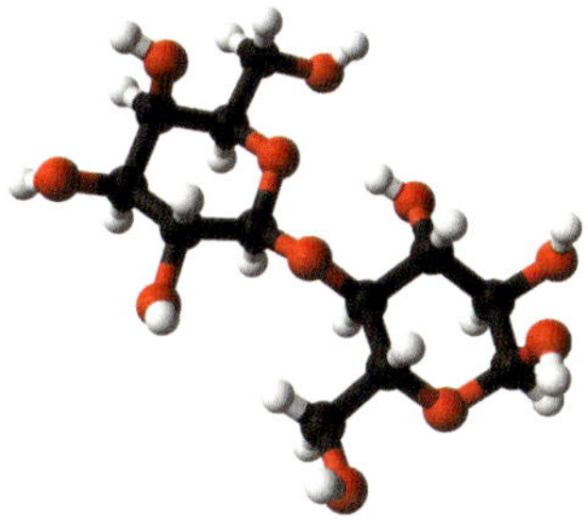

The molecular structure of α-lactose, as determined by X-ray crystallography.

Lactose is a disaccharide derived from the condensation of galactose and glucose, which form a β-1→4 glycosidic linkage. Its systematic name is β-D-galactopyranosyl-(1→4)-D-glucose. The glucose can be in either the α-pyranose form or the β-pyranose form, whereas the galactose can only have the β-pyranose form: hence α-lactose and β-lactose refer to the anomeric form of the glucopyranose ring alone.

Lactose is hydrolysed to glucose and galactose, isomerised in alkaline solution to lactulose, and catalytically hydrogenated to the corresponding polyhydric alcohol, lactitol. Lactose monohydrate crystals have a characteristic tomahawk shape that can be observed with a light microscope.

Isolation

Several million tons are produced annually as a by-product of the dairy industry. Whey is made up of 6.5% solids of which 4.8% is lactose that may be purified by crystallisation. Whey or milk plasma is the liquid remaining after milk is curdled and strained, for example in the production of cheese. Lactose makes up about 2–8% of milk by weight. Industrially, lactose is produced from whey permeate – that is whey filtrated for all major proteins. The protein fraction is used in infant nutrition and sport nutrition while the permeate can be evaporated to 60–65% solids and crystallized while cooling . Lactose can also be precipitated from whey using ethanol. Since it is insoluble in ethanol, lactose precipitates, in about 65% yield.

Metabolism

Infant mammals nurse on their mothers to drink milk, which is rich in lactose. The intestinal villi secrete the enzyme called lactase (β-D-galactosidase) to digest it. This enzyme cleaves the lactose molecule into its two subunits, the simple sugars glucose and galactose, which can be absorbed. Since lactose occurs mostly in milk, in most mammals, the production of lactase gradually decreases with maturity due to a lack of continuing consumption.

Many people with ancestry in Europe, West Asia, South Asia, the Sahel belt in West Africa, East Africa and a few other parts of Central Africa maintain lactase production into adulthood. In many of these areas, milk from mammals such as cattle, goats, and sheep is used as a large source of food. Hence, it was in these regions that genes for lifelong lactase production first evolved. The genes of adult lactose tolerance have evolved independently in various ethnic groups. By descent, more than 70% of western Europeans can drink milk as adults, compared with less than 30% of people from areas of Africa, eastern and south-eastern Asia and Oceania. In people who are lactose intolerant, lactose is not broken down and provides food for gas-producing gut flora, which can lead to diarrhea, bloating, flatulence, and other gastrointestinal symptoms.

Lactation

Lactose's synthesis is essentially one-way and is unable to cross the Golgi membrane. The resulting accumulation of lactose is the primary source of osmotic pressure, drawing in the bulk of the water content of milk.

Maltose

Maltose,also known as maltobiose or malt sugar, is a disaccharide formed from two units of glucose joined with an α(1→4) bond. In the isomer isomaltose, the two glucose molecules are joined with an α(1→6) bond. Maltose is the two-unit member of the amylose homologous series, the key structural motif of starch. When beta-amylase breaks down starch, it removes two glucose units at a time, producing maltose. An example of this reaction is found in germinating seeds, which is why it was named after malt. Unlike sucrose, it is a reducing sugar.

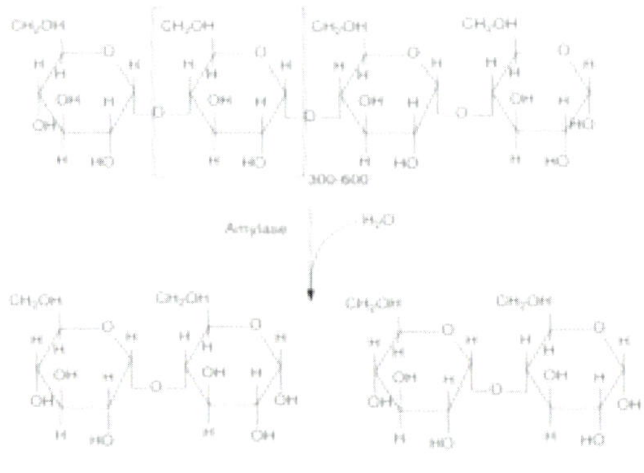

Amylase reaction consisting of hydrolyzing amylose, producing maltose

Structure and Nomenclature

Maltose is a disaccharide: the carbohydrates are generally divided into monosaccharides, oligosaccharides, and polysaccharides depending on the number of sugar subunits. Maltose, with two sugar units, is an oligosaccharide, specifically a disaccharide, because it consists of two glucose molecules. Glucose is a hexose: a monosaccharide containing six carbon atoms. The two glucose units are in the pyranose form and are joined by an O-glycosidic bond, with the first carbon (C_1) of the first glucose linked to the fourth carbon (C_4) of the second glucose, indicated as $(1 \rightarrow 4)$. The link is characterized as α because the glycosidic bond to the anomeric carbon (C_1) is in the opposite plane from the CH_2OH substituent in the same ring (C_6 of the first glucose). If the glycosidic bond to the anomeric carbon (C_1) were in the same plane as the CH_2OH substituent, it would be classified as a $\beta(1 \rightarrow 4)$ bond, and the resulting molecule would be cellobiose. The anomeric carbon (C_1) of the second glucose molecule, which is not involved in a glycosidic bond, could be either an α- or β-anomer depending on the bond direction of the attached hydroxyl group relative to the CH_2OH substituent of the same ring, resulting in either α-maltose or β-maltose.

An isomer of maltose is isomaltose. This is similar to maltose but instead of a bond in the $\alpha(1 \rightarrow 4)$ position, it is in the $\alpha(1 \rightarrow 6)$ position, the same bond that is found at the branch points of glycogen and amylopectin.

Properties

Like glucose, maltose is a reducing sugar, because the ring of one of the two glucose units can open to present a free aldehyde group; the other one cannot because of the nature of the glycosidic bond. Maltose can be broken down to glucose by the maltase enzyme, which catalyses the hydrolysis of the glycosidic bond.

Maltose in aqueous solution exhibits mutarotation, because the α and β isomers that are formed by the different conformations of the anomeric carbon have different specific rotations, and in aqueous solutions, these two forms are in equilibrium.

It has a sweet taste, but is only about 30-60% as sweet as sugar, depending on the concentration. A 10% solution of maltose is 35% as sweet as sucrose.

Oligosaccharide

An oligosaccharide is a saccharide polymer containing a small number (typically three to ten) of monosaccharides (simple sugars). Oligosaccharides can have many functions including cell recognition and cell binding. For example, glycolipids have an important role in the immune response.

They are normally present as glycans: oligosaccharide chains linked to lipids or to compatible amino acid side chains in proteins, by *N*- or *O*-glygosidic bonds. N-linked oligosaccharides are always pentasaccharides attached to asparagine via a beta linkage to the amine nitrogen of the side chain. Alternately, O-linked oligosaccharides are generally attached to threonine or serine on the alcohol group of the side chain. Not all natural oligosaccharides occur as components of glycoproteins or glycolipids. Some, such as the raffinose series, occur as storage or transport carbohydrates in plants. Others, such as maltodextrins or cellodextrins, result from the microbial breakdown of larger polysaccharides such as starch or cellulose.

Glycosylation

In biology, glycosylation is the process by which a carbohydrate is covalently attached to an organic molecule, creating structures such as glycoproteins and glycolipids.

N-linked Oligosaccharides

N-linked glycosylation involves oligosaccharide attachment to asparagine via a beta linkage to the amine nitrogen of the side chain. The process of N-linked glycosylation occurs cotranslationally, or concurrently while the proteins is being translated. Since it is added cotranslationally, it is believed that N-linked glycosylation helps determine the folding of polypeptides due to the hydrophilic nature of sugars. All N-linked Oligosaccharides are pentasaccharides: five monosaccharides long.

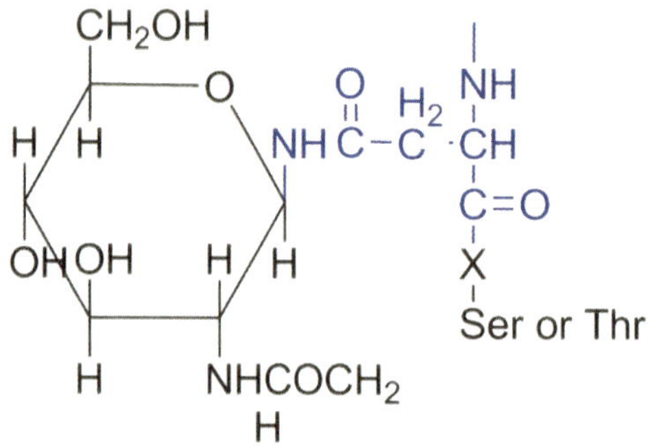

An example of N-linked oligosaccharide, shown here with GlcNAc. X is any amino acid except proline.

In N-glycosylation for eukaryotes, the oligosaccharide substrate is assembled right at the membrane of the endoplasmatic reticulum. For prokaryotes, this process occurs at the plasma membrane. In both cases, the acceptor substrate is an asparagine residue. The asparagine residue linked to an N-linked oligosaccharide usually occurs in the sequence Asn-X-Ser/Thr, where X can be any amino acid except for proline, although it is rare to see Asp, Glu, Leu, or Trp in this position.

O-linked Oligosaccharides

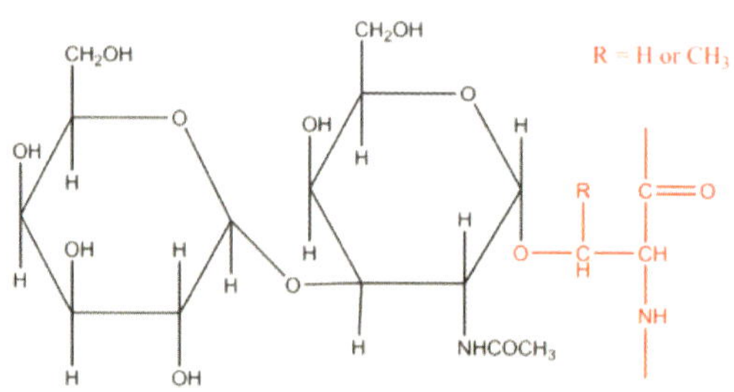

An example of O-Linked oligosaccharide with β-Galactosyl-(1n3)-α-N-acetylgalactosaminyl-Ser/ Thr.

Oligosaccharides that participate in O-linked glycosylation are attached to threonine or serine on the hydroxyl group of the side chain. O-linked glycosylation occurs in the golgi apparatus, where monosaccharide units are added to a complete polypeptide chain. Cell surface proteins and extracellular proteins are O-glycosylated. Glycosylation sites in O-linked oligosaccharides are determined by the secondary and tertiary structures of the polypeptide, which dictate where glycosyltransferases will add sugars.

Glycosylated Biomolecules

Glycoproteins and glycolipids are by definition covalently bonded to carbohydrates. They are very abundant on the surface of the cell, and their interactions contribute to the overall stability of the cell.

Glycoproteins

Glycoproteins have distinct Oligosaccharide structures which have significant effects on many of their properties, affecting critical functions such as antigenicity, solubility, and resistance to proteases. Glycoproteins are relevant as cell-surface receptors, cell-adhesion molecules, immunoglobulins, and tumor antigens.

Glycolipids

Glycolipids are important for cell recognition, and are important for modulating the function of membrane proteins that act as receptors. Glycolipids are lipid molecules bound to oligosaccharides, generally present in the lipid bilayer. Additionally, they can serve as receptors for cellular recognition and cell signaling. The head of the oligosaccharide serves as a binding partner in receptor activity. The binding mechanisms of receptors to the oligosaccharides depends on the composition of the oligosaccharides that are exposed or presented above the surface of the membrane. There is great diversity in the binding mechanisms of glycolipids, which is what makes them such an important target for pathogens as a site for interaction and entrance. For example, the chaperone activity of glycolipids has been studied for its relevance to HIV infection.

Functions

Cell Recognition

All cells are coated in either glycoproteins or glycolipids, both of which help determine cell types. Lectins, or proteins that bind carbohydrates, can recognize specific oligosaccharides and provide useful information for cell recognition based on oligosaccharide binding.

An important example of oligosaccharide cell recognition is the role of glycolipids in determining blood types. The various blood types are distinguished by the glycan modification present on the surface of blood cells. These can be visualized using mass spectrometry. The oligosaccharides found on the A, B, and H antigen occur on the non-reducing ends of the oligosaccharide. The H antigen (which indicates an O blood type) serves as a precursor for the A and B antigen. Therefore, a person with A blood type will have the A antigen and H antigen present on the glycolipids of the red blood cell plasma membrane. A person with B blood type will have the B and H antigen present. A person with AB blood type will have A, B, and H antigens present. And finally, a person with O blood type will only have the H antigen present. This means all blood types have the H antigen, which explains why the O blood type is known as the "universal donor".

Cell Adhesion

Many cells produce specific carbohydrate-binding proteins known as lectins, which mediate cell adhesion with oligosaccharides. Selectins - a family of lectins - mediate certain cell-cell adhesion processes, including those of leukocytes to endothelial cells. In an immune response, endothelial cells can express certain selectins transiently in response to damage or injury to the cells. In response, a reciprocal selectin-oligosaccharide interaction will occur between the two molecules which allows the white blood cell to help eliminate the infection or damage. Protein-Carbohydrate bonding is often mediated by hydrogen bonding and van der Waals forces.

Sources

Oligosaccharides are a component of fibre from plant tissue. FOS and inulin are present in Jerusalem artichoke, burdock, chicory, leeks, onions, and asparagus. Inulin is a significant part of the daily diet of most of the world's population. FOS can also be synthesized by enzymes of the fungus Aspergillus niger acting on sucrose. GOS is naturally found in soybeans and can be synthesized from lactose. FOS, GOS, and inulin are also sold as nutritional supplements.

Trisaccharide

Trisaccharides are oligosaccharides composed of three monosaccharides with two glycosidic bonds connecting them. Similar to the disaccharides, each glycosidic bond can be formed between any hydroxyl group on the component monosaccharides. Even if all three component sugars are the same (e.g., glucose), different bond combinations (regiochemistry) and stereochemistry (alpha- or beta-) result in trisaccharides that are diastereoisomers with different chemical and physical properties.

Raffinose

Raffinose is a trisaccharide composed of galactose, glucose, and fructose. It can be found in beans, cabbage, brussels sprouts, broccoli, asparagus, other vegetables, and whole grains. Raffinose can be hydrolyzed to D-galactose and sucrose by the enzyme α-galactosidase (α-GAL), an enzyme not found in the human digestive tract. α-GAL also hydrolyzes other α-galactosides such as stachyose, verbascose, and galactinol, if present. The enzyme does not cleave β-linked galactose, as in lactose.

The raffinose family of oligosaccharides (RFOs) are alpha-galactosyl derivatives of sucrose, and the most common are the trisaccharide raffinose, the tetrasaccharide stachyose, and the pentasaccharide verbascose. RFOs are almost ubiquitous in the plant kingdom, being found in a large variety of seeds from many different families, and they rank second only to sucrose in abundance as soluble carbohydrates.

Humans and other monogastric animals (pigs and poultry) do not possess the α-GAL enzyme to break down RFOs and these oligosaccharides pass undigested through the stomach and upper intestine. In the lower intestine, they are fermented by gas-producing bacteria that do possess the α-GAL enzyme and make carbon dioxide, methane or hydrogen—leading to the flatulence commonly associated with eating beans and other vegetables. α-GAL is present in digestive aids such as the product Beano.

Procedures concerning cryopreservation have used raffinose to provide hypertonicity for cell desiccation prior to freezing. Either raffinose or sucrose is used as a base substance for sucralose.

Polysaccharide

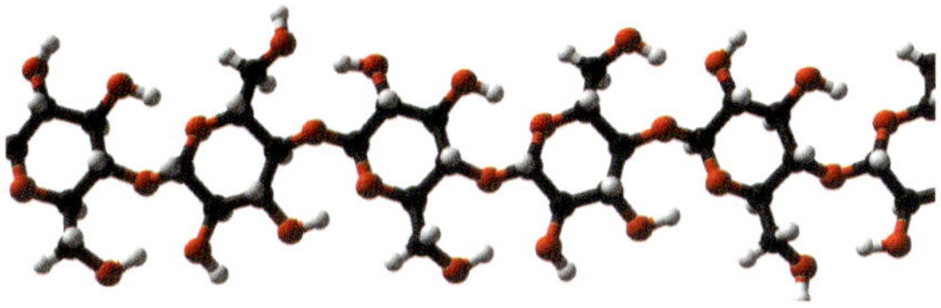

3D structure of cellulose, a beta-glucan polysaccharide.

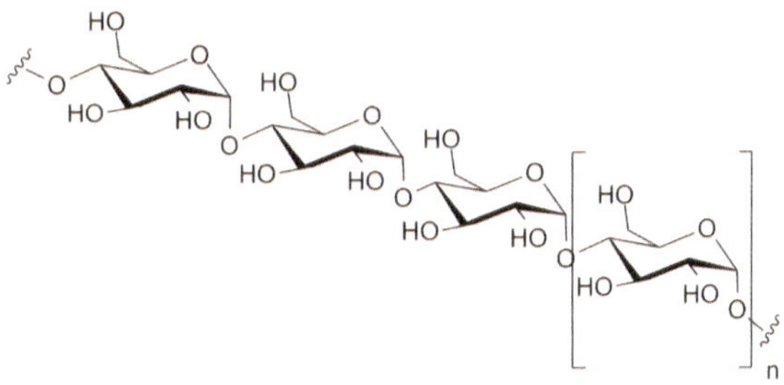

Amylose is a linear polymer of glucose mainly linked with α(1→4) bonds. It can be made of several thousands of glucose units. It is one of the two components of starch, the other being amylopectin.

Polysaccharides are polymeric carbohydrate molecules composed of long chains of monosaccharide units bound together by glycosidic linkages, and on hydrolysis give the constituent monosaccharides or oligosaccharides. They range in structure from linear to highly branched. Examples include storage polysaccharides such as starch and glycogen, and structural polysaccharides such as cellulose and chitin.

Polysaccharides are often quite heterogeneous, containing slight modifications of the repeating unit. Depending on the structure, these macromolecules can have distinct properties from their monosaccharide building blocks. They may be amorphous or even insoluble in water. When all the monosaccharides in a polysaccharide are the same type, the polysaccharide is called a *homopolysaccharide* or *homoglycan*, but when more than one type of monosaccharide is present they are called *heteropolysaccharides* or *heteroglycans*.

Natural saccharides are generally of simple carbohydrates called monosaccharides with general formula $(CH_2O)_n$ where n is three or more. Examples of monosaccharides are glucose, fructose, and glyceraldehyde. Polysaccharides, meanwhile, have a general formula of $C_x(H_2O)_y$ where x is usually a large number between 200 and 2500. When the repeating units in the polymer backbone are six-carbon monosaccharides, as is often the case, the general formula simplifies to $(C_6H_{10}O_5)_n$, where typically $40 \leq n \leq 3000$.

As a rule of thumb, polysaccharides contain more than ten monosaccharide units, whereas oligosaccharides contain three to ten monosaccharide units; but the precise cutoff varies somewhat according to convention. Polysaccharides are an important class of biological polymers. Their function in living organisms is usually either structure- or storage-related. Starch (a polymer of glucose) is used as a storage polysaccharide in plants, being found in the form of both amylose and the branched amylopectin. In animals, the structurally similar glucose polymer is the more densely branched glycogen, sometimes called "animal starch". Glycogen's properties allow it to be metabolized more quickly, which suits the active lives of moving animals.

Cellulose and chitin are examples of structural polysaccharides. Cellulose is used in the cell walls of plants and other organisms, and is said to be the most abundant organic molecule on Earth. It has many uses such as a significant role in the paper and textile industries, and is used as a feedstock for the production of rayon (via the viscose process), cellulose acetate, celluloid, and nitrocellulose. Chitin has a similar structure, but has nitrogen-containing side branches, increasing its strength. It is found in arthropod exoskeletons and in the cell walls of some fungi. It also has multiple uses, including surgical threads. Polysaccharides also include callose or laminarin, chrysolaminarin, xylan, arabinoxylan, mannan, fucoidan and galactomannan.

Function

Structure

Nutrition polysaccharides are common sources of energy. Many organisms can easily break down starches into glucose; however, most organisms cannot metabolize cellulose or other polysaccharides like chitin and arabinoxylans. These carbohydrate types can be metabolized by some bacteria and protists. Ruminants and termites, for example, use microorganisms to process cellulose.

Even though these complex polysaccharides are not very digestible, they provide important dietary elements for humans. Called dietary fiber, these carbohydrates enhance digestion among other benefits. The main action of dietary fiber is to change the nature of the contents of the gastrointestinal tract, and to change how other nutrients and chemicals are absorbed. Soluble fiber binds to bile acids in the small intestine, making them less likely to enter the body; this in turn lowers cholesterol levels in the blood. Soluble fiber also attenuates the absorption of sugar, reduces sugar response after eating, normalizes blood lipid levels and, once fermented in the colon, produces short-chain fatty acids as byproducts with wide-ranging physiological activities (discussion below). Although insoluble fiber is associated with reduced diabetes risk, the mechanism by which this occurs is unknown.

Not yet formally proposed as an essential macronutrient (as of 2005), dietary fiber is nevertheless regarded as important for the diet, with regulatory authorities in many developed countries recommending increases in fiber intake.

Storage Polysaccharides

Starch

Starch is a glucose polymer in which glucopyranose units are bonded by *alpha*-linkages. It is made up of a mixture of amylose (15–20%) and amylopectin (80–85%). Amylose consists of a linear chain of several hundred glucose molecules and Amylopectin is a branched molecule made of several thousand glucose units (every chain of 24–30 glucose units is one unit of Amylopectin). Starches are insoluble in water. They can be digested by breaking the *alpha*-linkages (glycosidic bonds). Both humans and animals have amylases, so they can digest starches. Potato, rice, wheat, and maize are major sources of starch in the human diet. The formations of starches are the ways that plants store glucose.

Glycogen

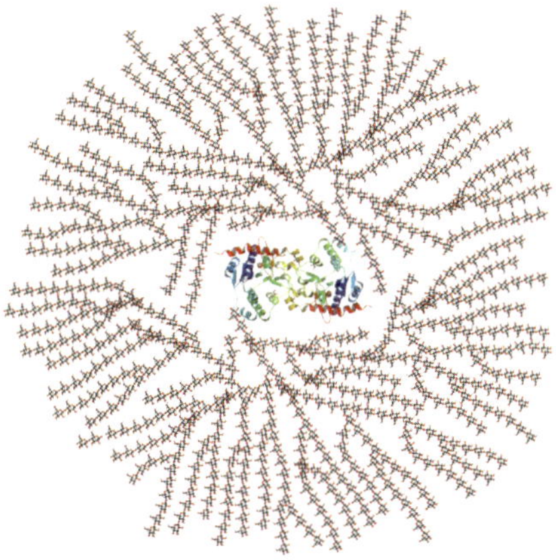

Schematic 2-D cross-sectional view of glycogen. A core protein of glycogenin is surrounded by branches of glucose units. The entire globular granule may contain approximately 30,000 glucose units.

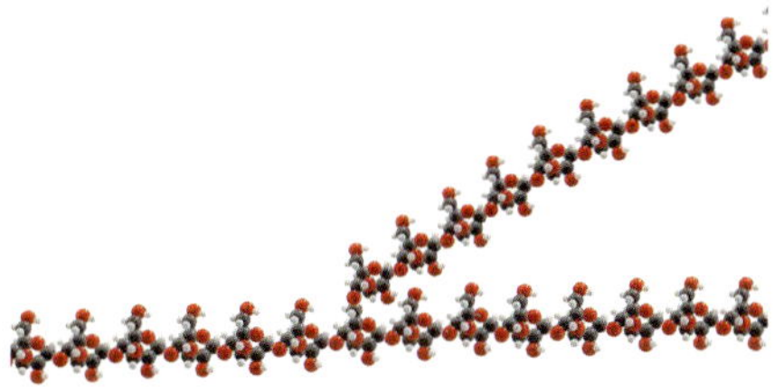

A view of the atomic structure of a single branched strand of glucose units in a glycogen molecule.

Glycogen serves as the secondary long-term energy storage in animal and fungal cells, with the primary energy stores being held in adipose tissue. Glycogen is made primarily by the liver and the muscles, but can also be made by glycogenesis within the brain and stomach. Glycogen is analogous to starch, a glucose polymer in plants, and is sometimes referred to as *animal starch*, having a similar structure to amylopectin but more extensively branched and compact than starch. Glycogen is a polymer of $\alpha(1\rightarrow4)$ glycosidic bonds linked, with $\alpha(1\rightarrow6)$-linked branches. Glycogen is found in the form of granules in the cytosol/cytoplasm in many cell types, and plays an important role in the glucose cycle. Glycogen forms an energy reserve that can be quickly mobilized to meet a sudden need for glucose, but one that is less compact and more immediately available as an energy reserve than triglycerides (lipids).

In the liver hepatocytes, glycogen can compose up to eight percent (100–120 g in an adult) of the fresh weight soon after a meal. Only the glycogen stored in the liver can be made accessible to other organs. In the muscles, glycogen is found in a low concentration of one to two percent of the muscle mass. The amount of glycogen stored in the body—especially within the muscles, liver, and red blood cells—varies with physical activity, basal metabolic rate, and eating habits such as intermittent fasting. Small amounts of glycogen are found in the kidneys, and even smaller amounts in certain glial cells in the brain and white blood cells. The uterus also stores glycogen during pregnancy, to nourish the embryo.

Glycogen is composed of a branched chain of glucose residues. It is stored in liver and muscles.

- It is an energy reserve for animals.
- It is the chief form of carbohydrate stored in animal body.
- It is insoluble in water. It turns brown-red when mixed with iodine.
- It also yields glucose on hydrolysis.

Structural Polysaccharides

Arabinoxylans

Arabinoxylans are found in both the primary and secondary cell walls of plants and are the copolymers of two sugars: arabinose and xylose.

Cellulose

The structural component of plants are formed primarily from cellulose. Wood is largely cellulose and lignin, while paper and cotton are nearly pure cellulose. Cellulose is a polymer made with repeated glucose units bonded together by *beta*-linkages. Humans and many animals lack an enzyme to break the *beta*-linkages, so they do not digest cellulose. Certain animals such as termites can digest cellulose, because bacteria possessing the enzyme are present in their gut. Cellulose is insoluble in water. It does not change color when mixed with iodine. On hydrolysis, it yields glucose. It is the most abundant carbohydrate in nature.

Chitin

Chitin is one of many naturally occurring polymers. It forms a structural component of many animals, such as exoskeletons. Over time it is bio-degradable in the natural environment. Its breakdown may be catalyzed by enzymes called chitinases, secreted by microorganisms such as bacteria and fungi, and produced by some plants. Some of these microorganisms have receptors to simple sugars from the decomposition of chitin. If chitin is detected, they then produce enzymes to digest it by cleaving the glycosidic bonds in order to convert it to simple sugars and ammonia.

Chemically, chitin is closely related to chitosan (a more water-soluble derivative of chitin). It is also closely related to cellulose in that it is a long unbranched chain of glucose derivatives. Both materials contribute structure and strength, protecting the organism.

Pectins

Pectins are a family of complex polysaccharides that contain 1,4-linked α-D-galactosyl uronic acid residues. They are present in most primary cell walls and in the non-woody parts of terrestrial plants.

Acidic Polysaccharides

Acidic polysaccharides are polysaccharides that contain carboxyl groups, phosphate groups and/or sulfuric ester groups.

Starch

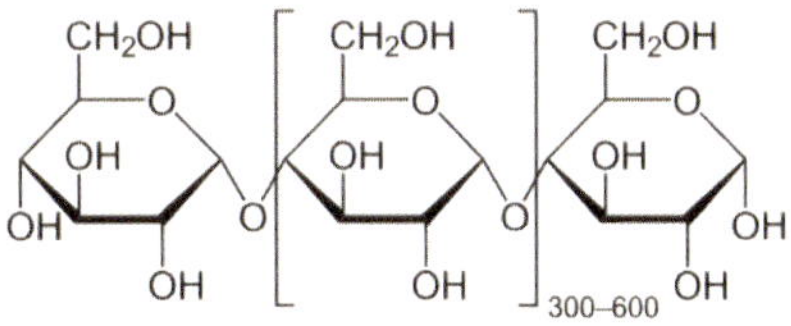

Structure of the amylose molecule

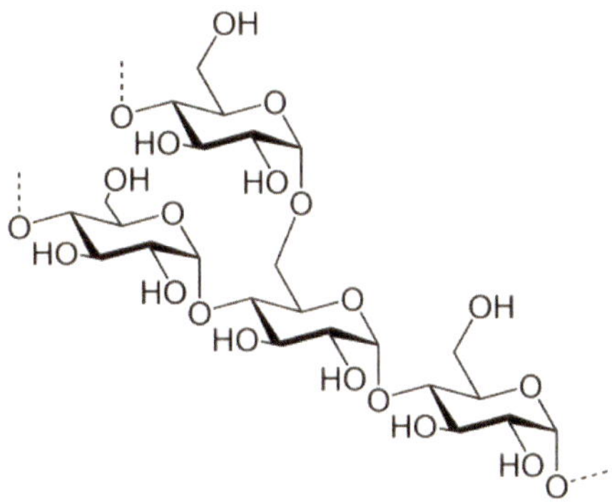

Structure of the amylopectin molecule

Starch or amylum is a polymeric carbohydrate consisting of a large number of glucose units joined by glycosidic bonds. This polysaccharide is produced by most green plants as energy storage. It is the most common carbohydrate in human diets and is contained in large amounts in staple foods like potatoes, wheat, maize (corn), rice, and cassava.

Pure starch is a white, tasteless and odorless powder that is insoluble in cold water or alcohol. It consists of two types of molecules: the linear and helical amylose and the branched amylopectin. Depending on the plant, starch generally contains 20 to 25% amylose and 75 to 80% amylopectin by weight. Glycogen, the glucose store of animals, is a more highly branched version of amylopectin.

In industry, starch is converted into sugars, for example by malting, and fermented to produce ethanol in the manufacture of beer, whisky and biofuel. It is processed to produce many of the sugars used in processed foods. Mixing most starches in warm water produces a paste, such as wheatpaste, which can be used as a thickening, stiffening or gluing agent. The biggest industrial non-food use of starch is as an adhesive in the papermaking process. Starch can be applied to parts of some garments before ironing, to stiffen them.

Properties

Structure

While amylose was thought to be completely unbranched, it is now known that some of its molecules contain a few branch points. Amylose is a much smaller molecule than amylopectin. About one quarter of the mass of starch granules in plants consist of amylose, although there are about 150 times more amylose than amylopectin molecules.

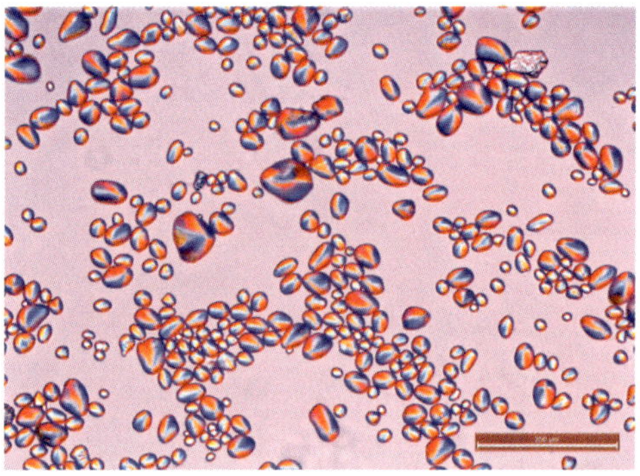

Starch, 100x magnified, under polarized light

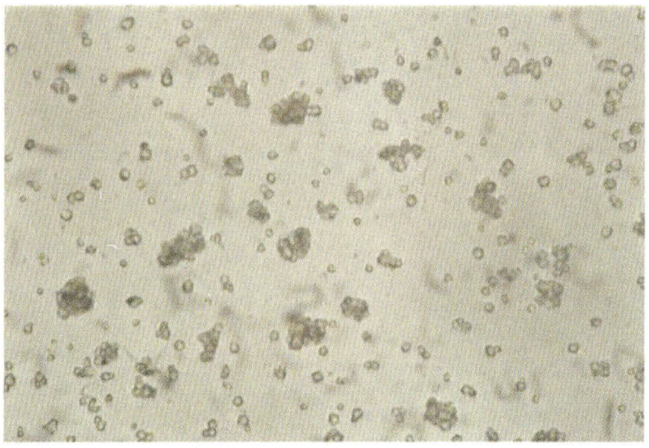

Rice starch seen on light microscope. Characteristic for the rice starch is that starch granules have an angular outline and some of them are attached to each other and form larger granules

Starch molecules arrange themselves in the plant in semi-crystalline granules. Each plant species has a unique starch granular size: rice starch is relatively small (about 2 μm) while potato starches have larger granules (up to 100 μm).

Starch becomes soluble in water when heated. The granules swell and burst, the semi-crystalline structure is lost and the smaller amylose molecules start leaching out of the granule, forming a network that holds water and increasing the mixture's viscosity. This process is called starch gelatinization. During cooking, the starch becomes a paste and increases further in viscosity. During cooling or prolonged storage of the paste, the semi-crystalline structure partially recovers and the starch paste thickens, expelling water. This is mainly caused by retrogradation of the amylose. This process is responsible for the hardening of bread or staling, and for the water layer on top of a starch gel (syneresis).

Some cultivated plant varieties have pure amylopectin starch without amylose, known as *waxy starches*. The most used is waxy maize, others are glutinous rice and waxy potato starch. Waxy starches have less retrogradation, resulting in a more stable paste. High amylose starch, amylomaize, is cultivated for the use of its gel strength and for use as a resistant starch (a starch that resists digestion) in food products.

Synthetic amylose made from cellulose has a well-controlled degree of polymerization. Therefore, it can be used as a potential drug deliver carrier.

Certain starches, when mixed with water, will produce a non-newtonian fluid sometimes nicknamed "oobleck".

Hydrolysis

The enzymes that break down or hydrolyze starch into the constituent sugars are known as amylases.

Alpha-amylases are found in plants and in animals. Human saliva is rich in amylase, and the pancreas also secretes the enzyme. Individuals from populations with a high-starch diet tend to have more amylase genes than those with low-starch diets.

Beta-amylase cuts starch into maltose units. This process is important in the digestion of starch and is also used in brewing, where amylase from the skin of seed grains is responsible for converting starch to maltose (Malting, Mashing).

Dextrinization

If starch is subjected to dry heat, it breaks down to form dextrins, also called "pyrodextrins" in this context. This break down process is known as dextrinization. (Pyro)dextrins are mainly yellow to brown in color and dextrinization is partially responsible for the browning of toasted bread.

Chemical Tests

A triiodide (I_3^-) solution formed by mixing iodine and iodide (usually from potassium iodide) is used to test for starch; a dark blue color indicates the presence of starch. The details of this reaction are not yet fully known, but it is thought that the iodine (I_3^- and I_5^- ions) fit inside the coils of amylose, the charge transfers between the iodine and the starch, and the energy level spacings in the resulting complex correspond to the absorption spectrum in the visible light region. The strength of the resulting blue color depends on the amount of amylose present. Waxy starches with little or no amylose present will color red. Benedict's test and Fehling's test is also done to indicate the presence of starch.

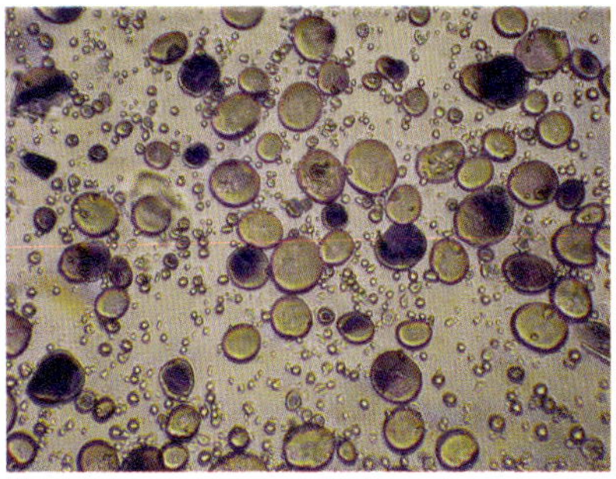

Granules of wheat starch, stained with iodine, photographed through a light microscope

Starch indicator solution consisting of water, starch and iodide is often used in redox titrations: in the presence of an oxidizing agent the solution turns blue, in the presence of reducing agent the blue color disappears because triiodide (I_3^-) ions break up into three iodide ions, disassembling the starch-iodine complex. A 0.3% w/w solution is the standard concentration for a starch indicator. It is made by adding 3 grams of soluble starch to 1 liter of heated water; the solution is cooled before use (starch-iodine complex becomes unstable at temperatures above 35 °C).

Each species of plant has a unique type of starch granules in granular size, shape and crystallization pattern. Under the microscope, starch grains stained with iodine illuminated from behind with polarized light show a distinctive Maltese cross effect (also known as extinction cross and birefringence).

Amylose

Amylose is a polysaccharide made of α-D-glucose units, bonded to each other through α(1→4) glycosidic bonds. It is one of the two components of starch, making up approximately 20-30%. The other component is amylopectin. Because of its tightly packed helical structure, amylose is more resistant to digestion than other starch molecules and is therefore an important form of resistant starch.

Structure

Amylose is made up of α(1→4) bound glucose molecules. The carbon atoms on glucose are numbered, starting at the aldehyde (C=O) carbon, so, in amylose, the 1-carbon on one glucose molecule is linked to the 4-carbon on the next glucose molecule (α(1→4) bonds). The structural formula of amylose is pictured at right. The number of repeated glucose subunits (n) is usually in the range of 300 to 3000, but can be many thousands.

There are three main forms of amylose chains can take. It can exist in a disordered amorphous conformation or two different helical forms. It can bind with itself in a double helix (A or B form), or it can bind with another hydrophobic guest molecule such as iodine, a fatty acid, or an aromatic compound. This is known as the V form and is how amylopectin binds to amylose to form starch. Within this group, there are many different variations. Each is notated with V and then a subscript indicating the number of glucose units per turn. The most common is the V_6 form, which has six glucose units a turn. V_8 and possibly V_7 forms exist as well. These provide an even larger space for the guest molecule to bind.

This linear structure can have some rotation around the phi and psi angles, but for the most part bound glucose ring oxygens lie on one side of the structure. The α(1→4) structure promotes the formation of a helix structure, making it possible for hydrogen bonds to form between the oxygen atoms bound at the 2-carbon of one glucose molecule and the 3-carbon of the next glucose molecule.

Physical Properties

Because the long linear chains of amylose more readily crystallize than amylopectin (which has short, highly branched chains), high-amylose starch is more resistant to digestion. Unlike amylopectin, amylose is not soluble in cold water. It also reduces the crystallinity of amylopectin and how easily water can infiltrate the starch. The higher the amylose content, the less expansion potential and the lower the gel strength for the same starch concentration. This can be countered partially by increasing the granule size.

Fiber X-ray diffraction analysis coupled with computer-based structure refinement has found A-, B-, and C- polymorphs of amylose. Each form corresponds to either the A-, the B-, or the C- starch forms. A- and B- structures have different helical crystal structures and water contents, whereas the C- structure is a mixture of A- and B- unit cells, resulting in an intermediate packing density between the two forms.

Function

Amylose is important in plant energy storage. It is less readily digested than amylopectin; however, because it is more linear than amylopectin, it takes up less space. As a result, it is the preferred starch for storage in plants. It makes up about 30% of the stored starch in plants, though the specific percentage varies by species. The digestive enzyme α-amylase is responsible for the breakdown of the starch molecule into maltotriose and maltose, which can be used as sources of energy.

Amylose is also an important thickener, water binder, emulsion stabilizer, and gelling agent in both industrial and food-based contexts. Loose helical amylose chains have a hydrophobic interior that can bind to hydrophobic molecules such as lipids and aromatic compounds. The one problem with this is that, when it crystallizes or associates, it can lose some stability, often releasing water in the process (syneresis). When amylose concentration is increased, gel stickiness decreases but gel firmness increases. When other things including amylopectin bind to amylose, the viscosity can be affected, but incorporating κ-carrageenan, alginate, xanthan gum, or low-molecular-weight sugars can reduce the loss in stability. The ability to bind water can add substance to food, possibly serving as a fat replacement. For example, amylose is responsible for causing white sauce to thicken, but, upon cooling, some separation between the solid and the water will occur.

In a laboratory setting, it can act as a marker. Iodine molecules fit neatly inside the helical structure of amylose, binding with the starch polymer that absorbs certain known wavelengths of light. Hence, a common test is the iodine test for starch. Mix starch with a small amount of yellow iodine solution. In the presence of amylose, a blue-black color will be observed. The intensity of the color can be tested with a colorimeter, using a red filter to discern the concentration of starch present in the solution. It is also possible to use starch as an indicator in titrations involving iodine reduction. It is also used in amylose magnetic beads and resin to separate maltose-binding protein.

Amylopectin

Amylopectin is a water-soluble polysaccharide and highly branched polymer of glucose found in plants. It is one of the two components of starch, the other being amylose.

Glucose units are linked in a linear way with α(1→4) glycosidic bonds. Branching takes place with α(1→6) bonds occurring every 24 to 30 glucose units, resulting in a soluble molecule that can be quickly degraded as it has many end points onto which enzymes can attach. In contrast, amylose contains very few α(1→6) bonds, or even none at all. This causes amylose to be hydrolyzed more slowly, but have higher density and be insoluble.

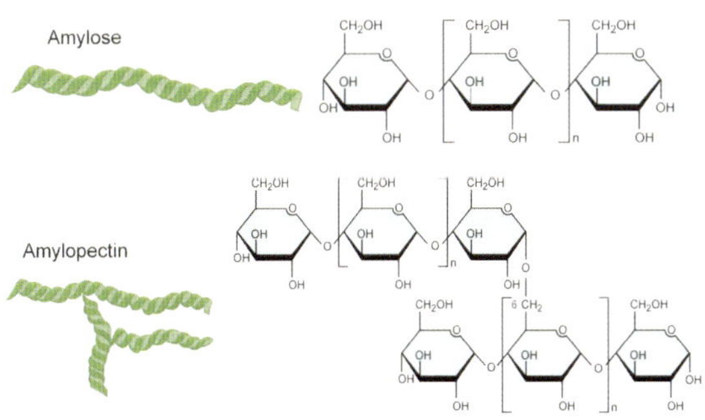

Structure of the amylose and amylopectin molecule

Its counterpart in animals is glycogen, which has the same composition and structure, but with more extensive branching that occurs every eight to 12 glucose units.

Plants store starch within specialized organelles called amyloplasts. When energy is needed for cell work, the plant hydrolyzes the starch, releasing the glucose subunits. Humans and other animals that eat plant foods also use amylase, an enzyme that assists in breaking down amylopectin.

Starch is made of about 70% amylopectin by weight, though it varies depending on the source (higher in medium-grain rice to 100% in glutinous rice, waxy potato starch, and waxy corn, and lower in long-grain rice, amylomaize, and russet potatoes, for example). Amylopectin is highly branched, being formed of 2,000 to 200,000 glucose units. Its inner chains are formed of 20-24 glucose subunits.

Dissolved amylopectin starch has a lower tendency of retrogradation (gelling) during storage and cooling. For this main reason, the waxy starches are used in different applications mainly as a thickening agent or stabilizer.

Maltodextrin

Maltodextrin is a polysaccharide that is used as a food additive. It is produced from starch by partial hydrolysis and is usually found as a white hygroscopic spray-dried powder. Maltodextrin is easily digestible, being absorbed as rapidly as glucose and might be either moderately sweet or almost flavorless (depending on the degree of polymerisation).

It is commonly used for the production of soft drinks and candy. It can also be found as an ingredient in a variety of other processed foods.

Structure

Maltodextrin consists of D-glucose units connected in chains of variable length. The glucose units are primarily linked with $\alpha(1\rightarrow4)$ glycosidic bonds. Maltodextrin is typically composed of a mixture of chains that vary from three to 17 glucose units long.

Maltodextrins are classified by DE (dextrose equivalent) and have a DE between 3 and 20. The higher the DE value, the shorter the glucose chains, the higher the sweetness, the higher the solubility, and the lower heat resistance. Above DE 20, the European Union's CN code calls it glucose syrup; at DE 10 or lower the customs CN code nomenclature classifies maltodextrins as dextrins.

Carbohydrate Metabolism

Carbohydrate metabolism denotes the various biochemical processes responsible for the formation, breakdown, and interconversion of carbohydrates in living organisms.

Carbohydrates are central to many essential metabolic pathways. Plants synthesize carbohydrates from carbon dioxide and water through photosynthesis, allowing them to store energy absorbed from sunlight internally. When animals and fungi consume plants, they use cellular respiration to break down these stored carbohydrates to make energy available to cells. Both animals and plants temporarily store the released energy in the form of high energy molecules, such as ATP, for use in various cellular processes.

Although humans consume a variety of carbohydrates, digestion breaks down complex carbohydrates into a few simple monomers for metabolism: glucose, fructose, and galactose. Glucose constitutes about 80% of the products, and is the primary structure that is distributed to cells in the tissues, where it is broken down or stored as glycogen. In aerobic respiration, the main form of cellular respiration used by humans, glucose and oxygen are metabolized to release energy, with carbon dioxide and water as byproducts. Most of the fructose and galactose travel to the liver, where they can be converted to glucose.

Some simple carbohydrates have their own enzymatic oxidation pathways, as do only a few of the more complex carbohydrates. The disaccharide lactose, for instance, requires the enzyme lactase to be broken into its monosaccharide components, glucose and galactose.

Metabolic Pathways

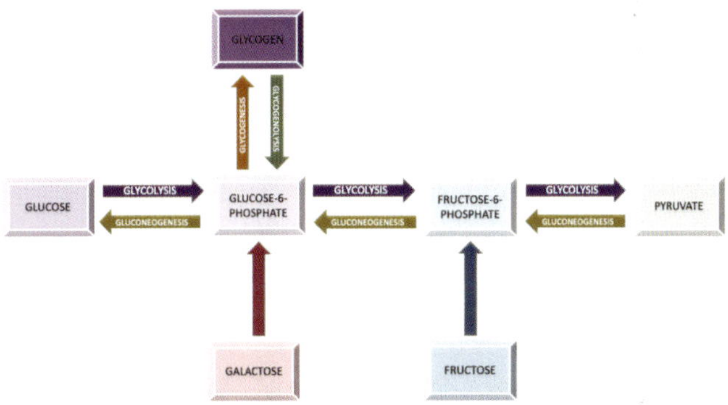

Overview of connections between metabolic processes.

Glycolysis

Glycolysis is the process of breaking down a glucose molecule into two pyruvate molecules, while storing energy released during this process as ATP and NADH. Nearly all organisms that break down glucose utilize glycolysis. Glucose regulation and product use are the primary categories in which these pathways differ between organisms. In some tissues and organisms, glycolysis is the sole method of energy production. This pathway is anaerobic, because it doesn't require oxygen.

Glycolysis consists of ten steps, split into two phases. During the first phase, it requires the breakdown of two ATP molecules. During the second phase, chemical energy from the intermediates is transferred into ATP and NADH. The breakdown of one molecule of glucose results in two molecules of pyruvate, which can be further oxidized to access more energy in later processes.

Glucogenesis

Gluconeogenesis is the reverse process of glycolysis. It involves the conversion of non-carbohydrate molecules into glucose. The non-carbohydrate molecules that are converted in this pathway include pyruvate, lactate, glycerol, alanine, and glutamine. This process occurs when there are lowered amounts of glucose. The production of glucose by this pathway is important to tissues that cannot use any other fuels, such as the brain. The liver is the primary location of gluconeogenesis, but some also occurs in the kidney.

This pathway is regulated by multiple different molecules. Glucagon, adrenocorticotropic hormone, and ATP encourage gluconeogenesis. Gluconeogenesis is inhibited by AMP, ADP, and insulin.

Glycogenolysis

Glycogenolysis refers to the breakdown of glycogen. In the liver, muscles, and the kidney, this process occurs to provide glucose when necessary. A single glucose molecule is cleaved from a branch of glycogen, and is transformed into glucose-1-phosphate during this process. This molecule can then be converted to glucose-6-phosphate, an intermediate in the glycolysis pathway.

Glucose-6-phosphate can then progress through glycolysis. Glycolysis only requires the input of one molecule of ATP when the glucose originates in glycogen. Alternatively, glucose-6-phosphate can be converted back into glucose in the liver and the kidneys, allowing it to raise blood glucose levels if necessary.

Glucagon in the liver stimulates glycogenolysis when the blood glucose is lowered, known as hypoglycemia. The glycogen in the liver can function as a backup source of glucose between meals. Adrenaline stimulates the breakdown of glycogen in the skeletal muscle during exercise. In the muscles, glycogen ensures a rapidly accessible energy source for movement.

Glycogenesis

Glycogenesis refers to the process of synthesizing glycogen. In humans, excess glucose is converted to glycogen via this process. Glycogen is a highly branched structure, consisting of glucose, in the form of glucose-6-phosphate, linked together. The branching of glycogen increases its solubility, and allows for a higher number of glucose molecules to be accessible for breakdown. Glycogenesis occurs primarily in the liver, skeletal muscles, and kidney.

Pentose Phosphate Pathway

The pentose phosphate pathway is an alternative method of oxidizing glucose. It occurs in the liver, adipose tissue, adrenal cortex, testis, milk glands, phagocyte cells, and red blood cells. It produces products that are used in other cell processes, while reducing NADP to NADPH. This pathway is regulated through changes in the activity of glucose-6-phosphate dehydrogenase.

Fructose Metabolism

Fructose must undergo certain extra steps in order to enter the glycolysis pathway. Enzymes located in certain tissues can add a phosphate group to fructose. This phosphorylation creates fructose-6-phosphate, an intermediate in the glycolysis pathway that can be broken down directly in those tissues. This pathway occurs in the muscles, adipose tissue, and kidney. In the liver, enzymes produce fructose-1-phosphate, which enters the glycolysis pathway and is later cleaved into glyceraldehyde and dihydroxyacetone phosphate.

Galactose Metabolism

Lactose, or milk sugar, consists of one molecule of glucose and one molecule of galactose. After separation from glucose, galactose travels to the liver for conversion to glucose. Galactokinase uses one molecule of ATP to phosphorylate galactose. The phosphorylated galactose is then converted to glucose-1-phosphate, and then eventually glucose-6-phosphate, which can be broken down in glycolysis.

Glycolysis

Glycolysis (from *glycose*, an older term for glucose + *-lysis* degradation) is the metabolic pathway that converts glucose $C_6H_{12}O_6$, into pyruvate, $CH_3COCOO^- + H^+$. The free energy released in this process is used to form the high-energy molecules ATP (adenosine triphosphate) and NADH (reduced nicotinamide adenine dinucleotide).

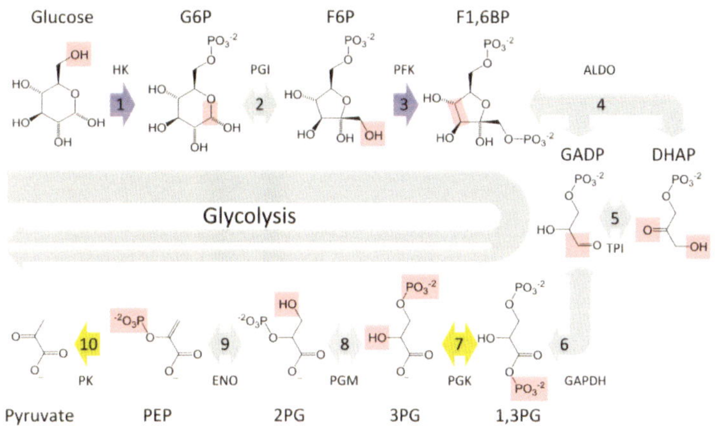

The metabolic pathway of glycolysis converts glucose to pyruvate by via a series of intermediate metabolites. Each chemical modification (red box) is performed by a different enzyme. Steps 1 and 3 consume ATP (blue) and steps 7 and 10 produce ATP (yellow). Since steps 6-10 occur twice per glucose molecule, this leads to a net production of ATP.

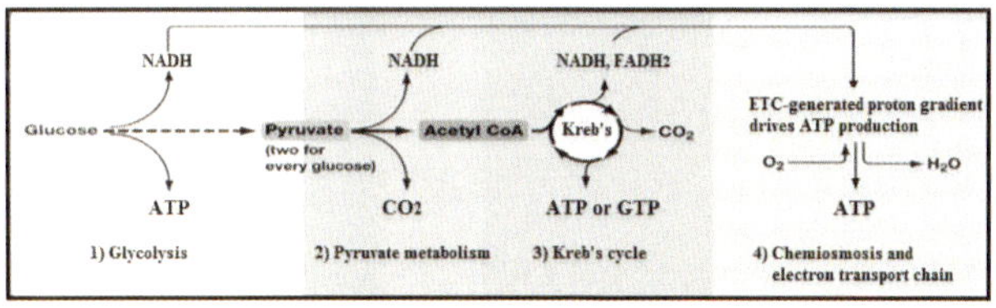

Summary of aerobic respiration

Glycolysis is a determined sequence of ten enzyme-catalyzed reactions. The intermediates provide entry points to glycolysis. For example, most monosaccharides, such as fructose and galactose, can be converted to one of these intermediates. The intermediates may also be directly useful. For example, the intermediate dihydroxyacetone phosphate (DHAP) is a source of the glycerol that combines with fatty acids to form fat.

Glycolysis is an oxygen independent metabolic pathway, meaning that it does not use molecular oxygen (i.e. atmospheric oxygen) for any of its reactions. However the products of glycolysis (pyruvate and NADH + H^+) are sometimes metabolized using atmospheric oxygen. When molecular oxygen is used for the metabolism of the products of glycolysis the process is usually referred to as aerobic, whereas if no oxygen is used the process is said to be anaerobic. Thus, glycolysis occurs, with variations, in nearly all organisms, both aerobic and anaerobic. The wide occurrence of glycolysis indicates that it is one of the most ancient metabolic pathways. Indeed, the reactions that constitute glycolysis and its parallel pathway, the pentose phosphate pathway, occur metal-catalyzed under the oxygen-free conditions of the Archean oceans, also in the absence of enzymes. Glycolysis could thus have originated from chemical constraints of the prebiotic world.

Glycolysis occurs in most organisms in the cytosol of the cell. The most common type of glycolysis is the *Embden–Meyerhof–Parnas (EMP pathway)*, which was discovered by Gustav Embden, Otto Meyerhof, and Jakub Karol Parnas. Glycolysis also refers to other pathways, such as the *Entner–Doudoroff pathway* and various heterofermentative and homofermentative pathways. However, the discussion here will be limited to the Embden–Meyerhof–Parnas pathway.

The entire glycolysis pathway can be separated into two phases:

1. The Preparatory/Investment Phase – wherein ATP is consumed.
2. The Pay Off Phase – wherein ATP is produced.

Overview

The overall reaction of glycolysis is:

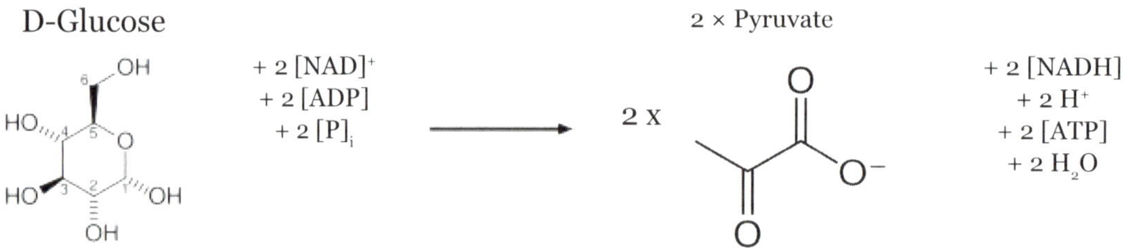

The use of symbols in this equation makes it appear unbalanced with respect to oxygen atoms, hydrogen atoms, and charges. Atom balance is maintained by the two phosphate (P_i) groups:

- Each exists in the form of a hydrogen phosphate anion (HPO_4^{2-}), dissociating to contribute 2 H^+ overall
- Each liberates an oxygen atom when it binds to an ADP (adenosine diphosphate) molecule, contributing 2 O overall

Charges are balanced by the difference between ADP and ATP. In the cellular environment, all three hydroxyl groups of ADP dissociate into $-O^-$ and H^+, giving ADP^{3-}, and this ion tends to exist in an ionic bond with Mg^{2+}, giving $ADPMg^-$. ATP behaves identically except that it has four hydroxyl groups, giving $ATPMg^{2-}$. When these differences along with the true charges on the two phosphate groups are considered together, the net charges of −4 on each side are balanced.

For simple fermentations, the metabolism of one molecule of glucose to two molecules of pyruvate has a net yield of two molecules of ATP. Most cells will then carry out further reactions to 'repay' the used NAD^+ and produce a final product of ethanol or lactic acid. Many bacteria use inorganic compounds as hydrogen acceptors to regenerate the NAD^+.

Cells performing aerobic respiration synthesize much more ATP, but not as part of glycolysis. These further aerobic reactions use pyruvate and NADH + H^+ from glycolysis. Eukaryotic aerobic respiration produces approximately 34 additional molecules of ATP for each glucose molecule, however most of these are produced by a vastly different mechanism to the substrate-level phosphorylation in glycolysis.

The lower-energy production, per glucose, of anaerobic respiration relative to aerobic respiration, results in greater flux through the pathway under hypoxic (low-oxygen) conditions, unless alternative sources of anaerobically oxidizable substrates, such as fatty acids, are found.

Metabolism of common monosaccharides, including glycolysis, gluconeogenesis, glycogenesis and glycogenolysis

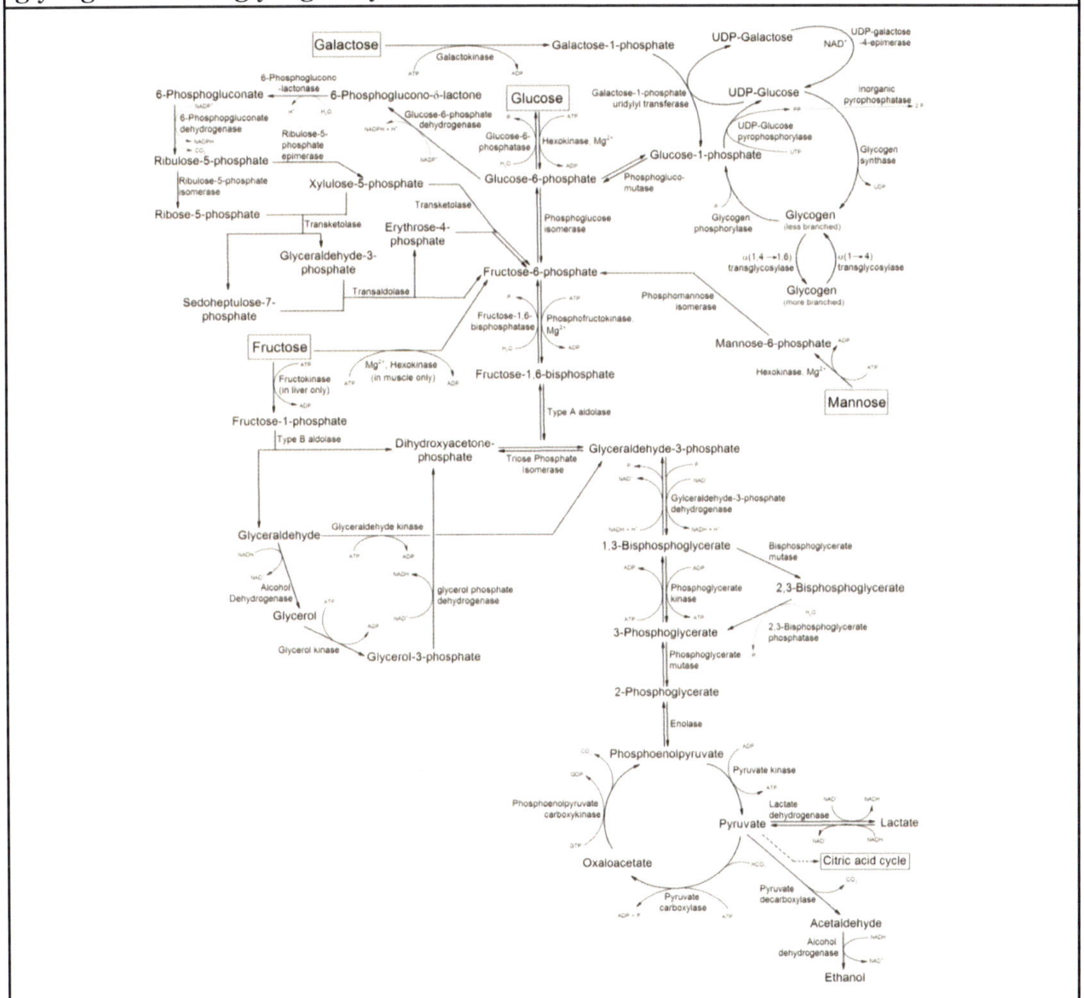

Post-glycolysis Processes

The overall process of glycolysis is:

$$\text{Glucose} + 2\ NAD^+ + 2\ ADP + 2\ P_i \rightarrow 2\ \text{Pyruvate} + 2\ NADH + 2\ H^+ + 2\ ATP + 2\ H_2O$$

If glycolysis were to continue indefinitely, all of the NAD^+ would be used up, and glycolysis would stop. To allow glycolysis to continue, organisms must be able to oxidize NADH back to NAD^+. How this is performed depends on which external electron acceptor is available.

Anoxic Regeneration of NAD+

One method of doing this is to simply have the pyruvate do the oxidation; in this process, pyruvate is converted to lactate (the conjugate base of lactic acid) in a process called lactic acid fermentation:

$$\text{Pyruvate} + \text{NADH} + H^+ \rightarrow \text{Lactate} + \text{NAD}^+$$

This process occurs in the bacteria involved in making yogurt (the lactic acid causes the milk to curdle). This process also occurs in animals under hypoxic (or partially anaerobic) conditions, found, for example, in overworked muscles that are starved of oxygen. In many tissues, this is a cellular last resort for energy; most animal tissue cannot tolerate anaerobic conditions for an extended period of time.

Some organisms, such as yeast, convert NADH back to NAD^+ in a process called ethanol fermentation. In this process, the pyruvate is converted first to acetaldehyde and carbon dioxide, and then to ethanol.

Lactic acid fermentation and ethanol fermentation can occur in the absence of oxygen. This anaerobic fermentation allows many single-cell organisms to use glycolysis as their only energy source.

Anoxic regeneration of NAD^+ is only an effective means of energy production during short, intense exercise in vertebrates, for a period ranging from 10 seconds to 2 minutes during a maximal effort in humans. (At lower exercise intensities it can sustain muscle activity in diving animals, such as seals, whales and other aquatic vertebrates, for very much longer periods of time.) Under these conditions NAD^+ is replenished by NADH donating its electrons to pyruvate to form lactate. This produces 2 ATP molecules per glucose molecule, or about 5% of glucose's energy potential (38 ATP molecules in bacteria). But the speed at which ATP is produced in this manner is about 100 times that of oxidative phosphorylation. The pH in the cytoplasm quickly drops when hydrogen ions accumulate in the muscle, eventually inhibiting the enzymes involved in glycolysis.

The burning sensation in muscles during hard exercise can be attributed to the release of hydrogen ions during the shift to glucose fermentation from glucose oxidation to carbon dioxide and water, when aerobic metabolism can no longer keep pace with the energy demands of the muscles. These hydrogen ions form a part of lactic acid. The body falls back on this less efficient but faster method of producing ATP under low oxygen conditions. This is thought to have been the primary means of energy production in earlier organisms before oxygen reached high concentrations in the atmosphere between 2000 and 2500 million years ago, and thus would represent a more ancient form of energy production than the aerobic replenishment of NAD^+ in cells.

The liver in mammals gets rid of this excess lactate by transforming it back into pyruvate under aerobic conditions.

Fermention of pyruvate to lactate is sometimes also called "anaerobic glycolysis", however, glycolysis ends with the production of pyruvate regardless of the presence or absence of oxygen.

In the above two examples of fermentation, NADH is oxidized by transferring two electrons to pyruvate. However, anaerobic bacteria use a wide variety of compounds as the terminal electron acceptors in cellular respiration: nitrogenous compounds, such as nitrates and nitrites; sulfur compounds, such as sulfates, sulfites, sulfur dioxide, and elemental sulfur; carbon dioxide; iron compounds; manganese compounds; cobalt compounds; and uranium compounds.

Aerobic Regeneration of NAD+, and Disposal of Pyruvate

In aerobic organisms, a complex mechanism has been developed to use the oxygen in air as the final electron acceptor.

- Firstly, the NADH + H^+ generated by glycolysis has to be transferred to the mitochondrion to be oxidized, and thus to regenerate the NAD^+ necessary for glycolysis to continue. However the inner mitochondrial membrane is impermeable to NADH and NAD^+. Use is therefore made of two "shuttles" to transport the electrons from NADH across the mitochondrial membrane. They are the malate-aspartate shuttle and the glycerol phosphate shuttle. In the former the electrons from NADH are transferred to cytosolic oxaloacetate to form malate. The malate then traverses the inner mitochondrial membrane into the mitochondrial matrix, where it is reoxidized by NAD^+ forming intra-mitochondrial oxaloacetate and NADH. The oxaloacetate is then re-cycled to the cytosol via its conversion to aspartate which is readily transported out of the mitochondrion. In the glycerol phosphate shuttle electrons from cytosolic NADH are transferred to dihydroxyacetone to form glycerol-3-phosphate which readily traverses the outer mitochondrial membrane. Glycerol-3-phosphate is then reoxidized to dihydroxyacetone, donating its electrons to FAD instead of NAD^+. This reaction takes place on the inner mitochondrial membrane, allowing $FADH_2$ to donate its electrons directly to coenzyme Q (ubiquinone) which is part of the electron transport chain which ultimately transfers electrons to molecular oxygen (O_2), with the formation of water, and the release of energy eventually captured in the form of ATP.
- The glycolytic end-product, pyruvate (plus NAD^+) is converted to acetyl-CoA, CO_2 and NADH + H^+ within the mitochondria in a process called pyruvate decarboxylation.
- The resulting acetyl-CoA enters the citric acid cycle (or Krebs Cycle), where the acetyl group of the acetyl-CoA is converted into carbon dioxide by two decarboxylation reactions with the formation of yet more intra-mitochondrial NADH + H^+.
- The intra-mitochondrial NADH + H^+ is oxidized to NAD^+ by the electron transport chain, using oxygen as the final electron acceptor to form water. The energy released during this process is used to create a hydrogen ion (or proton) gradient across the inner membrane of the mitochondrion.
- Finally, the proton gradient is used to produce about 2.5 ATP for every NADH + H^+ oxidized in a process called oxidative phosphorylation.

Conversion of Carbohydrates into Fatty Acids and Cholesterol

The pyruvate produced by glycolysis is an important intermediary in the conversion of carbohydrates into fatty acids and cholesterol. This occurs via the conversion of pyruvate into acetyl-CoA in the mitochondrion. However, this acetyl CoA needs to be transported into cytosol where the synthesis of fatty acids and cholesterol occurs. This cannot occur directly. To obtain cytosolic acetyl-CoA, citrate (produced by the condensation of acetyl CoA with oxaloacetate) is removed from the citric acid cycle and carried across the inner mitochondrial membrane into the cytosol. There it is cleaved by ATP citrate lyase into acetyl-CoA and oxaloacetate. The oxaloacetate is returned to mitochondrion as malate (and then back into oxaloacetate to transfer more acetyl-CoA out of the mitochondrion). The cytosolic acetyl-CoA can be carboxylated by acetyl-CoA carboxylase into malonyl CoA, the first committed step in the synthesis of fatty acids, or it can be combined with acetoacetyl-CoA to form 3-hydroxy-3-methylglutaryl-CoA (HMG-CoA) which is the rate limiting step controlling the synthesis of cholesterol. Cholesterol can be used as is, as a structural component of cellular membranes, or it can be used to synthesize the steroid hormones, bile salts, and vitamin D.

Conversion of Pyruvate into Oxaloacetate for the Citric Acid Cycle

Pyruvate molecules produced by glycolysis are actively transported across the inner mitochondrial membrane, and into the matrix where they can either be oxidized and combined with coenzyme A to form CO_2, acetyl-CoA, and NADH, or they can be carboxylated (by pyruvate carboxylase) to form oxaloacetate. This latter reaction "fills up" the amount of oxaloacetate in the citric acid cycle, and is therefore an anaplerotic reaction, increasing the cycle's capacity to metabolize acetyl-CoA when the tissue's energy needs (e.g. in heart and skeletal muscle) are suddenly increased by activity. In the citric acid cycle all the intermediates (e.g. citrate, iso-citrate, alpha-ketoglutarate, succinate, fumarate, malate and oxaloacetate) are regenerated during each turn of the cycle. Adding more of any of these intermediates to the mitochondrion therefore means that that additional amount is retained within the cycle, increasing all the other intermediates as one is converted into the other. Hence the addition of oxaloacetate greatly increases the amounts of all the citric acid intermediates, thereby increasing the cycle's capacity to metabolize acetyl CoA, converting its acetate component into CO_2 and water, with the release of enough energy to form 11 ATP and 1 GTP molecule for each additional molecule of acetyl CoA that combines with oxaloacetate in the cycle.

To cataplerotically remove oxaloacetate from the citric cycle, malate can be transported from the mitochondrion into the cytoplasm, decreasing the amount of oxaloacetate that can be regenerated. Furthermore, citric acid intermediates are constantly used to form a variety of substances such as the purines, pyrimidines and porphyrins.

Gluconeogenesis

Gluconeogenesis (GNG) is a metabolic pathway that results in the generation of glucose from certain non-carbohydrate carbon substrates. From breakdown of proteins, these substrates include glucogenic amino acids (although not ketogenic amino acids); from breakdown of lipids (such as triglycerides), they include glycerol (although not fatty acids); and from other steps in metabolism they include pyruvate and lactate.

Gluconeogenesis is one of several main mechanisms used by humans and many other animals to maintain blood glucose levels, avoiding low levels (hypoglycemia). Other means include the degradation of glycogen (glycogenolysis) and fatty acid catabolism.

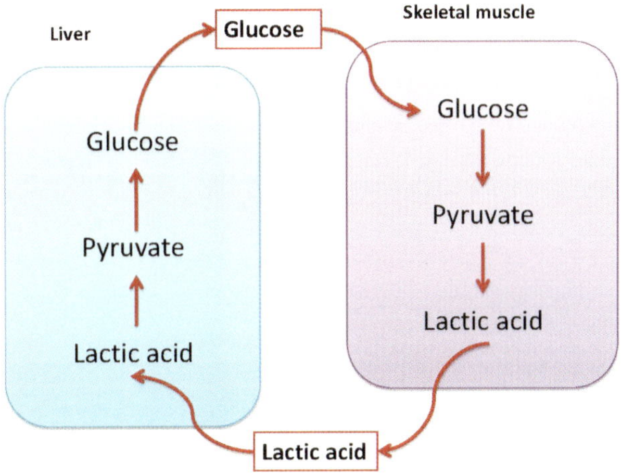

Simplified Gluconeogenesis Pathway

Gluconeogenesis is a ubiquitous process, present in plants, animals, fungi, bacteria, and other microorganisms. In vertebrates, gluconeogenesis takes place mainly in the liver and, to a lesser extent, in the cortex of the kidneys. In ruminants, this tends to be a continuous process. In many other animals, the process occurs during periods of fasting, starvation, low-carbohydrate diets, or intense exercise. The process is highly endergonic until it is coupled to the hydrolysis of ATP or GTP, effectively making the process exergonic. For example, the pathway leading from pyruvate to glucose-6-phosphate requires 4 molecules of ATP and 2 molecules of GTP to proceed spontaneously. Gluconeogenesis is often associated with ketosis. Gluconeogenesis is also a target of therapy for type 2 diabetes, such as the antidiabetic drug, metformin, which inhibits glucose formation and stimulates glucose uptake by cells. In ruminants, because dietary carbohydrates tend to be metabolized by rumen organisms, gluconeogenesis occurs regardless of fasting, low-carbohydrate diets, exercise, etc.

Precursors

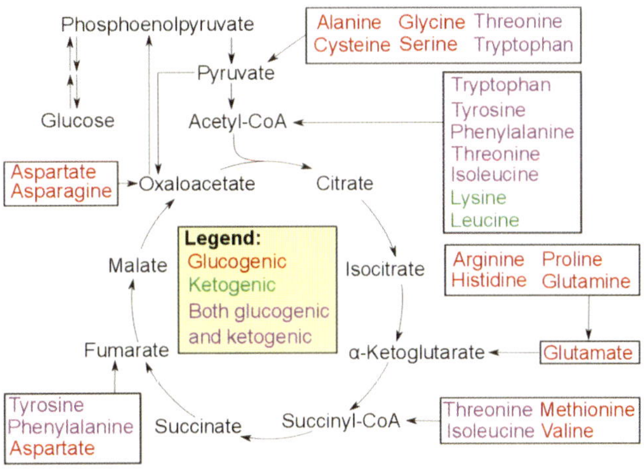

Catabolism of proteinogenic amino acids. Amino acids are classified according to the abilities of their products to enter gluconeogenesis:

- Glucogenic amino acids have this ability.
- Ketogenic amino acids do not. These products may still be used for ketogenesis or lipid synthesis.
- Some amino acids are catabolized into both glucogenic and ketogenic products.

In humans the main gluconeogenic precursors are lactate, glycerol (which is a part of the triacylglycerol molecule), alanine and glutamine. Altogether, they account for over 90% of the overall gluconeogenesis. Other glucogenic amino acids as well as all citric acid cycle intermediates, the latter through conversion to oxaloacetate, can also function as substrates for gluconeogenesis. In ruminants, propionate is the principal gluconeogenic substrate. Generally, consumption of gluconeogenic substrates in food does not result in increased gluconeogenesis.

Lactate is transported back to the liver where it is converted into pyruvate by the Cori cycle using the enzyme lactate dehydrogenase. Pyruvate, the first designated substrate of the gluconeogenic pathway, can then be used to generate glucose. Transamination or deamination of amino acids facilitates entering of their carbon skeleton into the cycle directly (as pyruvate or oxaloacetate), or indirectly via the citric acid cycle. The contribution of Cori cycle lactate to overall glucose production increases with fasting duration. Specifically, after 12, 20, and 40 hours of fasting by human volunteers, the contribution of Cori cycle lactate to gluconeogenesis was 41%, 71%, and 92%, respectively.

Whether even-chain fatty acids can be converted into glucose in animals has been a longstanding question in biochemistry. It is known that odd-chain fatty acids can be oxidized to yield propionyl-CoA, a precursor for succinyl-CoA, which can be converted to pyruvate and enter into gluconeogenesis. In plants, specifically seedlings, the glyoxylate cycle can be used to convert fatty acids (acetate) into the primary carbon source of the organism. The glyoxylate cycle produces four-carbon dicarboxylic acids that can enter gluconeogenesis.

In 1995, researchers identified the glyoxylate cycle in nematodes. In addition, the glyoxylate enzymes malate synthase and isocitrate lyase have been found in animal tissues. Genes coding for malate synthase have been identified in other metazoans including arthropods, echinoderms, and even some vertebrates. Mammals found to possess these genes include monotremes (platypus) and marsupials (opossum) but not placental mammals. Genes for isocitrate lyase are found only in nematodes, in which, it is apparent, they originated in horizontal gene transfer from bacteria.

The existence of glyoxylate cycles in humans has not been established, and it is widely held that fatty acids cannot be converted to glucose in humans directly. However, carbon-14 has been shown to end up in glucose when it is supplied in fatty acids. Despite these findings, it is considered unlikely that the 2-carbon acetyl-CoA derived from the oxidation of fatty acids would produce a net yield of glucose via the citric acid cycle – however, acetyl-CoA can be converted into pyruvate and lactate through the ketogenic pathway. Put simply, acetic acid (in the form of acetyl-CoA) is used to partially produce glucose; acetyl groups can only form part of the glucose molecules (not the 5th carbon atom) and require extra substrates (such as pyruvate) in order to form the rest of the glucose molecule. But a roundabout pathway *does* lead from acetyl-coA to pyruvate, via acetoacetate, acetone, hydroxyacetone (acetol) and then either propylene glycol or methylglyoxal.

Location

In mammals, gluconeogenesis has been believed to be restricted to the liver, the kidney, the intestine, and muscle, but recent evidence indicates gluconeogenesis occurring in astrocytes of the brain. These organs use somewhat different gluconeogenic precursors. The liver preferentially uses lactate, alanine and glycerol (especially alanine) while the kidney preferentially uses lactate, glutamine and glycerol (especially glutamine). Lactate from the Cori cycle, is quantitatively the largest source of substrate for gluconeogenesis, especially for the kidney. The liver uses both glycogenolysis and gluconeogenesis to produce glucose, whereas the kidney only uses gluconeogenesis. After a meal, the liver shifts to glycogen synthesis, whereas the kidney increases gluconeogenesis. The intestine uses mostly glutamine and glycerol.

Propionate is the principal substrate for gluconeogenesis in the ruminant liver, and the ruminant liver may make increased use of gluconeogenic amino acids, e.g. alanine, when glucose demand is increased. The capacity of liver cells to use lactate for gluconeogenesis declines from the preruminant stage to the ruminant stage in calves and lambs. In sheep kidney tissue, very high rates of gluconeogenesis from propionate have been observed.

In all species, the formation of oxaloacetate from pyruvate and TCA cycle intermediates is restricted to the mitochondrion, and the enzymes that convert Phosphoenolpyruvic acid (PEP) to glucose are found in the cytosol. The location of the enzyme that links these two parts of gluconeogenesis by converting oxaloacetate to PEP – PEP carboxykinase (PEPCK) – is variable by species: it can be found entirely within the mitochondria, entirely within the cytosol, or dispersed evenly between the two, as it is in humans. Transport of PEP across the mitochondrial membrane is accomplished by dedicated transport proteins; however no such proteins exist for oxaloacetate. Therefore, in species that lack intra-mitochondrial PEPCK, oxaloacetate must be converted into malate or aspartate, exported from the mitochondrion, and converted back into oxaloacetate in order to allow gluconeogenesis to continue.

Pathway

Gluconeogenesis is a pathway consisting of a series of eleven enzyme-catalyzed reactions. The pathway will begin in either the liver or kidney, in the mitochondria or cytoplasm of those cells, this being dependent on the substrate being used. Many of the reactions are the reverse of steps found in glycolysis.

- Gluconeogenesis begins in the mitochondria with the formation of oxaloacetate by the carboxylation of pyruvate. This reaction also requires one molecule of ATP, and is catalyzed by pyruvate carboxylase. This enzyme is stimulated by high levels of acetyl-CoA (produced in β-oxidation in the liver) and inhibited by high levels of ADP and glucose.
- Oxaloacetate is reduced to malate using NADH, a step required for its transportation out of the mitochondria.
- Malate is oxidized to oxaloacetate using NAD^+ in the cytosol, where the remaining steps of gluconeogenesis take place.
- Oxaloacetate is decarboxylated and then phosphorylated to form phosphoenolpyruvate using the enzyme PEPCK. A molecule of GTP is hydrolyzed to GDP during this reaction.
- The next steps in the reaction are the same as reversed glycolysis. However, fructose 1,6-bisphosphatase converts fructose 1,6-bisphosphate to fructose 6-phosphate, using

one water molecule and releasing one phosphate (in glycolysis, phosphofructokinase 1 converts F6P and ATP to F1,6BP and ADP). This is also the rate-limiting step of gluconeogenesis.

- Glucose-6-phosphate is formed from fructose 6-phosphate by phosphoglucoisomerase (the reverse of step 2 in glycolysis). Glucose-6-phosphate can be used in other metabolic pathways or dephosphorylated to free glucose. Whereas free glucose can easily diffuse in and out of the cell, the phosphorylated form (glucose-6-phosphate) is locked in the cell, a mechanism by which intracellular glucose levels are controlled by cells.
- The final reaction of gluconeogenesis, the formation of glucose, occurs in the lumen of the endoplasmic reticulum, where glucose-6-phosphate is hydrolyzed by glucose-6-phosphatase to produce glucose and release an inorganic phosphate. Like two steps prior, this step is not a simple reversal of glycolysis, in which hexokinase catalyzes the conversion of glucose and ATP into G6P and ADP. Glucose is shuttled into the cytoplasm by glucose transporters located in the endoplasmic reticulum's membrane.

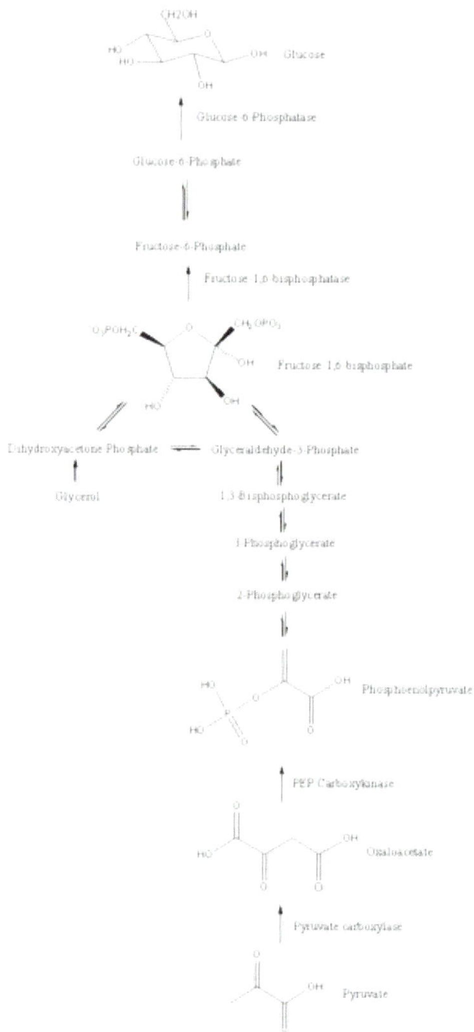

Gluconeogenesis pathway with key molecules and enzymes. Many steps are the opposite of those found in the glycolysis.

Metabolism of common monosaccharides, including glycolysis, gluconeogenesis, glycogenesis and glycogenolysis

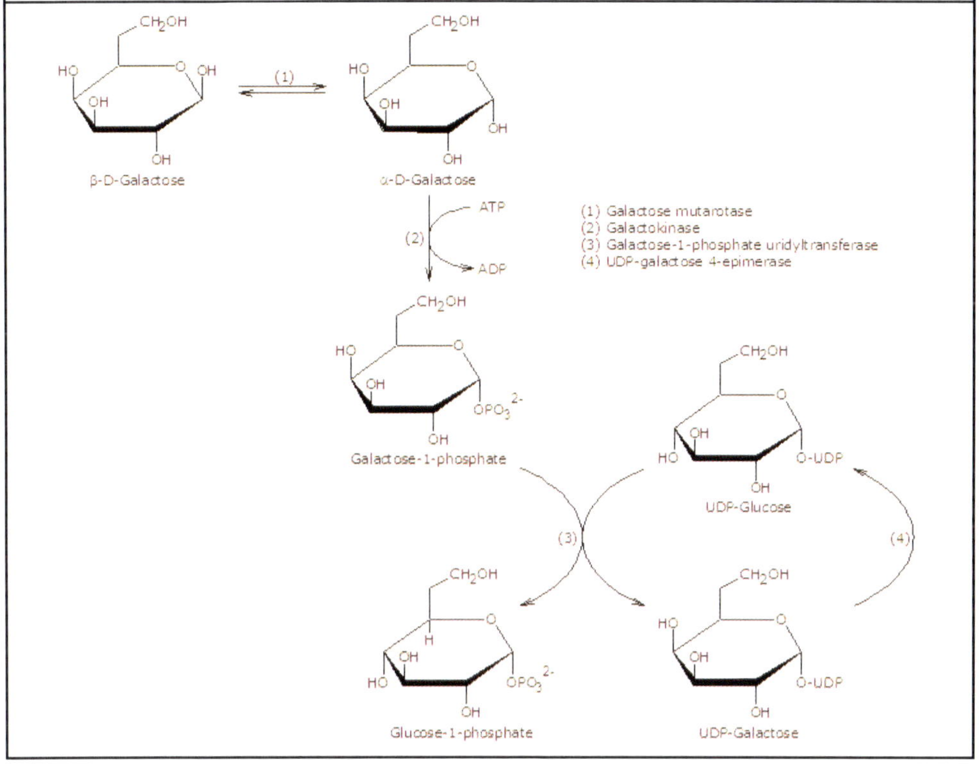

Regulation

While most steps in gluconeogenesis are the reverse of those found in glycolysis, three regulated and strongly endergonic reactions are replaced with more kinetically favorable reactions. Hexokinase/glucokinase, phosphofructokinase, and pyruvate kinase enzymes of glycolysis are replaced with glucose-6-phosphatase, fructose-1,6-bisphosphatase, and PEP carboxykinase/ pyruvate carboxylase. These enzymes are typically regulated by similar molecules, but with opposite results. For example, acetyl CoA and citrate activate gluconeogenesis enzymes (pyruvate carboxylase and fructose-1,6-bisphosphatase, respectively), while at the same time inhibiting the glycolytic enzyme pyruvate kinase. This system of reciprocal control allow glycolysis and gluconeogenesis to inhibit each other and prevents a futile cycle of synthesizing glucose to only break it down.

The majority of the enzymes responsible for gluconeogenesis are found in the cytosol; the exceptions are mitochondrial pyruvate carboxylase and, in animals, phosphoenolpyruvate carboxykinase. The latter exists as an isozyme located in both the mitochondrion and the cytosol. The rate of gluconeogenesis is ultimately controlled by the action of a key enzyme, fructose-1,6-bisphosphatase, which is also regulated through signal transduction by cAMP and its phosphorylation.

Global control of gluconeogenesis is mediated by glucagon (*released when blood glucose is low*); it triggers phosphorylation of enzymes and regulatory proteins by Protein Kinase A (a cyclic AMP regulated kinase) resulting in inhibition of glycolysis and stimulation of gluconeogenesis. Recent studies have shown that the absence of hepatic glucose production has no major effect on the control of fasting plasma glucose concentration. Compensatory induction of gluconeogenesis occurs in the kidneys and intestine, driven by glucagon, glucocorticoids, and acidosis.

References

- Maton, Anthea; Jean Hopkins; Charles William McLaughlin; Susan Johnson; Maryanna Quon Warner; David LaHart; Jill D. Wright (1993). Human Biology and Health. Englewood Cliffs, New Jersey: Prentice Hall. pp. 52–59. ISBN 0-13-981176-1
- "Heptanoate as a neural fuel: energetic and neurotransmitter precursors in normal and glucose transporter I-deficient (G1D) brain". Journal of Cerebral Blood Flow. 33: 175–82. doi:10.1038/jcbfm.2012.151
- Malik, Vasanti S.; Hu, Frank B. (2015). "Fructose and Cardiometabolic Health". Journal of the American College of Cardiology. 66 (14): 1615–1624. doi:10.1016/j.jacc.2015.08.025. ISSN 0735-1097. PMC 4592517 . PMID 26429086
- Campbell, Neil A.; Brad Williamson; Robin J. Heyden (2006). Biology: Exploring Life. Boston, Massachusetts: Pearson Prentice Hall. ISBN 0-13-250882-6
- Flitsch, Sabine L.; Ulijn, Rein V (2003). "Sugars tied to the spot". Nature. 421(6920): 219–20. Bibcode:-2003Natur.421..219F. doi:10.1038/421219a. PMID 12529622
- Laurence, K (1994). A Question of Labour: Indentured Immigration Into Trinidad & British Guiana, 1875–1917. St Martin's Press. ISBN 978-0-312-12172-3
- Westwater ML; Fletcher PC; Ziauddeen H (November 2016). "Sugar addiction: the state of the science". European Journal of Nutrition. 55 (Suppl 2): 55–69. doi:10.1007/s00394-016-1229-6. PMC 5174153 . PMID 27372453
- Food and Nutrition Board (2002/2005). Dietary Reference Intakes for Energy, Carbohydrate, Fiber, Fat, Fatty Acids, Cholesterol, Protein and Amino Acids. Washington, D.C.: The National Academies Press. Page 769. ISBN 0-309-08537-3
- Jenkins, David; Alexandra L. Jenkins; Thomas M.S. Woleve; Lilian H. Thompson; A. Venkat Rao (February 1986). "Simple and Complex Carbohydrates". Nutrition Reviews. 44 (2): 44–49. doi:10.1111/j.1753-4887.1986.tb07585.x
- Welsh, Jean A.; Andrea J. Sharma; Lisa Grellinger; Miriam B. Vos (2011). "Consumption of added sugars is decreasing in the United States". American Journal of Clinical Nutrition. American Society for Nutrition. 94 (3): 726–34. doi:10.3945/ajcn.111.018366. Retrieved January 18, 2014
- Clarke, S. F; Foster, J. R (2012). "A history of blood glucose meters and their role in self-monitoring of diabetes mellitus". British journal of biomedical science. 69 (2): 83–93. PMID 22872934

Chapter 3

Lipids: Types, Functions and Metabolism

Lipids are biological molecules which help in forming cell membranes. They also perform other processes such as energy storage and signaling. This all-inclusive chapter delves into the types of lipids, their metabolism and other biological functions facilitated by them.

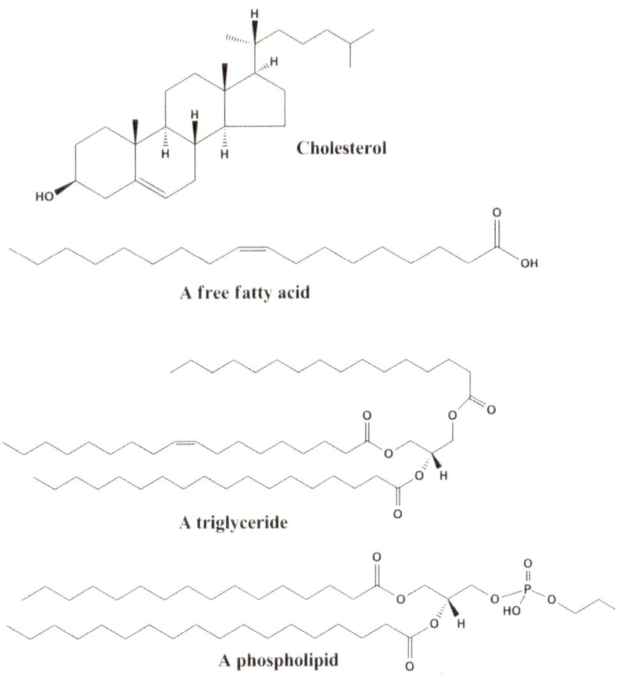

Structures of some common lipids. At the top are cholesterol and oleic acid. The middle structure is a triglyceride composed of oleoyl, stearoyl, and palmitoyl chains attached to a glycerol backbone. At the bottom is the common phospholipid phosphatidylcholine.

In biology, a lipid is a substance of biological origin that is soluble in nonpolar solvents. It comprises a group of naturally occurring molecules that include fats, waxes, sterols, fat-soluble vitamins (such as vitamins A, D, E, and K), monoglycerides, diglycerides, triglycerides, and phospholipids. The main biological functions of lipids include storing energy, signaling, and acting as structural components of cell membranes. Lipids have applications in the cosmetic and food industries as well as in nanotechnology.

Scientists sometimes broadly define lipids as hydrophobic or amphiphilic small molecules; the amphiphilic nature of some lipids allows them to form structures such as vesicles, multilamellar/unilamellar liposomes, or membranes in an aqueous environment. Biological lipids originate entirely or in part from two distinct types of biochemical subunits or "building-blocks": ketoacyl and isoprene groups. Using this approach, lipids may be divided into eight categories: fatty acids, glycerolipids, glycerophospholipids, sphingolipids, saccharolipids, and polyketides (derived from condensation of ketoacyl subunits); and sterol lipids and prenol lipids (derived from condensation of isoprene subunits).

Although the term "lipid" is sometimes used as a synonym for fats, fats are a subgroup of lipids called triglycerides. Lipids also encompass molecules such as fatty acids and their derivatives (including tri-, di-, monoglycerides, and phospholipids), as well as other sterol-containing metabolites such as cholesterol. Although humans and other mammals use various biosynthetic pathways both to break down and to synthesize lipids, some essential lipids cannot be made this way and must be obtained from the diet.

Phospholipid

Phospholipids are a class of lipids that are a major component of all cell membranes. They can form lipid bilayers because of their amphiphilic characteristic. The structure of the phospholipid molecule generally consists of two hydrophobic fatty acid "tails" and a hydrophilic "head" consisting of a phosphate group. The two components are joined together by a glycerol molecule. The phosphate groups can be modified with simple organic molecules such as choline.

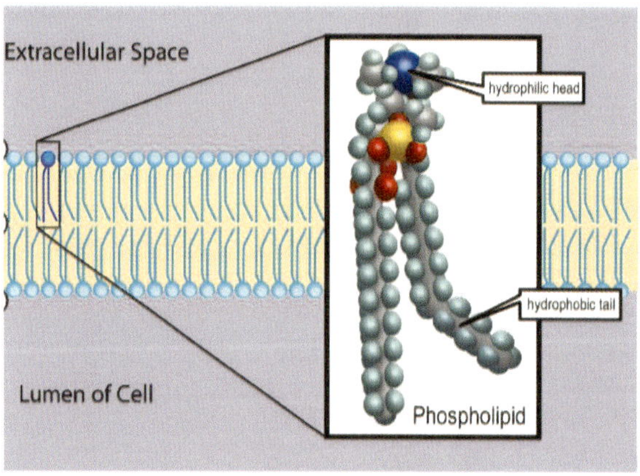

Phospholipid

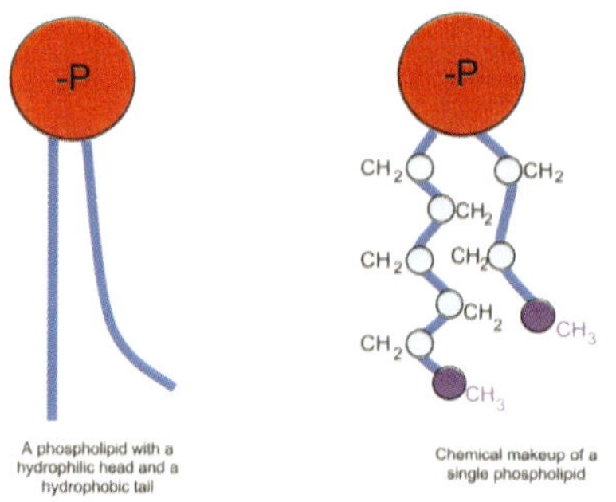

The left image shows a phospholipid, and the right image shows the chemical makeup.

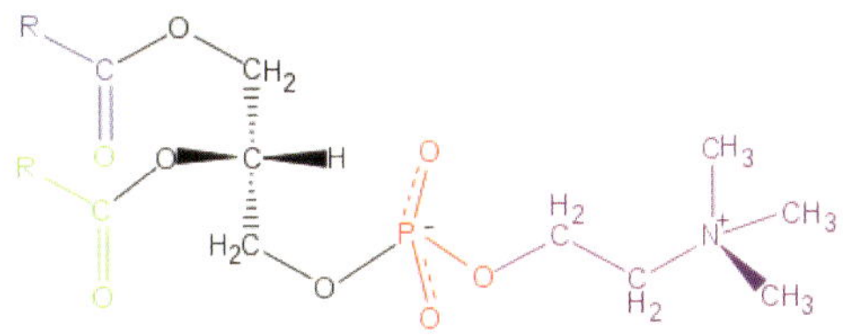

Phosphatidyl choline is the major component of lecithin. It is also a source for choline in the synthesis of acetylcholine in cholinergic neurons.

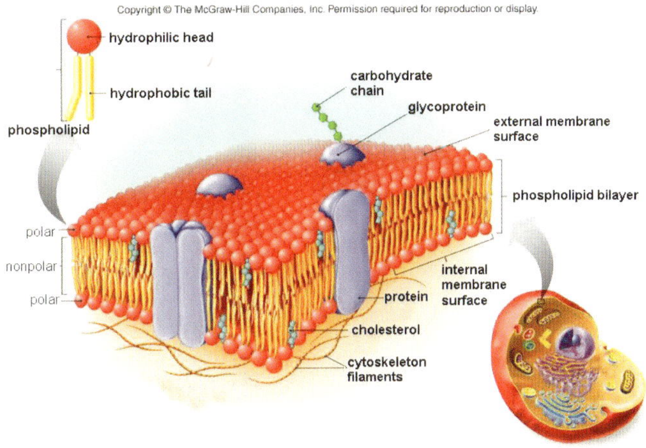

Cell membranes consist of phospholipid bilayers

The first phospholipid identified in 1847 as such in biological tissues was lecithin, or phosphatidylcholine, in the egg yolk of chickens by the French chemist and pharmacist, Theodore Nicolas Gobley. Biological membranes in eukaryotes also contain another class of lipid, sterol, interspersed among the phospholipids and together they provide membrane fluidity and mechanical strength. Purified phospholipids are produced commercially and have found applications in nanotechnology and materials science.

Amphiphilic Character

An amphiphile is a term describing a chemical compound possessing both hydrophilic (water-loving, polar) and lipophilic (fat-loving) properties. The phospholipid head contains a negatively charged phosphate group and glycerol; it is hydrophilic (attracted to water). The phospholipid tails usually consist of 2 long fatty acid chains; they are hydrophobic and avoid interactions with water. When placed in aqueous solutions, phospholipids are driven by hydrophobic interactions that result in the fatty acid tails aggregating to minimize interactions with water molecules. These specific properties allow phospholipids to play an important role in the phospholipid bilayer. In biological systems, the phospholipids often occur with other molecules (e.g., proteins, glycolipids, sterols) in a bilayer such as a cell membrane. Lipid bilayers occur when hydrophobic tails line up against one another, forming a membrane of hydrophilic heads on both sides facing the water.

Such movement can be described by the fluid mosaic model, that describes the membrane as a mosaic of lipid molecules that act as a solvent for all the substances and proteins within it, so proteins and lipid molecules are then free to diffuse laterally through the lipid matrix and migrate over the membrane. Sterols contribute to membrane fluidity by hindering the packing together of phospholipids. However, this model has now been superseded, as through the study of lipid polymorphism it is now known that the behaviour of lipids under physiological (and other) conditions is not simple.

Diacylglyceride Structures

- Phosphatidic acid (phosphatidate) (PA)
- Phosphatidylethanolamine (cephalin) (PE)
- Phosphatidylcholine (lecithin) (PC)
- Phosphatidylserine (PS)
- Phosphoinositides:
    - Phosphatidylinositol (PI)
    - Phosphatidylinositol phosphate (PIP)
    - Phosphatidylinositol bisphosphate (PIP2) and
    - Phosphatidylinositol trisphosphate (PIP3)

Phosphosphingolipids

- Ceramide phosphorylcholine (Sphingomyelin) (SPH)
- Ceramide phosphorylethanolamine (Sphingomyelin) (Cer-PE)
- Ceramide phosphoryllipid

Applications

Phospholipids have been widely used to prepare liposomal, ethosomal and other nanoformulations of topical, oral and parenteral drugs for differing reasons like improved bio-availability, reduced toxicity and increased permeability across membranes. Liposomes are often composed of phosphatidylcholine-enriched phospholipids and may also contain mixed Phospholipid chains with surfactant properties. The ethosomal formulation of ketoconazole using phospholipids is a promising option for transdermal delivery in fungal infections.

Simulations

Computational simulations of phospholipids are often performed using molecular dynamics with force fields such as GROMOS, CHARMM, or AMBER.

Characterization

Phospholipids are optically highly birefringent, i.e. their refractive index is different along their axis as opposed to perpendicular to it. Measurement of birefringence can be achieved using cross polarisers in a microscope to obtain an image of e.g. vesicle walls or using techniques such as dual polarisation interferometry to quantify lipid order or disruption in supported bilayers.

Analysis

There are no simple methods available for analysis of phospholipids since the close range of polarity between different phospholipid species makes detection difficult. Oil chemists often use spectroscopy to determine total Phosphorus abundance and then calculate approximate mass of phospholipids based on molecular weight of expected fatty acid species. Modern lipid profiling employs more absolute methods of analysis, with nuclear magnetic resonance spectroscopy (NMR), particularly ^{31}P-NMR, while HPLC-ELSD provides relative values.

Phospholipid Synthesis

Phospholipid synthesis occurs in the cytosole adjacent to ER membrane that is studded with proteins that act in synthesis (GPAT and LPAAT acyl transferases, phosphatase and choline phosphotransferase) and allocation (flippase and floppase). Eventually a vesicle will bud off from the ER containing phospholipids destined for the cytoplasmic cellular membrane on its exterior leaflet and phospholipids destined for the exoplasmic cellular membrane on its inner leaflet.

Sources

Common sources of industrially produced phospholipids are soya, rapeseed, sunflower, chicken eggs, bovine milk, fish eggs etc. Each source has a unique profile of individual phospholipid species and consequently differing applications in food, nutrition, pharmaceuticals, cosmetics and drug delivery.

Market

The key companies operating in the global Phospholipid market include Avanti Polar Lipids, Lipoid GmbH, VAV Life Sciences Pvt. Ltd. and NOF Corporation.

In Signal Transduction

Some types of phospholipid can be split to produce products that function as second messengers in signal transduction. Examples include phosphatidylinositol (4,5)-bisphosphate (PIP_2), that can be split by the enzyme Phospholipase C into inositol triphosphate (IP_3) and diacylglycerol (DAG), which both carry out the functions of the G_q type of G protein in response to various stimuli and intervene in various processes from long term depression in neurons to leukocyte signal pathways started by chemokine receptors.

Phospholipids also intervene in prostaglandin signal pathways as the raw material used by lipase enzymes to produce the prostaglandin precursors. In plants they serve as the raw material to produce Jasmonic acid, a plant hormone similar in structure to prostaglandins that mediates defensive responses against pathogens.

Food Technology

Phospholipids can act as emulsifiers, enabling oils to form a colloid with water. Phospholipids are one of the components of lecithin which is found in egg-yolks, as well as being extracted from soy beans, and is used as a food additive in many products, and can be purchased as a dietary supplement. Lysolecithins are typically used for water-oil emulsions like margarine, due to their higher HLB ratio.

Phospholipid Derivatives

- Natural phospholipid derivates:

  egg PC (Egg lecithin), egg PG, soy PC, hydrogenated soy PC, sphingomyelin as natural phospholipids.

- Synthetic phospholipid derivates:
  - Phosphatidic acid (DMPA, DPPA, DSPA)
  - Phosphatidylcholine (DDPC, DLPC, DMPC, DPPC, DSPC, DOPC, POPC, DEPC)
  - Phosphatidylglycerol (DMPG, DPPG, DSPG, POPG)
  - Phosphatidylethanolamine (DMPE, DPPE, DSPE DOPE)
  - Phosphatidylserine (DOPS)
  - PEG phospholipid (mPEG-phospholipid, polyglycerin-phospholipid, funcitional-ized-phospholipid, terminal activated-phospholipid)

Cardiolipin

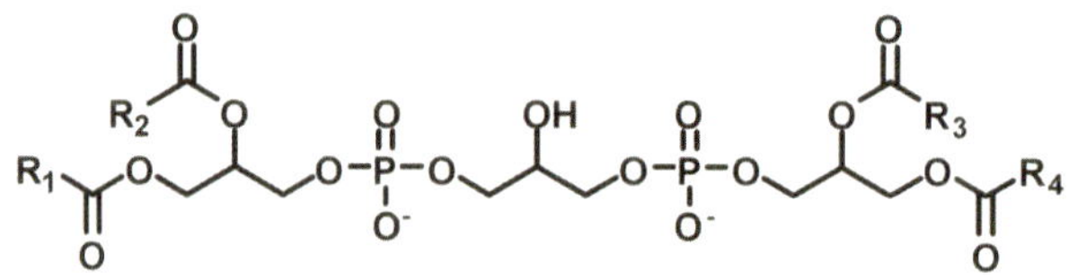

Cardiolipin

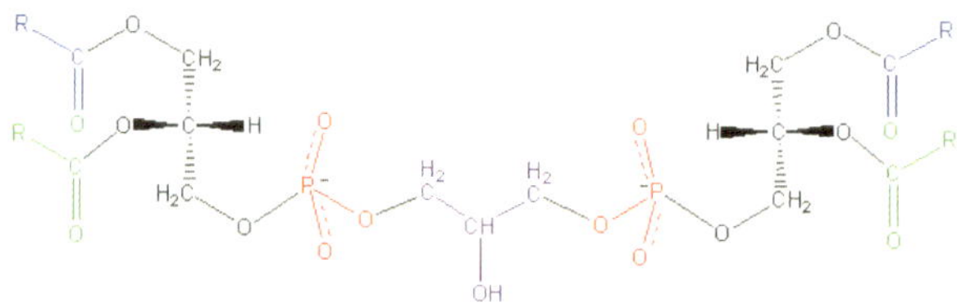

Cardiolipin (alternate image)

Cardiolipin (IUPAC name "1,3-bis(*sn*-3'-phosphatidyl)-*sn*-glycerol") is an important component of the inner mitochondrial membrane, where it constitutes about 20% of the total lipid composition. It can also be found in the membranes of most bacteria. The name 'cardiolipin' is derived from the fact that it was first found in animal hearts. It was first isolated from beef heart in the early 1940s. In mammalian cells, but also in plant cells, cardiolipin (CL) is found almost exclusively in the inner mitochondrial membrane where it is essential for the optimal function of numerous enzymes that are involved in mitochondrial energy metabolism.

Structure

Cardiolipin (CL) is a kind of diphosphatidylglycerol lipid. Two phosphatidic acid moieties connect with a glycerol backbone in the center to form a dimeric structure. So it has four alkyl groups and potentially carries two negative charges. As there are four distinct alkyl chains in cardiolipin, the potential for complexity of this molecule species is enormous. However, in most animal tissues, cardiolipin contains

18-carbon fatty alkyl chains with 2 unsaturated bonds on each of them. It has been proposed that the (18:2)4 acyl chain configuration is an important structural requirement for the high affinity of CL to inner membrane proteins in mammalian mitochondria. However, studies with isolated enzyme preparations indicate that its importance may vary depending on the protein examined.

Source	Molecular species				
	Fatty acid R1	Fatty acid R2	Fatty acid R3	Fatty acid R4	Amount (Mol%)
Bovine heart	18:2	18:2	18:2	18:2	48
	18:3	18:2	18:2	18:2	21
	18:2	18:3	18:2	18:2	
	18:2	18:2	18:2	18:3	
	18:2	18:2	18:3	18:2	
	18:2	18:1	18:2	18:2	15
	18:2	18:2	18:2	18:1	
Rat liver	18:2	18:2	18:2	18:2	57
	18:2	18:1	18:2	18:2	35
	18:2	18:2	18:2	18:1	

Cardiolipin in animal tissues

Since there are two phosphates in the molecule, each of them can catch one proton. Although it has a symmetric structure, ionizing one phosphate happens at a very different levels of acidity than ionizing both: $pK_1 = 3$ and $pK_2 > 7.5$. So under normal physiological conditions (wherein pH is around 7), the molecule may carry only one negative charge. The hydroxyl groups (–OH and $-O^-$) on phosphate would form a stable intramolecular hydrogen bond with the centered glycerol's hydroxyl group, thus forming a bicyclic resonance structure. This structure traps one proton, which is quite helpful for oxidative phosphorylation.

As the head group forms such compact bicycle structure, the head group area is quite small relative to the big tail region consisting of 4 acyl chains. Based on this special structure, the fluorescent mitochondrial indicator, nonyl acridine orange (NAO) was introduced in 1982, and was later found to target mitochondria by binding to CL. NAO has a very large head and small tail structure which can compensate with cardiolipin's small head large tail structure, and arrange in a highly ordered way. Several studies were published utilizing NAO both as a quantitative mitochondrial indicator and an indicator of CL content in mitochondria. However, NAO is influenced by membrane potential and/or the spatial arrangement of CL, so it's not proper to use NAO for CL or mitochondria quantitative studies of intact respiring mitochondria. But NAO still represents a simple method of assessing CL content.

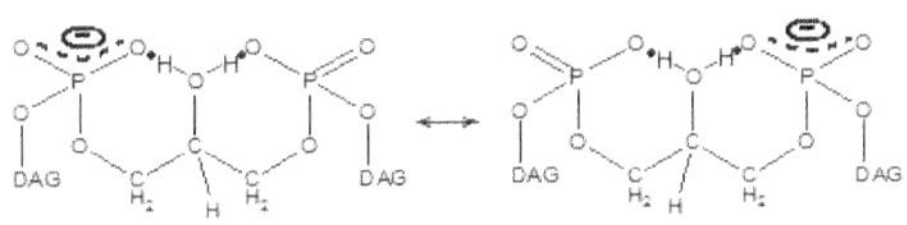

Bicyclic structure of Cardiolipin

Cardiolipin bicyclic structure

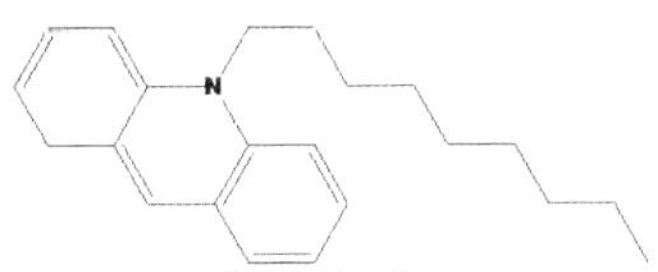

10-N-Nonyl acridine

Structure of NAO

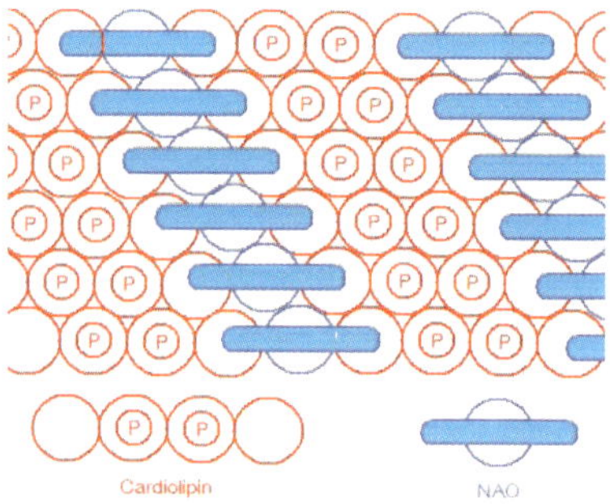

NAO & CL arranged in a highly ordered way

Metabolism and Catabolism

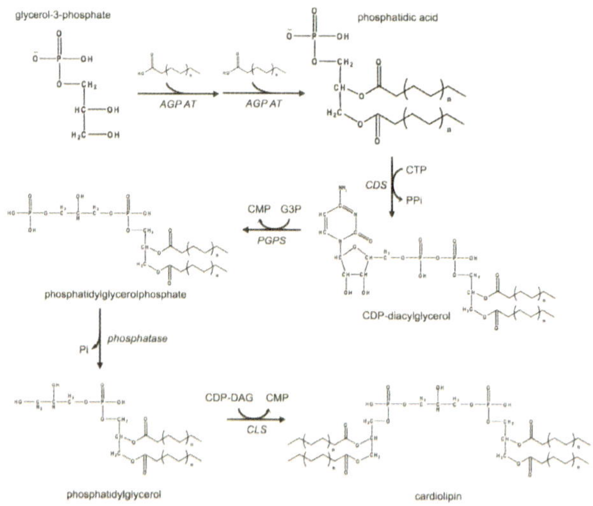

Cardiolipin synthesis in eukaryotes

Metabolism

Eukaryotic Pathway

In eukaryotes such as yeasts, plants and animals, the synthesis processes are believed to happen in mitochondria. The first step is the acylation of glycerol-3-phosphate by a glycerol-3-phosphate acyltransferase. Then acylglycerol-3-phosphate can be once more acylated to form a phosphatidic acid (PA). With the help of the enzyme CDP-DAG synthase (CDS) (phosphatidate cytidylyltransferase), PA is converted into cytidinediphosphate-diacylglycerol (CDP-DAG). The following step is conversion of CDP-DAG to phosphatidylglycerol phosphate (PGP) by the enzyme PGP synthase, followed by dephosphorylation by PTPMT1 to form PG. Finally, a molecule of CDP-DAG is bound to PG to form one molecule of cardiolipin, catalyzed by the mitochondria-localized enzyme cardiolipin synthase (CLS).

Prokaryotic Pathway

In prokaryotes such as bacteria, diphosphatidylglycerol synthase catalyses a transfer of the phosphatidyl moiety of one phosphatidylglycerol to the free 3'-hydroxyl group of another, with the elimination of one molecule of glycerol, via the action of an enzyme related to phospholipase D. The enzyme can operate in reverse under some physiological conditions to remove cardiolipin.

Catabolism

Catabolism of cardiolipin may happen by the catalysis of phospholipase A2 (PLA) to remove fatty acyl groups. Phospholipase D (PLD) in the mitochondrion hydrolyses cardiolipin to phosphatidic acid.

Functions

Regulates Aggregate Structures

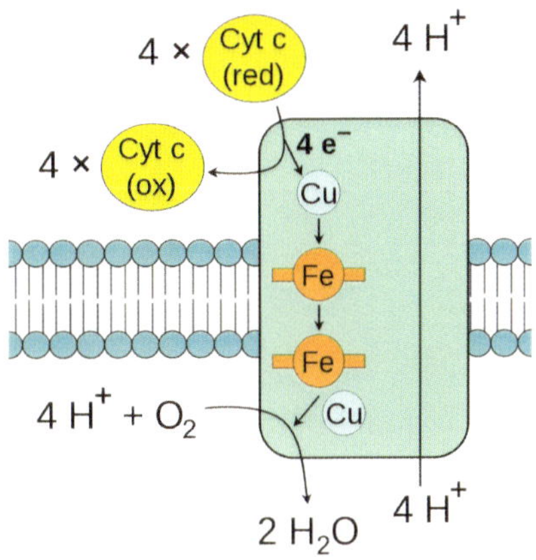

Respiratory electron transfer of Complex IV

Because of cardiolipin's unique structure, a change in pH and the presence of divalent cations can induce a structural change. CL shows a great variety of forms of aggregates. It is found that in the presence of Ca^{2+} or other divalent cations, CL can be induced to have a lamellar-to-hexagonal (L_a-H_{II}) phase transition. And it is believed to have a close connection with membrane fusion.

Helps to Build Quaternary Structure

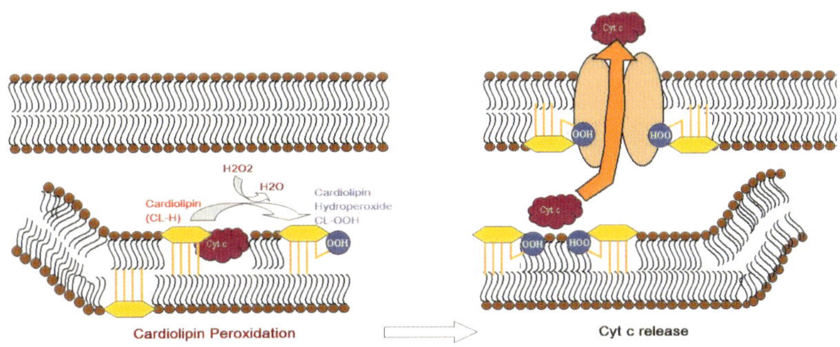

The mechanism whereby CL triggers apoptosis

The enzyme cytochrome c oxidase or Complex IV is a large transmembrane protein complex found in bacteria and the mitochondrion. It is the last enzyme in the respiratory electron transport chain of mitochondria (or bacteria) located in the mitochondrial (or bacterial) membrane. It receives an electron from each of four cytochrome c molecules, and transfers them to one oxygen molecule, converting molecular oxygen to two molecules of water. Complex IV has been shown to require two associated CL molecules in order to maintain its full enzymatic function. Cytochrome bc1(Complex III) also needs cardiolipin to maintain its quaternary structure and functional role. Complex V of the oxidative phosphorylation machinery also displays high binding affinity for CL, binding four molecules of CL per molecule of complex V.

Triggers Apoptosis

Cardiolipin distribution to the outer mitochondrial membrane would lead to apoptosis of the cells, as evidenced by cytochrome c (cyt c) release, Caspase-8 activation, MOMP induction and NLRP3 inflammasome activation. During apoptosis, cyt c is released from the intermembrane spaces of mitochondria into the cytosol. Cyt c can then bind to the IP3 receptor on ER, stimulating calcium release, which then reacts back to cause the release of cyt c. When the calcium concentration reaches a toxic level, this causes cell death. Cytochrome c is thought to play a role in apoptosis via the release of apoptotic factors from the mitochondria. A cardiolipin-specific oxygenase produces CL hydroperoxides which can result in the conformation change of the lipid. The oxidized CL transfers from the inner membrane to the outer membrane, and then helps to form a permeable pore which releases cyt c.

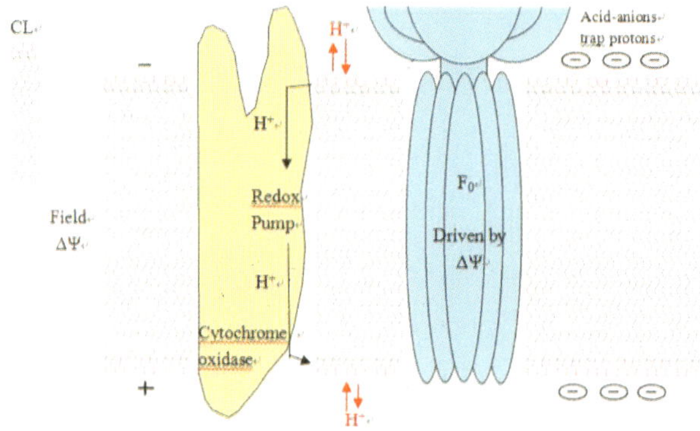

CL serves as a proton trap in oxidative phosphorylation

Serves as Proton Trap for Oxidative Phosphorylation

During the oxidative phosphorylation process catalyzed by Complex IV, large quantities of protons are transferred from one side of the membrane to another side causing a large pH change. CL is suggested to function as a proton trap within the mitochondrial membranes, thereby strictly localizing the proton pool and minimizing the changes in pH in the mitochondrial intermembrane space.

This function is due to CL's unique structure. As stated above, CL can trap a proton within the bicyclic structure while carrying a negative charge. Thus, this bicyclic structure can serve as an electron buffer pool to release or absorb protons to maintain the pH near the membranes.

Other Functions

- Cholesterol translocation from outer to the inner membrane of mitochondrial.
- Activates mitochondrial cholesterol side-chain cleavage.
- Import protein into mitochondrial matrix.
- Anticoagulant function.

Glycolipid

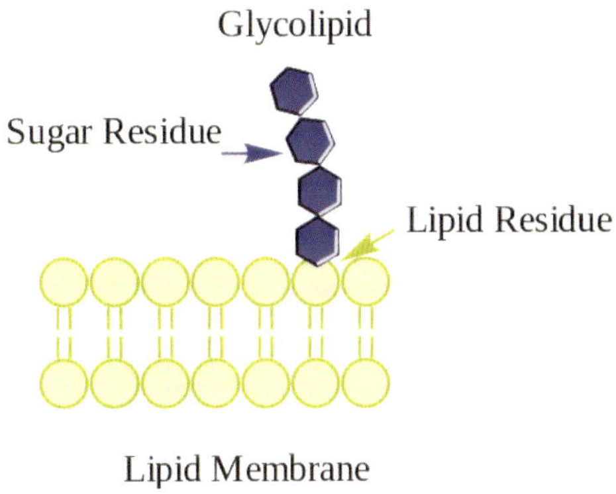

Glycolipid

Glycolipids are lipids with a carbohydrate attached by a glycosidic bond. Their role is to maintain the stability of the cell membrane and to facilitate cellular recognition, which is crucial to the immune response and in the connections that allow cells to connect to one another to form tissues. Glycolipids are found on the surface of all eukaryotic cell membranes, where they extend from the phospholipid bilayer into the extracellular environment.

Structure

The essential feature of a glycolipid is the presence of a monosaccharide or oligosaccharide bound to a lipid moiety. The most common lipids in cellular membranes are glycerolipids and sphingolipids, which have glycerol or a sphingosine backbones, respectively. Fatty acids are connected to this backbone, so that the lipid as a whole has a polar head and a non-polar tail. The lipid bilayer of the cell membrane consists of two layers of lipids, with the inner and outer surfaces of the membrane made up of the polar head groups, and the inner part of the membrane made up of the non-polar fatty acid tails.

The saccharides that are attached to the polar head groups on the outside of the cell are the ligand components of glycolipids, and are likewise polar, allowing them to be soluble in the aqueous environment surrounding the cell. The lipid and the saccharide form a glycoconjugate through a glycosidic bond, which is a covalent bond. The anomeric carbon of the sugar binds to a free hydroxyl group on the lipid backbone. The structure of these saccharides varies depending on the structure of the molecules to which they bind.

Metabolism

Glycosyltransferases

Enzymes called glycosyltransferases link the saccharide to the lipid molecule, and also play a role in assembling the correct oligosaccharide so that the right receptor can be activated on the cell which responds to the presence of the glycolipid on the surface of the cell. The glycolipid is assembled in the Golgi apparatus and embedded in the surface of a vesicle which is then transported to the cell membrane. The vesicle merges with the cell membrane so that the glycolipid can be presented on the cell's outside surface.

Glycoside Hydrolases

Glycoside hydrolases catalyze the breakage of glycosidic bonds. They are used to modify the oligosaccharide structure of the glycan after it has been added onto the lipid. They can also remove glycans from glycolipids to turn them back into unmodified lipids.

Defects in Metabolism

Sphingolipidoses are a group of diseases that are associated with the accumulation of sphingolipids which have not been degraded correctly, normally due to a defect in a glycoside hydrolase enzyme. Sphingolipidoses are typically inherited, and their effects depend on which enzyme is affected, and the degree of impairment. One notable example is Niemann–Pick disease which can cause pain and damage to neural networks, and is usually fatal in early infancy.

Function

Cell–cell Interactions

The main function of glycolipids in the body is to serve as recognition sites for cell–cell interactions. The saccharide of the glycolipid will bind to a specific complementary carbohydrate or to a lectin (carbohydrate-binding protein), of a neighboring cell. The interaction of these cell surface markers is the basis of cell recognitions, and initiates cellular responses that contribute to activities such as regulation, growth, and apoptosis.

Immune Responses

An example of how glycolipids function within the body is the interaction between leukocytes and endothelial cells during inflammation. Selectins, a class of lectins found on the surface of leukocytes and endothelial cells bind to the carbohydrates attached to glycolipids to initiate the immune response. This binding causes leukocytes to leave circulation and congregate near the site of inflammation. This is the initial binding mechanism, which is followed by the expression of integrins which form stronger bonds and allow leukocytes to migrate toward the site of inflammation. Glycolipids are also responsible for other responses, notably the recognition of host cells by viruses.

Blood Types

Blood types are an example of how glycolipids on cell membranes mediate cell interactions with the surrounding environment. The four main human blood types (A, B, AB, O) are determined by the oligosaccharide attached to a specific glycolipid on the surface of red blood cells, which acts as an antigen. The unmodified antigen, called the H antigen, is the characteristic of type O, and is present on red blood cells of all blood types. Blood type A has an N-acetylgalactosamine added as the main determining structure, type B has a galactose, and type AB has all three of these antigens. Antigens which are not present in an individual's blood will cause antibodies to be produced, which will bind to the foreign glycolipids. For this reason, people with blood type AB can receive transfusions from all blood types (the universal acceptor), and people with blood type O can act as donors to all blood types (the universal donor).

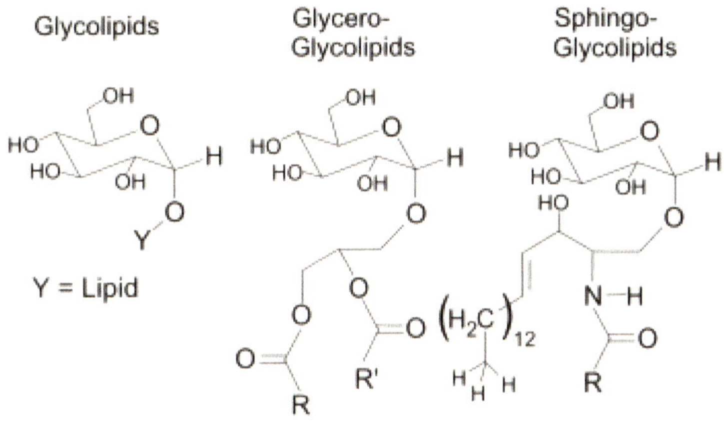

Chemical structure of glycolipids

Types of Glycolipids

This list is incomplete; you can help by expanding it.

- Glyceroglycolipids: a sub-group of glycolipids characterized by an acetylated or non-acetylated glycerol with at least one fatty acid as the lipid complex. Glyceroglycolipids are often associated with photosynthetic membranes and their functions. The subcategories of glyceroglycolipids depend on the carbohydrate attached.
    - Galactolipids: defined by a galactose sugar attached to a glycerol lipid molecule. They are found in chloroplast membranes and are associated with photosynthetic properties.
    - Sulfolipids: have a sulfur-containing functional group in the sugar moiety attached to a lipid. An important group is the sulfoquinovosyl diacylglycerols which are associated with the sulfur cycle in plants.
- Glycosphingolipids: a sub-group of glycolipids based on sphingolipids. Glycosphingolipids are mostly located in nervous tissue and are responsible for cell signaling.
    - Cerebrosides: a group glycosphingolipids involved in nerve cell membranes.
        - Galactocerebrosides: a type of cerebroseide with galactose as the saccharide moiety.
        - Glucocerebrosides: a type of cerebroside with glucose as the saccharide moiety; often found in non-neural tissue.
        - Sulfatides: a class of glycolipids containing a sulfate group in the carbohydrate with a ceramide lipid backbone. They are involved in numerous biological functions ranging from immune response to nervous system signaling.
    - Gangliosides: the most complex animal glycolipids. They contain negatively charged oligosacchrides with one or more sialic acid residues; more than 200 different gangliosides have been identified. They are most abundant in nerve cells.
    - Globosides: glycosphingolipids with more than one sugar as part of the carbohydrate complex. They have a variety of functions; failure to degrade these molecules leads to Fabry disease.

- Glycophosphosphingolipids: complex glycophospholipids from fungi, yeasts, and plants, where they were originally called "phytoglycolipids". They may be as complicated a set of compounds as the negatively charged gangliosides in animals.
- Glycophosphatidylinositols: a sub-group of glycolipids defined by a phosphatidylinositol lipid moiety bound to a carbohydrate complex. They can be bound to the C-terminus of a protein and have various functions associated with the different proteins they can be bound to.

Sophorolipid

A sophorolipid is a surface-active glycolipid compound that can be synthesized by a selected number of non-pathogenic yeast species. The first report on sophorolipids was published in 1961, but in the last two decades, because of growing environmental awareness, they have gained increased attention as potential bio-surfactants due to their biodegradability and low eco-toxicity. Today sophorolipids are considered promising bio-surfactants.

Biosurfactants

The term "biosurfactant" can be applied to a surfactant that is obtained directly from a natural source (from plant, animal or microbial cells) by some kind of separation procedure such as extraction, precipitation or distillation. These surfactants have recently gained appeal as an alternative to surfactants organically synthesized from petrochemicals. While many bio-surfactants prove to be acceptable substitutes for their synthetic counterparts, often minor chemical processing can alter their properties to meet specific needs. There is debate regarding how much organic synthesis should be involved while still being able to call a given substance a bio-surfactant, but currently the term is not protected and often "bio-based surfactant" is used interchangeably with "bio-surfactant".

Since surface active agents are abundant in nature, the factor that limits the production of bio-surfactants is definitely not their presence, but likely the technological cost of scaled-up production, separation, and purification. In most instances this makes the cost of bio-based surfactants much higher than the cost of their synthetic equivalents. For example, most biological membranes are made up of a phospholipid bilayer which is composed of surfactant molecules having a polar head and non-polar or hydrophobic tail. However while these phospholipids are ubiquitous in nature, their intracellular location and interaction with other cellular components make their recovery and purification difficult and costly.

Other than sophorolipids, common biosurfactants include rhamnolipids which can be produced by some species of Pseudomonas, and Lecithin which can be obtained either from soybean or from egg yolk and is a common food ingredient. One of interesting surfactants producing by yeast are cellobiose lipids consist of a residue of cellobiose, the disaccharide composed of two glucose residues linked by a 1,4'-beta-glycoside bond, and fatty acid residue as an aglycone . Cellobiose lipids are effective fungicides against many species of ascomycetous and basidiomycetous yeast and mycelial fungi, including human pathogens, Candida albicans and Cryptococcus neoformans.

Microbial Biosurfactant

Usually the term bio-surfactant is used for the surfactants of microbial origin. This is because the shorter generation time of microbial biomass compared to biomass of animal or plant origin, makes the production of bio-surfactants using bacteria and fungi more industrially relevant. Furthermore, most of the bio-surfactants produced by microbes are synthesized extracellularly and many microbes are known to produce bio-surfactants in large relative quantities. Both these factors (i.e. extracellular location and high concentration) can make the process of recovery, concentration and purification of microbial bio-surfactants simple and cost effective. Recent research has meant sophorolipids can be recovered during a fermentation using a gravity separator in a loop with the bioreactor, enabling the production of >770 g/l sophorolipid at a productivity 4.24 g/l/h, some of the highest values seen in a fermentation process.

Structure and Properties

Honey like viscous, oily, acidic sophorolipids produced by solid state fermentation (SSF) of *starmerella bombicola.*

Sophorolipids are glycolipid class of microbial biosurfactants which consist of a hydrophobic fatty acid tail of 16 or 18 carbon atoms and a hydrophilic carbohydrate head, sophorose. which is a glucose di-saccharide with an unusual β-1,2 bond and can be acetylated on the 6′- and/or 6′′- positions. One terminal or sub terminal hydroxylated fatty acid is β-glycosidically linked to the sophorose molecule. The carboxylic end of this fatty acid is either free (acidic or open form) or internally esterified at the 4′′ or in some rare cases at the 6′- or 6′′-position (lactonic form). The hydroxy fatty acid itself counts in general 16 or 18 carbon atoms and can have one or more unsaturated bonds. The physicochemical and biological properties of sophorolipids are significantly influenced by the distribution of the lactone vs. acidic forms produced in the fermentative broth. In general, lactone sophorolipids are more efficient in reducing surface tension and are better antimicrobial agents, whereas acidic sophorolipids display better foaming properties. Acetyl groups can also lower the hydrophilicity of sophorolipids and enhance their antiviral and cytokine stimulating effects.

Biosurfactant Properties

Two desirable properties of biosurfactants are biodegradability and low toxicity. While molecules of biological origin are almost always bio-degradable, the rate at which they are degraded can vary significantly as does the degree to which they can be considered toxic. Though many biosurfactants are easily biodegradable and exhibit low toxicity, it is inappropriate to assume that all bio-surfactants possess these same traits under all conditions. Furthermore, even though a bio-surfactant may not be toxic itself, the producing micro-organism could be pathogenic and hence the production of such bio-surfactants may lead to environment concerns. For example, Pseudomonads,including the opportunistic pathogen *Pseudomonas aeruginosa*, can be used to produce the bio-surfactant rhamnolipid. Similar concerns have been expressed over the widespread production of citric acid by the opportunistic pathogen *Aspergillus niger* but were limited to ensuring that the final product was sufficiently pure and that the production process was contained. In the case of rhamnolipids, there are many non-pathogenic strains of Pseudomonas which can be used to produce that molecule. Sophorolipids are synthesised by various non pathogenic yeast species such as *Candida apicola*, *Rhodotorula bogoriensis*, *Wickerhamiella domercqiae*, and *Starmerella bombicola*.

Besides biodegradibility, low toxicity, and high production potential, sophorolipids also have a high surface and interfacial activity compared to many chemically synthesized surfactants. These properties help justify the higher costs of production and purification associated with large scale production. Generally a surfactant is considered as a good surfactant if it can lower surface tension (ST) of water from 72 to 30-35 mN/m and the interfacial tension (IT) water/hexadecane from 40 to 1 mN/m. Sophorolipids are reported to reduce ST up to 29 mN/m and IT to 5 mN/m, at the much lower concentrations than the critical micelle concentration(CMC) of many of the synthetic surfactants. In addition to this, sophorolipids are reported to function under wide ranges of temperatures, pressures and ionic strengths; and they also possess a number of other useful biological activities including Antimicrobial, virucidal, Anticancer, Immuno-modulatory properties.

Research

A detailed and comprehensive literature review on the various aspects of sophorolipids production (e.g. producing micro-organisms, bio-synthetic pathway, effect of medium components and other fermentation conditions and downstream process of sophorolipids is available in the published work of Van Bogaert et al. This work also discusses potential application of sophorolipids (and their derivatives) as well as the potential for genetic engineering strains to enhance sophorolipid yields. Researchers have focused on optimization of sophorolipid production in submerged fermentation, but some efforts have also investigated the possibility of sophorololipid production using solid state fermentation (SSF). The production process can be significantly impacted by the specific properties of the carbon and oil substrates used; and several inexpensive alternatives to more traditional substrates have been investigated. These potential substrates include: biodiesel by-product streams, waste frying oil, restaurant waste oil, industrial fatty acid residues, mango seed fat, and soybean dark oil. The use of most of these substrates have resulted in lower yields compared to traditional fermentation substrates.

Limitations

There no doubt that the sophorolipids are the most promising microbial biosurfactant from microbial origin. However even sophorolipids have few shortcomings. For example, sophorolipids are not readily soluble in aqueous medium at acidic pH. Their solubility increases above pH 5.0 and are freely soluble above pH 6.0. However sophorolipids shows instability at pH values higher than 7.0 to 7.5 for long term storage: beyond this point, irreversible hydrolysis of the acetyl groups and ester bonds is observed. Sophorolipids are low foaming surfactants even at high concentrations; these limitations excludes them from a number of applications (for example: preparation of soaps), but makes them suitable for non-foam requiring purposes. (for example: cleansing solution for glass and metallic surfaces, as food additives). Further, even though sophorolipids have excellent antimicrobial, anticancer and antiviral properties, its oral administration is almost impossible, since it gets readily hydrolyzed in stomach as well as intestine, thus they can either be useful for topical preparations such as antimicrobial creams, or may require development of parenteral or novel dosage forms, if to be used for systemic infections.

Chemical Modifications of Sophorolipids and Polysophorolipids

To enhance the performance of surfactant properties of natural sophorolipids, chemical modification methods have been actively pursued. Recently, researchers demonstrated the possibility of applying sophorolipids as building blocks via ring-opening metathesis polymerization for a new type of polymers, known as polysophorolipids which show promising potentials in biomaterials applications.

Sphingolipid

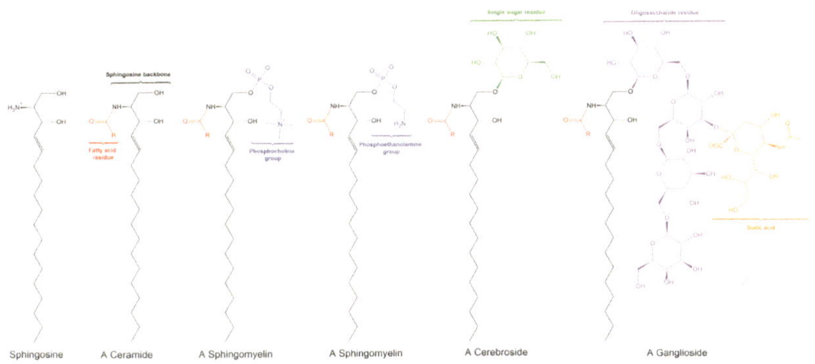

General structures of sphingolipids

Sphingolipids, or glycosylceramides, are a class of lipids containing a backbone of sphingoid bases, a set of aliphatic amino alcohols that includes sphingosine. They were discovered in brain extracts in the 1870s and were named after the mythological Sphinx because of their enigmatic nature. These compounds play important roles in signal transmission and cell recognition. Sphingolipidoses, or disorders of sphingolipid metabolism, have particular impact on neural tissue. A sphingolipid with an R group consisting of a hydrogen atom only is a ceramide. Other common R groups include phosphocholine, yielding a sphingomyelin, and various sugar monomers or dimers, yielding cerebrosides and globosides, respectively. Cerebrosides and globosides are collectively known as glycosphingolipids.

Structure

The long-chain bases, sometimes simply known as sphingoid bases, are the first non-transient products of *de novo* sphingolipid synthesis in both yeast and mammals. These compounds, specifically known as phytosphingosine and dihydrosphingosine (also known as sphinganine, although this term is less common), are mainly C_{18} compounds, with somewhat lower levels of C_{20} bases. Ceramides and glycosphingolipids are *N*-acyl derivatives of these compounds.

The sphingosine backbone is O-linked to a (usually) charged head group such as ethanolamine, serine, or choline.

The backbone is also amide-linked to an acyl group, such as a fatty acid.

Types

Simple sphingolipids, which include the sphingoid bases and ceramides, make up the early products of the sphingolipid synthetic pathways.

- Sphingoid bases are the fundamental building blocks of all sphingolipids. The main mammalian sphingoid bases are dihydrosphingosine and sphingosine, while dihydrosphingosine and phytosphingosine are the principle sphingoid bases in yeast. Sphingosine, dihydrosphingosine, and phytosphingosine may be phosphorylated.
- Ceramides, as a general class, are *N*-acylated sphingoid bases lacking additional head groups.
    - Dihydroceramide is produced by *N*-acylation of dihydrosphingosine. Dihydroceramide is found in both yeast and mammalian systems.
    - Ceramide is produced in mammalian systems by desaturation of dihydroceramide by dihydroceramide desaturase 1 (DES1). This highly bioactive molecule may also be phosphorylated to form ceramide-1-phosphate.
    - Phytoceramide is produced in yeast by hydroxylation of dihydroceramide at C-4.

Complex sphingolipids may be formed by addition of head groups to ceramide or phytoceramide:

- Sphingomyelins have a phosphocholine or phosphoethanolamine molecule with an ester linkage to the 1-hydroxy group of a ceramide.
- Glycosphingolipids are ceramides with one or more sugar residues joined in a β-glycosidic linkage at the 1-hydroxyl position.
    - Cerebrosides have a single glucose or galactose at the 1-hydroxy position.
        - Sulfatides are sulfated cerebrosides.
    - Gangliosides have at least three sugars, one of which must be sialic acid.
- Inositol-containing ceramides, which are derived from phytoceramide, are produced in yeast. These include inositol phosphorylceramide, mannose inositol phosphorylceramide, and mannose diinositol phosphorylceramide.

Mammalian Sphingolipid Metabolism

De novo sphingolipid synthesis begins with formation of 3-keto-dihydrosphingosine by serine palmitoyltransferase. The preferred substrates for this reaction are palmitoyl-CoA and serine. However, studies have demonstrated that serine palmitoyltransferase has some activity toward other species of fatty acyl-CoA and alternative amino acids, and the diversity of sphingoid bases has recently been reviewed. Next, 3-keto-dihydrosphingosine is reduced to form dihydrosphingosine. Dihydrosphingosine is acylated by a (dihydro)-ceramide synthase, such as Lass1p or Lass2p (also termed as CerS), to form dihydroceramide. This is desaturated to form ceramide.

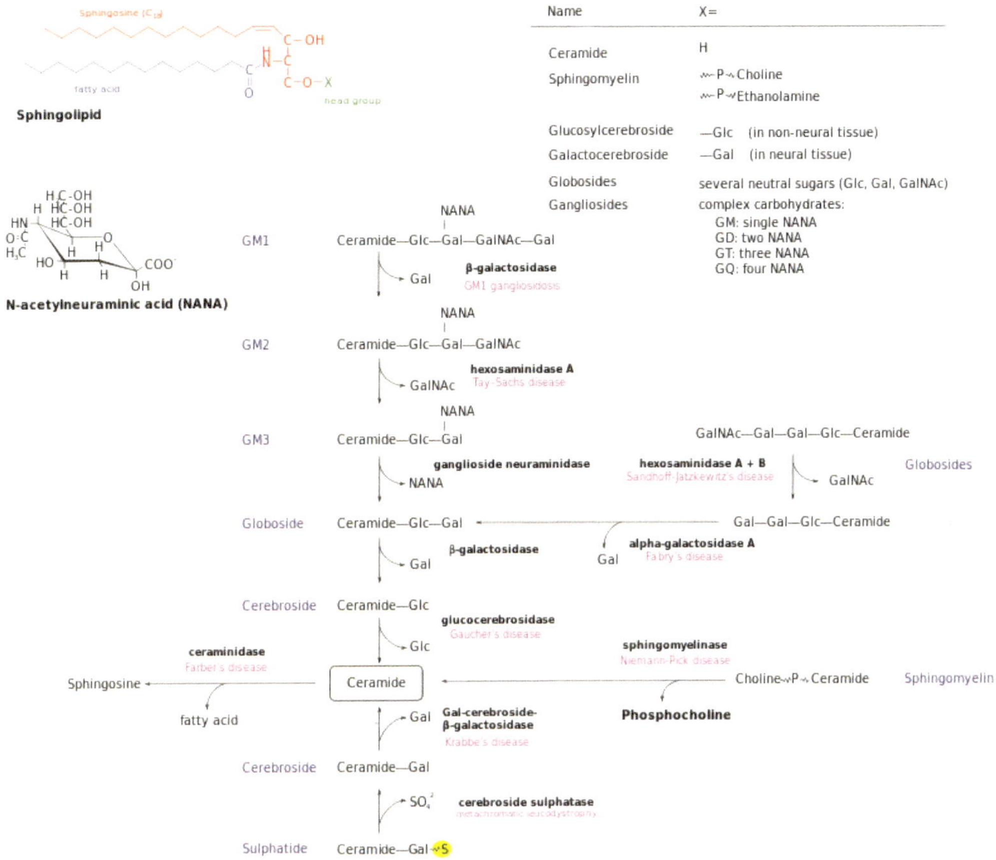

Metabolic pathways of various forms of sphingolipids. Sphingolipidoses are labeled at corresponding stages that are deficient.

Ceramide may subsequently have several fates. It may be phosphorylated by ceramide kinase to form ceramide-1-phosphate. Alternatively, it may be glycosylated by glucosylceramide synthase or galactosylceramide synthase. Additionally, it can be converted to sphingomyelin by the addition of a phosphorylcholine headgroup by sphingomyelin synthase. Diacylglycerol is generated by this process. Finally, ceramide may be broken down by a ceramidase to form sphingosine. Sphingosine may be phosphorylated to form sphingosine-1-phosphate. This may be dephosphorylated to reform sphingosine.

Breakdown pathways allow the reversion of these metabolites to ceramide. The complex glycosphingolipids are hydrolyzed to glucosylceramide and galactosylceramide. These lipids are then hydrolyzed by beta-glucosidases and beta-galactosidases to regenerate ceramide. Similarly, sphingomyelin may be broken down by sphingomyelinase to form ceramide.

The only route by which sphingolipids are converted to non-sphingolipids is through sphingosine-1-phosphate lyase. This forms ethanolamine phosphate and hexadecenal.

Functions of Mammalian Sphingolipids

Sphingolipids are commonly believed to protect the cell surface against harmful environmental factors by forming a mechanically stable and chemically resistant outer leaflet of the plasma membrane lipid bilayer. Certain complex glycosphingolipids were found to be involved in specific functions, such as cell recognition and signaling. Cell recognition depends mainly on the physical properties of the sphingolipids, whereas signaling involves specific interactions of the glycan structures of glycosphingolipids with similar lipids present on neighboring cells or with proteins.

Recently, simple sphingolipid metabolites, such as ceramide and sphingosine-1-phosphate, have been shown to be important mediators in the signaling cascades involved in apoptosis, proliferation, stress responses, necrosis, inflammation, autophagy, senescence, and differentiation. Ceramide-based lipids self-aggregate in cell membranes and form separate phases less fluid than the bulk phospholipids. These sphingolipid-based microdomains, or "lipid rafts" were originally proposed to sort membrane proteins along the cellular pathways of membrane transport. At present, most research focuses on the organizing function during signal transduction.

Sphingolipids are synthesized in a pathway that begins in the ER and is completed in the Golgi apparatus, but these lipids are enriched in the plasma membrane and in endosomes, where they perform many of their functions. Transport occurs via vesicles and monomeric transport in the cytosol. Sphingolipids are virtually absent from mitochondria and the ER, but constitute a 20-35 molar fraction of plasma membrane lipids.

In experimental animals, feeding sphingolipids inhibits colon carcinogenesis, reduces LDL cholesterol and elevates HDL cholesterol.

Yeast Sphingolipids

Because of the incredible complexity of mammalian systems, yeast are often used as a model organism for working out new pathways. These single-celled organisms are often more genetically tractable than mammalian cells, and strain libraries are available to supply strains harboring almost any non-lethal open reading frame single deletion. The two most commonly used yeasts are *Saccharomyces cerevisiae* and *Schizosaccharomyces pombe*, although research is also done in the pathogenic yeast *Candida albicans*.

In addition to the important structural functions of complex sphingolipids (inositol phosphorylceramide and its mannosylated derivatives), the sphingoid bases phytosphingosine and dihydrosphingosine (sphinganine) play vital signaling roles in *S. cerevisiae*. These effects include regulation of endocytosis, ubiquitin-dependent proteolysis (and, thus, regulation of nutrient uptake), cytoskeletal dynamics, the cell cycle, translation, posttranslational protein modification, and the heat stress response. Additionally, modulation of sphingolipid metabolism by phosphatidylinositol (4,5)-bisphosphate signaling *via* Slm1p and Slm2p and calcineurin has recently been described. Additionally, a substrate-level interaction has been shown between complex sphingolipid synthesis and cycling of phosphatidylinositol 4-phosphate by the phosphatidylinositol kinase Stt4p and the lipid phosphatase Sac1p.

Plant Sphingolipids

Higher plants contain a wider variety of sphingolipids than animals and fungi.

Disorders

There are several disorders of sphingolipid metabolism, known as sphingolipidoses. The main members of this group are Niemann-Pick disease, Fabry disease, Krabbe disease, Gaucher disease, Tay-Sachs disease and Metachromatic leukodystrophy. They are generally inherited in an autosomal recessive fashion, but notably Fabry disease is X-linked. Taken together, sphingolipidoses have an incidence of approximately 1 in 10,000, but substantially more in certain populations such as Ashkenazi Jews. Enzyme replacement therapy is available to treat mainly Fabry disease and Gaucher disease, and people with these types of sphingolipidoses may live well into adulthood. The other types are generally fatal by age 1 to 5 years for infantile forms, but progression may be mild for juvenile- or adult-onset forms.

Sphingolipids have also been implicated with the frataxin protein (Fxn), the deficiency of which is associated with Friedreich's ataxia (FRDA). Loss of Fxn in the nervous system in mice also activates an iron/sphingolipid/PDK1/Mef2 pathway, indicating that the mechanism is evolutionarily conserved. Furthermore, sphingolipid levels and PDK1 activity are also increased in hearts of FRDA patients, suggesting that a similar pathway is affected in FRDA. Other research has demonstrated that iron accumulation in the nervous systems of flies enhances the synthesis of sphingolipids, which in turn activates 3-phosphoinositide dependent protein kinase-1 (Pdk1) and myocyte enhancer factor-2 (Mef2) to trigger neurodegeneration of adult photoreceptors.

Lipid Bilayer

The lipid bilayer (or phospholipid bilayer) is a thin polar membrane made of two layers of lipid molecules. These membranes are flat sheets that form a continuous barrier around all cells. The cell membranes of almost all living organisms and many viruses are made of a lipid bilayer, as are the membranes surrounding the cell nucleus and other sub-cellular structures. The lipid bilayer is the barrier that keeps ions, proteins and other molecules where they are needed and prevents them from diffusing into areas where they should not be. Lipid bilayers are ideally suited to this role, even though they are only a few nanometers in width, they are impermeable to most water-soluble (hydrophilic) molecules. Bilayers are particularly impermeable to ions, which allows cells to regulate salt concentrations and pH by transporting ions across their membranes using proteins called ion pumps.

Biological bilayers are usually composed of amphiphilic phospholipids that have a hydrophilic phosphate head and a hydrophobic tail consisting of two fatty acid chains. Phospholipids with certain head groups can alter the surface chemistry of a bilayer and can, for example, serve as signals as well as "anchors" for other molecules in the membranes of cells. Just like the heads, the tails of lipids can also affect membrane properties, for instance by determining the phase of the bilayer. The bilayer can adopt a solid gel phase state at lower temperatures but undergo phase transition to a fluid state at higher temperatures, and the chemical properties of the lipids' tails influence at which temperature this happens. The packing of lipids within the bilayer also affects its mechanical properties, including its resistance to stretching and bending. Many of these properties have been studied with the use of artificial "model" bilayers produced in a lab. Vesicles made by model bilayers have also been used clinically to deliver drugs.

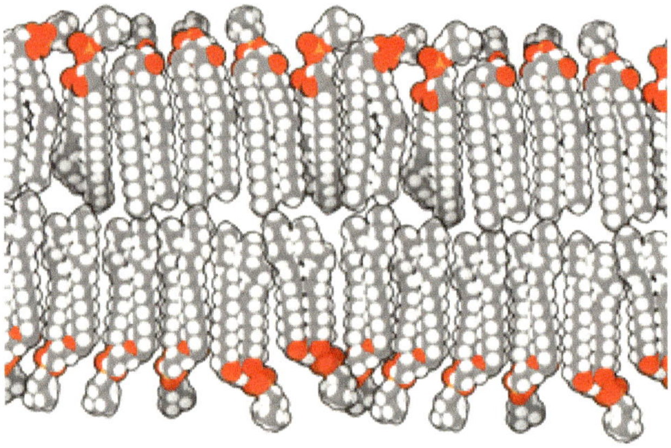

This fluid lipid bilayer cross section is made up entirely of phosphatidylcholine.

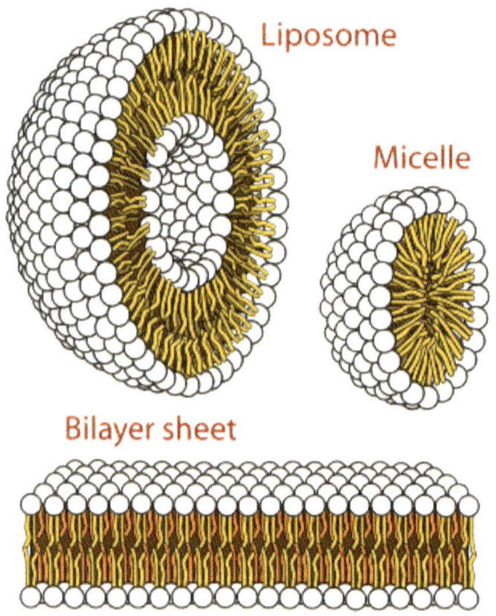

The three main structures phospholipids form in solution; the liposome (a closed bilayer), the micelle and the bilayer.

Biological membranes typically include several types of molecules other than phospholipids. A particularly important example in animal cells is cholesterol, which helps strengthen the bilayer and decrease its permeability. Cholesterol also helps regulate the activity of certain integral membrane proteins. Integral membrane proteins function when incorporated into a lipid bilayer, and they are held tightly to lipid bilayer with the help of an annular lipid shell. Because bilayers define the boundaries of the cell and its compartments, these membrane proteins are involved in many intra- and inter-cellular signaling processes. Certain kinds of membrane proteins are involved in the process of fusing two bilayers together. This fusion allows the joining of two distinct structures as in the fertilization of an egg by sperm or the entry of a virus into a cell. Because lipid bilayers are quite fragile and invisible in a traditional microscope, they are a challenge to study. Experiments on bilayers often require advanced techniques like electron microscopy and atomic force microscopy.

Structure and Organization

When phospholipids are exposed to water, they self-assemble into a two-layered sheet with the hydrophobic tails pointing toward the center of the sheet. This arrangement results in two "leaflets" that are each a single molecular layer. The center of this bilayer contains almost no water and excludes molecules like sugars or salts that dissolve in water. The assembly process is driven by interactions between hydrophobic molecules (also called the hydrophobic effect). An increase in interactions between hydrophobic molecules (causing clustering of hydrophobic regions) allows water molecules to bond more freely with each other, increasing the entropy of the system. This complex process includes non-covalent interactions such as van der Waals forces, electrostatic and hydrogen bonds.

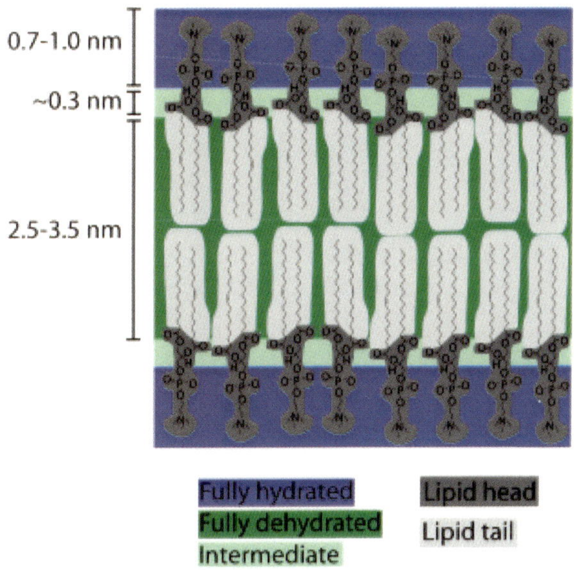

Schematic cross sectional profile of a typical lipid bilayer. There are three distinct regions: the fully hydrated headgroups, the fully dehydrated alkane core and a short intermediate region with partial hydration. Although the head groups are neutral, they have significant dipole moments that influence the molecular arrangement.

Cross Section Analysis

The lipid bilayer is very thin compared to its lateral dimensions. If a typical mammalian cell (diameter ~10 micrometers) were magnified to the size of a watermelon (~1 ft/30 cm), the lipid bilayer making up the plasma membrane would be about as thick as a piece of office paper. Despite being only a few nanometers thick, the bilayer is composed of several distinct chemical regions across its cross-section. These regions and their interactions with the surrounding water have been characterized over the past several decades with x-ray reflectometry, neutron scattering and nuclear magnetic resonance techniques.

The first region on either side of the bilayer is the hydrophilic headgroup. This portion of the membrane is completely hydrated and is typically around 0.8-0.9 nm thick. In phospholipid bilayers the phosphate group is located within this hydrated region, approximately 0.5 nm outside the hydrophobic core. In some cases, the hydrated region can extend much further, for instance in lipids with a large protein or long sugar chain grafted to the head. One common example of such a modification in nature is the lipopolysaccharide coat on a bacterial outer membrane, which helps retain a water layer around the bacterium to prevent dehydration.

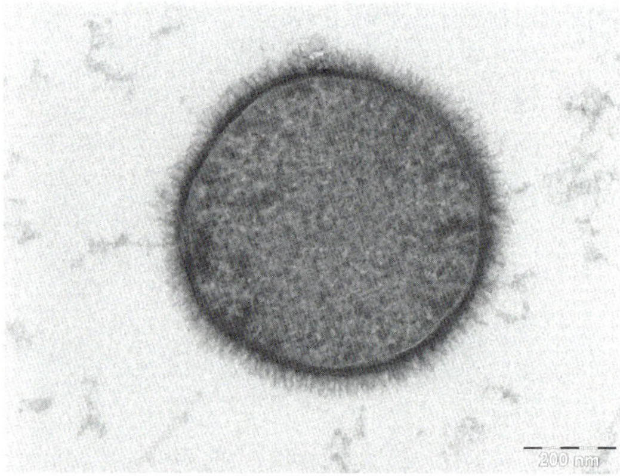

TEM image of a bacterium. The furry appearance on the outside is due to a coat of long-chain sugars attached to the cell membrane. This coating helps trap water to prevent the bacterium from becoming dehydrated.

Next to the hydrated region is an intermediate region that is only partially hydrated. This boundary layer is approximately 0.3 nm thick. Within this short distance, the water concentration drops from 2M on the headgroup side to nearly zero on the tail (core) side. The hydrophobic core of the bilayer is typically 3-4 nm thick, but this value varies with chain length and chemistry. Core thickness also varies significantly with temperature, in particular near a phase transition.

Asymmetry

In many naturally occurring bilayers, the compositions of the inner and outer membrane leaflets are different. In human red blood cells, the inner (cytoplasmic) leaflet is composed mostly of phosphatidylethanolamine, phosphatidylserine and phosphatidylinositol and its phosphorylated derivatives. By contrast, the outer (extracellular) leaflet is based on phosphatidylcholine, sphingomyelin and a variety of glycolipids, In some cases, this asymmetry is based on where the lipids are made in the cell and reflects their initial orientation. The biological functions of lipid asymmetry are imperfectly understood, although it is clear that it is used in several different situations. For example, when a cell undergoes apoptosis, the phosphatidylserine — normally localised to the cytoplasmic leaflet — is transferred to the outer surface: There, it is recognised by a macrophage that then actively scavenges the dying cell.

Lipid asymmetry arises, at least in part, from the fact that most phospholipids are synthesised and initially inserted into the inner monolayer: those that constitute the outer monolayer are then transported from the inner monolayer by a class of enzymes called flippases. Other lipids, such as sphingomyelin, appear to be synthesised at the external leaflet. Flippases are members of a larger family of lipid transport molecules that also includes floppases, which transfer lipids in the opposite direction, and scramblases, which randomize lipid distribution across lipid bilayers (as in apoptotic cells). In any case, once lipid asymmetry is established, it does not normally dissipate quickly because spontaneous flip-flop of lipids between leaflets is extremely slow.

It is possible to mimic this asymmetry in the laboratory in model bilayer systems. Certain types of very small artificial vesicle will automatically make themselves slightly asymmetric, although the mechanism by which this asymmetry is generated is very different from that in cells. By utilizing two different monolayers in Langmuir-Blodgett deposition or a combination of Langmuir-Blodgett and vesicle rupture deposition it is also possible to synthesize an asymmetric planar bilayer. This asymmetry may be lost over time as lipids in supported bilayers can be prone to flip-flop.

Phases and Phase Transitions

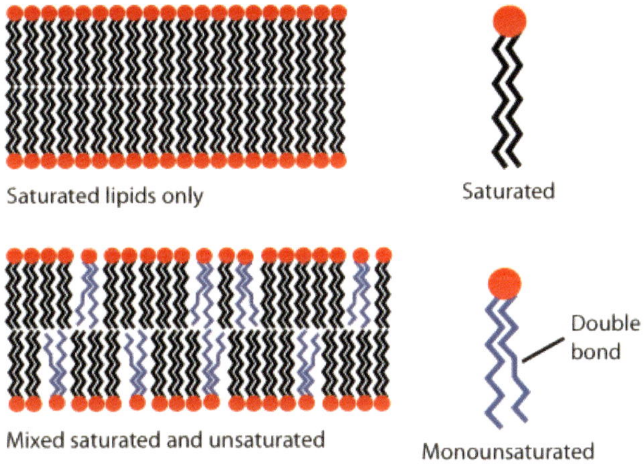

Diagram showing the effect of unsaturated lipids on a bilayer. The lipids with an unsaturated tail (blue) disrupt the packing of those with only saturated tails (black). The resulting bilayer has more free space and is, as a consequence, more permeable to water and other small molecules.

At a given temperature a lipid bilayer can exist in either a liquid or a gel (solid) phase. All lipids have a characteristic temperature at which they transition (melt) from the gel to liquid phase. In both phases the lipid molecules are prevented from flip-flopping across the bilayer, but in liquid phase bilayers a given lipid will exchange locations with its neighbor millions of times a second. This random walk exchange allows lipid to diffuse and thus wander across the surface of the membrane. Unlike liquid phase bilayers, the lipids in a gel phase bilayer have less mobility.

The phase behavior of lipid bilayers is determined largely by the strength of the attractive Van der Waals interactions between adjacent lipid molecules. Longer-tailed lipids have more area over which to interact, increasing the strength of this interaction and, as a consequence, decreasing the lipid mobility. Thus, at a given temperature, a short-tailed lipid will be more fluid than an otherwise identical long-tailed lipid. Transition temperature can also be affected by the degree of unsaturation of the lipid tails. An unsaturated double bond can produce a kink in the alkane chain, disrupting the lipid packing. This disruption creates extra free space within the bilayer that allows additional flexibility in the adjacent chains. An example of this effect can be noted in everyday life as butter, which has a large percentage saturated fats, is solid at room temperature while vegetable oil, which is mostly unsaturated, is liquid.

Most natural membranes are a complex mixture of different lipid molecules. If some of the components are liquid at a given temperature while others are in the gel phase, the two phases can coexist in spatially separated regions, rather like an iceberg floating in the ocean. This phase separation plays a critical role in biochemical phenomena because membrane components such as proteins can partition into one or the other phase and thus be locally concentrated or activated. One particularly important component of many mixed phase systems is cholesterol, which modulates bilayer permeability, mechanical strength, and biochemical interactions.

Surface Chemistry

While lipid tails primarily modulate bilayer phase behavior, it is the headgroup that determines the bilayer surface chemistry. Most natural bilayers are composed primarily of phospholipids, but sphingolipids and sterols such as cholesterol are also important components. Of the phospholipids, the most common headgroup is phosphatidylcholine (PC), accounting for about half the phospholipids in most mammalian cells. PC is a zwitterionic headgroup, as it has a negative charge on the phosphate group and a positive charge on the amine but, because these local charges balance, no net charge.

Other headgroups are also present to varying degrees and can include phosphatidylserine (PS) phosphatidylethanolamine (PE) and phosphatidylglycerol (PG). These alternate headgroups often confer specific biological functionality that is highly context-dependent. For instance, PS presence on the extracellular membrane face of erythrocytes is a marker of cell apoptosis, whereas PS in growth plate vesicles is necessary for the nucleation of hydroxyapatite crystals and subsequent bone mineralization. Unlike PC, some of the other headgroups carry a net charge, which can alter the electrostatic interactions of small molecules with the bilayer.

Biological Roles

Containment and Separation

The primary role of the lipid bilayer in biology is to separate aqueous compartments from their surroundings. Without some form of barrier delineating "self" from "non-self," it is difficult to even define the concept of an organism or of life. This barrier takes the form of a lipid bilayer in all known life forms except for a few species of archaea that utilize a specially adapted lipid monolayer. It has even been proposed that the very first form of life may have been a simple lipid vesicle with virtually its sole biosynthetic capability being the production of more phospholipids. The partitioning ability of the lipid bilayer is based on the fact that hydrophilic molecules cannot easily cross the hydrophobic bilayer core, as discussed in Transport across the bilayer below. The nucleus, mitochondria and chloroplasts have two lipid bilayers, while other sub-cellular structures are surrounded by a single lipid bilayer (such as the plasma membrane, endoplasmic reticula, Golgi apparatus and lysosomes).

Prokaryotes have only one lipid bilayer- the cell membrane (also known as the plasma membrane). Many prokaryotes also have a cell wall, but the cell wall is composed of proteins or long chain carbohydrates, not lipids. In contrast, eukaryotes have a range of organelles including the nucleus, mitochondria, lysosomes and endoplasmic reticulum. All of these sub-cellular compartments are surrounded by one or more lipid bilayers and, together, typically comprise the majority of the bilayer area present in the cell. In liver hepatocytes for example, the plasma membrane accounts for only two percent of the total bilayer area of the cell, whereas the endoplasmic reticulum contains more than fifty percent and the mitochondria a further thirty percent.

Signaling

Probably the most familiar form of cellular signaling is synaptic transmission, whereby a nerve impulse that has reached the end of one neuron is conveyed to an adjacent neuron via the release of neurotransmitters. This transmission is made possible by the action of synaptic vesicles loaded with the neurotransmitters to be released. These vesicles fuse with the cell membrane at the pre-synaptic terminal and release its contents to the exterior of the cell. The contents then diffuse across the synapse to the post-synaptic terminal.

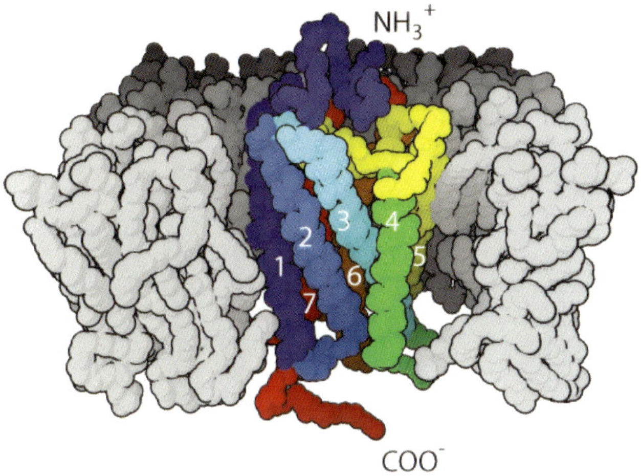

Illustration of a GPCR signaling protein. In response to a molecule such as a hormone binding to the exterior domain (blue) the GPCR changes shape and catalyzes a chemical reaction on the interior domain (red). The gray feature is the surrounding bilayer.

Lipid bilayers are also involved in signal transduction through their role as the home of integral membrane proteins. This is an extremely broad and important class of biomolecule. It is estimated that up to a third of the human proteome may be membrane proteins. Some of these proteins are linked to the exterior of the cell membrane. An example of this is the CD59 protein, which identifies cells as "self" and thus inhibits their destruction by the immune system. The HIV virus evades the immune system in part by grafting these proteins from the host membrane onto its own surface. Alternatively, some membrane proteins penetrate all the way through the bilayer and serve to relay individual signal events from the outside to the inside of the cell. The most common class of this type of protein is the G protein-coupled receptor (GPCR). GPCRs are responsible for much of the cell's ability to sense its surroundings and, because of this important role, approximately 40% of all modern drugs are targeted at GPCRs.

In addition to protein- and solution-mediated processes, it is also possible for lipid bilayers to participate directly in signaling. A classic example of this is phosphatidylserine-triggered phagocytosis. Normally, phosphatidylserine is asymmetrically distributed in the cell membrane and is present only on the interior side. During programmed cell death a protein called a scramblase equilibrates this distribution, displaying phosphatidylserine on the extracellular bilayer face. The presence of phosphatidylserine then triggers phagocytosis to remove the dead or dying cell.

Characterization Methods

The lipid bilayer is a very difficult structure to study because it is so thin and fragile. In spite of these limitations dozens of techniques have been developed over the last seventy years to allow investigations of its structure and function.

Electrical measurements are a straightforward way to characterize an important function of a bilayer: its ability to segregate and prevent the flow of ions in solution. By applying a voltage across the bilayer and measuring the resulting current, the resistance of the bilayer is determined. This resistance is typically quite high (10^8 Ohm-cm^2 or more) since the hydrophobic core is impermeable to charged species. The presence of even a few nanometer-scale holes results in a dramatic increase in current. The sensitivity of this system is such that even the activity of single ion channels can be resolved.

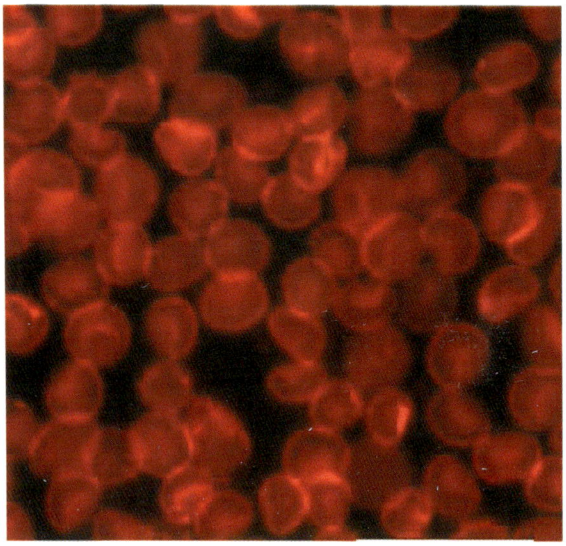

Human red blood cells viewed through a fluorescence microscope. The cell membrane has been stained with a fluorescent dye. Scale bar is 20μm.

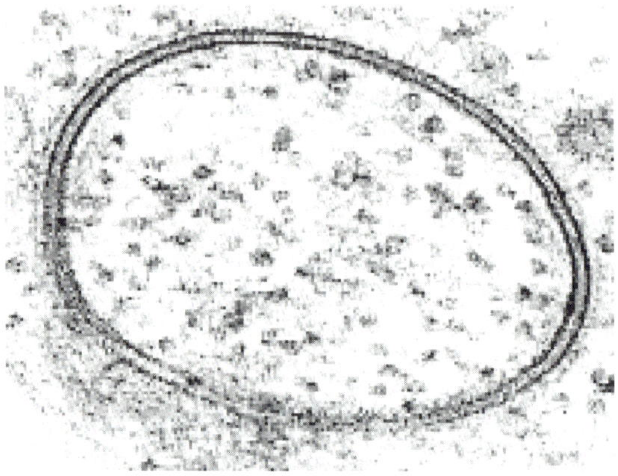

Transmission Electron Microscope (TEM) image of a lipid vesicle. The two dark bands around the edge are the two leaflets of the bilayer. Historically, similar images confirmed that the cell membrane is a bilayer

Electrical measurements do not provide an actual picture like imaging with a microscope can. Lipid bilayers cannot be seen in a traditional microscope because they are too thin. In order to see bilayers, researchers often use fluorescence microscopy. A sample is excited with one wavelength of light and observed in a different wavelength, so that only fluorescent molecules with a matching excitation and emission profile will be seen. Natural lipid bilayers are not fluorescent, so a dye is used that attaches to the desired molecules in the bilayer. Resolution is usually limited to a few hundred nanometers, much smaller than a typical cell but much larger than the thickness of a lipid bilayer.

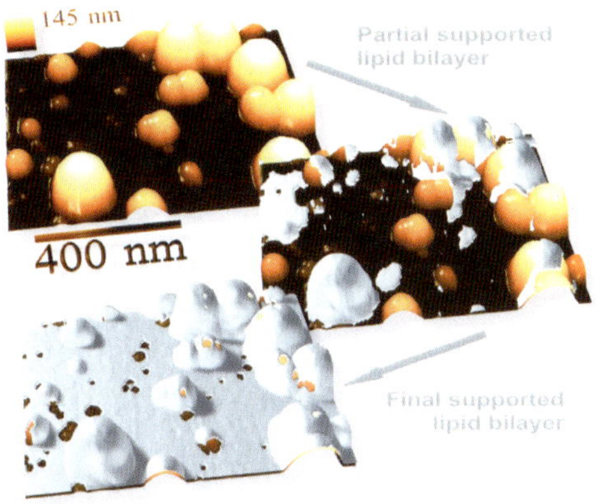

3d-Adapted AFM images showing formation of transmembrane pores (holes) in supported lipid bilayer

Electron microscopy offers a higher resolution image. In an electron microscope, a beam of focused electrons interacts with the sample rather than a beam of light as in traditional microscopy. In conjunction with rapid freezing techniques, electron microscopy has also been used to study the mechanisms of inter- and intracellular transport, for instance in demonstrating that exocytotic vesicles are the means of chemical release at synapses.

^{31}P-NMR(nuclear magnetic resonance) spectroscopy is widely used for studies of phospholipid bilayers and biological membranes in native conditions. The analysis of ^{31}P-NMR spectra of lipids could provide a wide range of information about lipid bilayer packing, phase transitions (gel phase, physiological liquid crystal phase, ripple phases, non bilayer phases), lipid head group orientation/dynamics, and elastic properties of pure lipid bilayer and as a result of binding of proteins and other biomolecules.

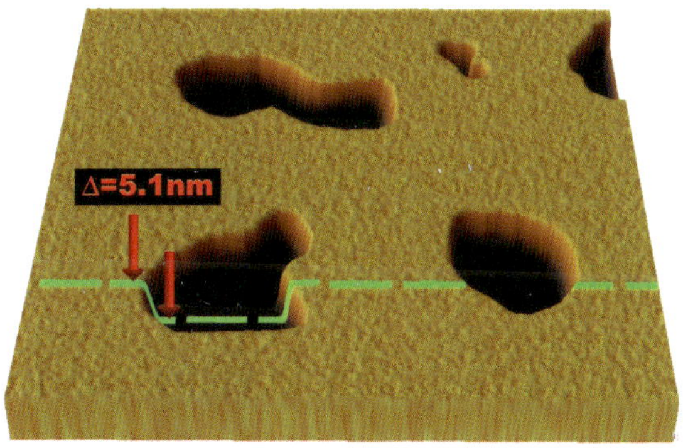

Illustration of a typical AFM scan of a supported lipid bilayer. The pits are defects in the bilayer, exposing the smooth surface of the substrate underneath.

A new method to study lipid bilayers is Atomic force microscopy (AFM). Rather than using a beam of light or particles, a very small sharpened tip scans the surface by making physical contact with the bilayer and moving across it, like a record player needle. AFM is a promising technique because it has the potential to image with nanometer resolution at room temperature and even under water or physiological buffer, conditions necessary for natural bilayer behavior. Utilizing this capability, AFM has been used to examine dynamic bilayer behavior including the formation of transmembrane pores (holes) and phase transitions in supported bilayers. Another advantage is that AFM does not require fluorescent or isotopic labeling of the lipids, since the probe tip interacts mechanically with the bilayer surface. Because of this, the same scan can image both lipids and associated proteins, sometimes even with single-molecule resolution. AFM can also probe the mechanical nature of lipid bilayers.

Lipid bilayers exhibit high levels of birefringence where the refractive index in the plane of the bilayer differs from that perpendicular by as much as 0.1 refractive index units. This has been used to characterise the degree of order and disruption in bilayers using dual polarisation interferometry to understand mechanisms of protein interaction.

Lipid bilayers are complicated molecular systems with many degrees of freedom. Thus atomistic simulation of membrane and in particular ab initio calculations of its properties is difficult and computationally expensive. Quantum chemical calculations has recently been successfully performed to estimate dipole and quadrupole moments of lipid membranes.

Transport Across the Bilayer

Passive Diffusion

Most polar molecules have low solubility in the hydrocarbon core of a lipid bilayer and, as a consequence, have low permeability coefficients across the bilayer. This effect is particularly pronounced for charged species, which have even lower permeability coefficients than neutral polar molecules. Anions typically have a higher rate of diffusion through bilayers than cations. Compared to ions, water molecules actually have a relatively large permeability through the bilayer, as evidenced by osmotic swelling. When a cell or vesicle with a high interior salt concentration is placed in a solution with a low salt concentration it will swell and eventually burst. Such a result would not be observed unless water was able to pass through the bilayer with relative ease. The anomalously large permeability of water through bilayers is still not completely understood and continues to be the subject of active debate. Small uncharged apolar molecules diffuse through lipid bilayers many orders of magnitude faster than ions or water. This applies both to fats and organic solvents like chloroform and ether. Regardless of their polar character larger molecules diffuse more slowly across lipid bilayers than small molecules.

Ion Pumps and Channels

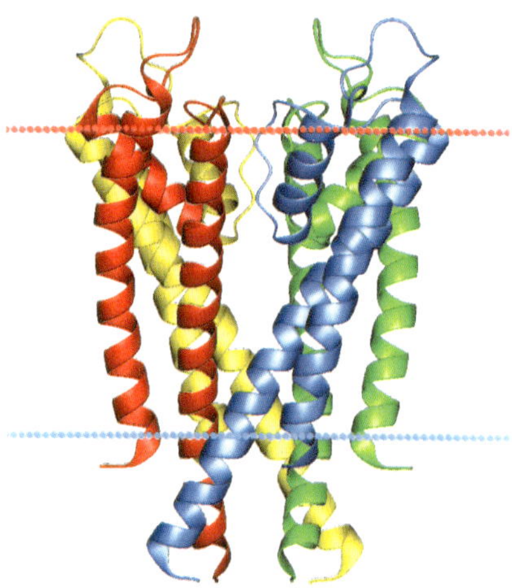

Structure of a potassium ion channel. The alpha helices penetrate the bilayer (boundaries indicated by red and blue lines), opening a hole through which potassium ions can flow

Two special classes of protein deal with the ionic gradients found across cellular and sub-cellular membranes in nature- ion channels and ion pumps. Both pumps and channels are integral membrane proteins that pass through the bilayer, but their roles are quite different. Ion pumps are the proteins that build and maintain the chemical gradients by utilizing an external energy source to move ions against the concentration gradient to an area of higher chemical potential. The energy source can be ATP, as is the case for the Na^+-K^+ ATPase. Alternatively, the energy source can be another chemical gradient already in place, as in the Ca^{2+}/Na^+ antiporter. It is through the action of ion pumps that cells are able to regulate pH via the pumping of protons.

In contrast to ion pumps, ion channels do not build chemical gradients but rather dissipate them in order to perform work or send a signal. Probably the most familiar and best studied example is the voltage-gated Na^+ channel, which allows conduction of an action potential along neurons. All ion pumps have some sort of trigger or "gating" mechanism. In the previous example it was electrical bias, but other channels can be activated by binding a molecular agonist or through a conformational change in another nearby protein.

Endocytosis and Exocytosis

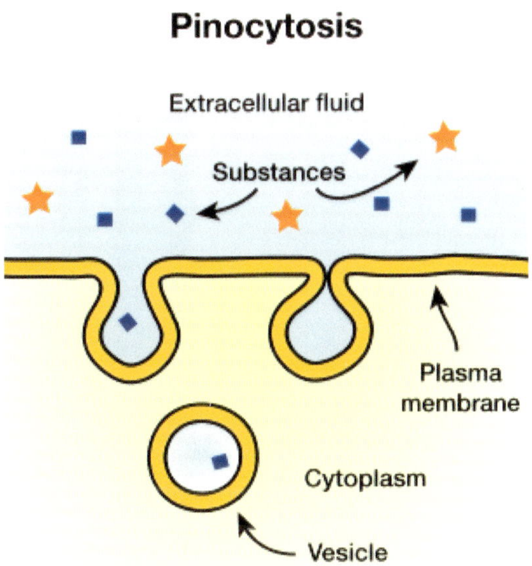

Schematic illustration of pinocytosis, a type of endocytosis

Some molecules or particles are too large or too hydrophilic to pass through a lipid bilayer. Other molecules could pass through the bilayer but must be transported rapidly in such large numbers that channel-type transport is impractical. In both cases, these types of cargo can be moved across the cell membrane through fusion or budding of vesicles. When a vesicle is produced inside the cell and fuses with the plasma membrane to release its contents into the extracellular space, this process is known as exocytosis. In the reverse process, a region of the cell membrane will dimple inwards and eventually pinch off, enclosing a portion of the extracellular fluid to transport it into the cell. Endocytosis and exocytosis rely on very different molecular machinery to function, but the two processes are intimately linked and could not work without each other. The primary mechanism of this interdependence is the sheer volume of lipid material involved. In a typical cell, an area of bilayer equivalent to the entire plasma membrane will travel through the endocytosis/exocytosis cycle in about half an hour. If these two processes were not balancing each other, the cell would either balloon outward to an unmanageable size or completely deplete its plasma membrane within a matter of minutes.

Exocytosis in prokaryotes: Membrane vesicular exocytosis, popularly known as membrane vesicle trafficking, a Nobel prize-winning (year, 2013) process, is traditionally regarded as a prerogative of eukaryotic cells. This *myth* was however broken with the revelation that nanovesicles, popularly known as bacterial outer membrane vesicles, released by gram-negative microbes, translocate bacterial signal molecules to host or target cells to carry out multiple processes in favour of the secreting microbe e.g., in *host cell invasion* and microbe-environment interactions, in general.

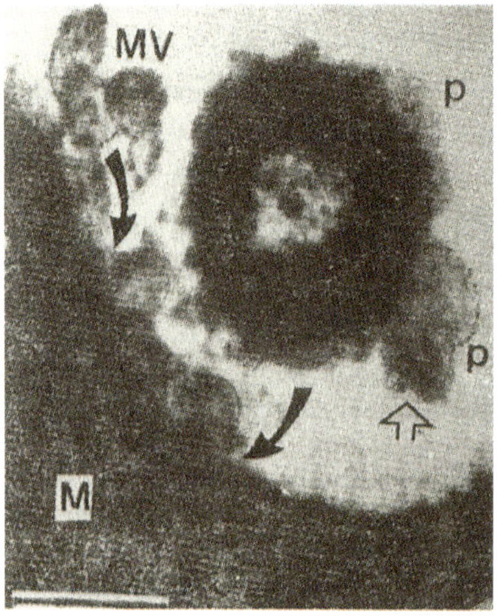

Exocytosis of outer membrane vesicles (MV) liberated from inflated periplasmic pockets (p) on surface of human *Salmonella* 3,10:r:- pathogens docking on plasma membrane of macrophage cells (M) in chicken ileum, for host-pathogen signaling *in vivo*.

Electroporation

Electroporation is the rapid increase in bilayer permeability induced by the application of a large artificial electric field across the membrane. Experimentally, electroporation is used to introduce hydrophilic molecules into cells. It is a particularly useful technique for large highly charged molecules such as DNA, which would never passively diffuse across the hydrophobic bilayer core. Because of this, electroporation is one of the key methods of transfection as well as bacterial transformation. It has even been proposed that electroporation resulting from lightning strikes could be a mechanism of natural horizontal gene transfer.

This increase in permeability primarily affects transport of ions and other hydrated species, indicating that the mechanism is the creation of nm-scale water-filled holes in the membrane. Although electroporation and dielectric breakdown both result from application of an electric field, the mechanisms involved are fundamentally different. In dielectric breakdown the barrier material is ionized, creating a conductive pathway. The material alteration is thus chemical in nature. In contrast, during electroporation the lipid molecules are not chemically altered but simply shift position, opening up a pore that acts as the conductive pathway through the bilayer as it is filled with water.

Mechanics

Lipid bilayers are large enough structures to have some of the mechanical properties of liquids or solids. The area compression modulus K_a, bending modulus K_b, and edge energy , can be used to describe them. Solid lipid bilayers also have a shear modulus, but like any liquid, the shear modulus is zero for fluid bilayers. These mechanical properties affect how the membrane functions. K_a and K_b affect the ability of proteins and small molecules to insert into the bilayer, and bilayer mechanical properties have been shown to alter the function of mechanically activated ion channels. Bilayer mechanical properties also govern what types of stress a cell can withstand without tearing. Although lipid bilayers can easily bend, most cannot stretch more than a few percent before rupturing.

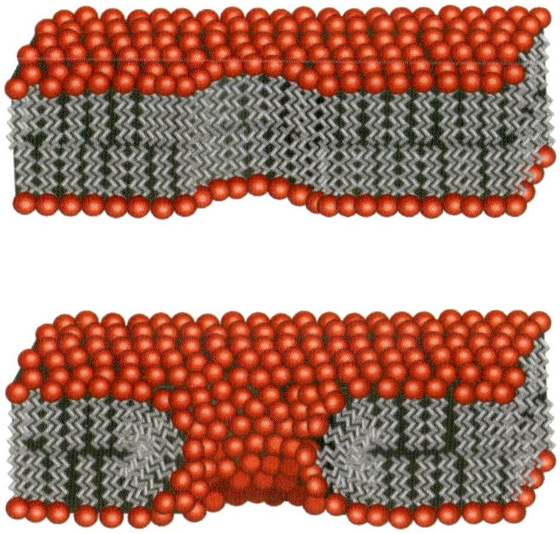

Schematic showing two possible conformations of the lipids at the edge of a pore. In the top image the lipids have not rearranged, so the pore wall is hydrophobic. In the bottom image some of the lipid heads have bent over, so the pore wall is hydrophilic.

As discussed in the Structure and organization section, the hydrophobic attraction of lipid tails in water is the primary force holding lipid bilayers together. Thus, the elastic modulus of the bilayer is primarily determined by how much extra area is exposed to water when the lipid molecules are stretched apart. It is not surprising given this understanding of the forces involved that studies have shown that K_a varies strongly with osmotic pressure but only weakly with tail length and unsaturation. Because the forces involved are so small, it is difficult to experimentally determine K_a. Most techniques require sophisticated microscopy and very sensitive measurement equipment.

In contrast to K_a, which is a measure of how much energy is needed to stretch the bilayer, K_b is a measure of how much energy is needed to bend or flex the bilayer. Formally, bending modulus is defined as the energy required to deform a membrane from its intrinsic curvature to some other curvature. Intrinsic curvature is defined by the ratio of the diameter of the head group to that of the tail group. For two-tailed PC lipids, this ratio is nearly one so the intrinsic curvature is nearly zero. If a particular lipid has too large a deviation from zero intrinsic curvature it will not form a bilayer and will instead form other phases such as micelles or inverted micelles. Addition of *small hydrophilic molecules* like *sucrose* into mixed lipid *lamellar liposomes* made from galactolipid-rich thylakoid membranes destabilises bilayers into micellar phase. Typically, K_b is not measured experimentally but rather is calculated from measurements of K_a and bilayer thickness, since the three parameters are related.

Λ is a measure of how much energy it takes to expose a bilayer edge to water by tearing the bilayer or creating a hole in it. The origin of this energy is the fact that creating such an interface exposes some of the lipid tails to water, but the exact orientation of these border lipids is unknown. There is some evidence that both hydrophobic (tails straight) and hydrophilic (heads curved around) pores can coexist.

Fusion

Fusion is the process by which two lipid bilayers merge, resulting in one connected structure. If this fusion proceeds completely through both leaflets of both bilayers, a water-filled bridge is formed and the solutions contained by the bilayers can mix. Alternatively, if only one leaflet from each bilayer is involved in the fusion process, the bilayers are said to be hemifused. Fusion is involved in many cellular processes, in particular in eukaryotes, since the eukaryotic cell is extensively sub-divided by lipid bilayer membranes. Exocytosis, fertilization of an egg by sperm and transport of waste products to the lysozome are a few of the many eukaryotic processes that rely on some form of fusion. Even the entry of pathogens can be governed by fusion, as many bilayer-coated viruses have dedicated fusion proteins to gain entry into the host cell.

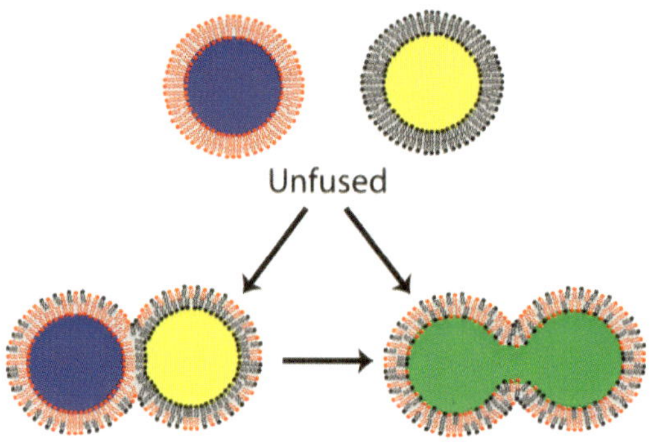

Illustration of lipid vesicles fusing showing two possible outcomes: hemifusion and full fusion. In hemifusion, only the outer bilayer leaflets mix. In full fusion both leaflets as well as the internal contents mix.

There are four fundamental steps in the fusion process. First, the involved membranes must aggregate, approaching each other to within several nanometers. Second, the two bilayers must come into very close contact (within a few angstroms). To achieve this close contact, the two surfaces must become at least partially dehydrated, as the bound surface water normally present causes bilayers to strongly repel. The presence of ions, in particular divalent cations like magnesium and calcium, strongly affects this step. One of the critical roles of calcium in the body is regulating membrane fusion. Third, a destabilization must form at one point between the two bilayers, locally distorting their structures. The exact nature of this distortion is not known. One theory is that a highly curved "stalk" must form between the two bilayers. Proponents of this theory believe that it explains why phosphatidylethanolamine, a highly curved lipid, promotes fusion. Finally, in the last step of fusion, this point defect grows and the components of the two bilayers mix and diffuse away from the site of contact.

Schematic illustration of the process of fusion through stalk formation.

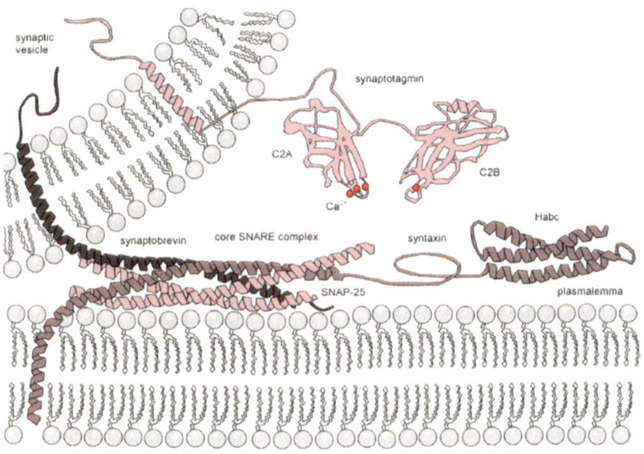

Diagram of the action of SNARE proteins docking a vesicle for exocytosis. Complementary versions of the protein on the vesicle and the target membrane bind and wrap around each other, drawing the two bilayers close together in the process.

The situation is further complicated when considering fusion *in vivo* since biological fusion is almost always regulated by the action of membrane-associated proteins. The first of these proteins to be studied were the viral fusion proteins, which allow an enveloped virus to insert its genetic material into the host cell (enveloped viruses are those surrounded by a lipid bilayer; some others have only a protein coat). Eukaryotic cells also use fusion proteins, the best-studied of which are the SNAREs. SNARE proteins are used to direct all vesicular intracellular trafficking. Despite years of study, much is still unknown about the function of this protein class. In fact, there is still an active debate regarding whether SNAREs are linked to early docking or participate later in the fusion process by facilitating hemifusion.

In studies of molecular and cellular biology it is often desirable to artificially induce fusion. The addition of polyethylene glycol (PEG) causes fusion without significant aggregation or biochemical disruption. This procedure is now used extensively, for example by fusing B-cells with myeloma cells. The resulting "hybridoma" from this combination expresses a desired antibody as determined by the B-cell involved, but is immortalized due to the melanoma component. Fusion can also be artificially induced through electroporation in a process known as electrofusion. It is believed that this phenomenon results from the energetically active edges formed during electroporation, which can act as the local defect point to nucleate stalk growth between two bilayers.

Model Systems

Lipid bilayers can be created artificially in the lab to allow researchers to perform experiments that cannot be done with natural bilayers. These synthetic systems are called model lipid bilayers. There are many different types of model bilayers, each having experimental advantages and disadvantages. They can be made with either synthetic or natural lipids. Among the most common model systems are:

- Black lipid membranes (BLM)
- Supported lipid bilayers (SLB)
- Tethered Bilayer Lipid Membranes (t-BLM)
- Vesicles

Liposome

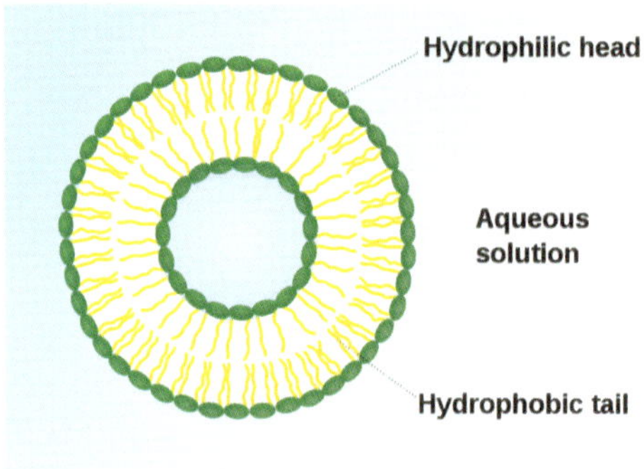

Scheme of a liposome formed by phospholipids in an aqueous solution.

A liposome is a spherical vesicle having at least one lipid bilayer. The liposome can be used as a vehicle for administration of nutrients and pharmaceutical drugs. Liposomes can be prepared by disrupting biological membranes (such as by sonication).

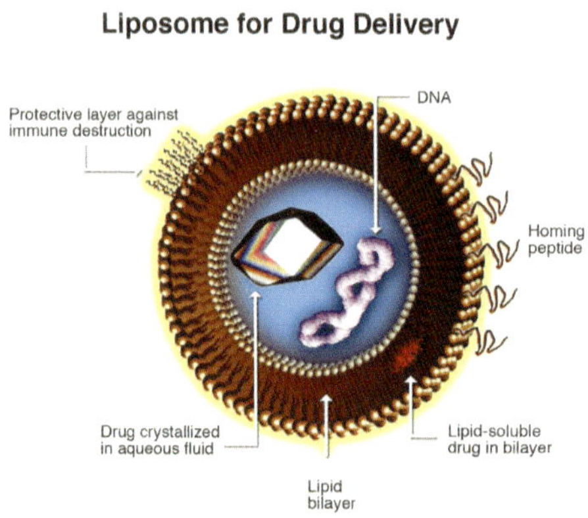

Liposomes are composite structures made of phospholipids and may contain small amounts of other molecules. Though liposomes can vary in size from low micrometer range to tens of micrometers, unilamellar liposomes, as pictured here, are typically in the lower size range with various targeting ligands attached to their surface allowing for their surface-attachment and accumulation in pathological areas for treatment of disease.

Liposomes are most often composed of phospholipids, especially phosphatidylcholine, but may also include other lipids, such as egg phosphatidylethanolamine, so long as they are compatible with lipid bilayer structure. A liposome design may employ surface ligands for attaching to unhealthy tissue.

The major types of liposomes are the multilamellar vesicle (MLV, with several lamellar phase lipid bilayers), the small unilamellar liposome vesicle (SUV, with one lipid bilayer), the large unilamellar vesicle (LUV), and the cochleate vesicle. A less desirable form are multivesicular liposomes in which one vesicle contains one or more smaller vesicles.

Liposomes should not be confused with lysosomes, or with micelles and reverse micelles composed of monolayers.

Discovery

The word *liposome* derives it is so named because its composition is primarily of phospholipid.

Liposomes were first described by British haematologist Alec D Bangham in 1961 (published 1964), at the Babraham Institute, in Cambridge. They were discovered when Bangham and R. W. Horne were testing the institute's new electron microscope by adding negative stain to dry phospholipids. The resemblance to the plasmalemma was obvious, and the microscope pictures served as the first evidence for the cell membrane being a bilayer lipid structure. Their integrity as a closed, bilayer structure, that could release its contents after detergent treatment (structure-linked latency) was established by Bangham, Standish and Weissmann in the next year. Weissmann - during a Cambridge pub discussion with Bangham - first named the structures "liposomes" after the lysosome, which his laboratory had been studying: a simple organelle the structure-linked latency of which could be disrupted by detergents and streptolysins. Liposomes can be easily distinguished from micelles and hexagonal lipid phases by negative staining transmission electron microscopy.

Alec Douglas Bangham with colleagues Jeff Watkins and Malcolm Standish wrote the 1965 paper that effectively launched the liposome "industry". Around this time he was joined at Babraham by Gerald Weissmann, an American physician with an interest in lysosomes. Now an emeritus professor at New York University School of Medicine, Weissmann recalls the two of them sitting in a Cambridge pub and reflecting on the role of lipid sheets in separating the interior of the cell from the exterior milieu. This insight, they felt, was to cell function what the discovery of the double helix had been to genetics. Bangham had called his lipid structures "multilamellar smectic mesophases" or sometimes "Banghasomes". It was Weissmann who proposed the more user-friendly term liposome.

Mechanism

A liposome has an aqueous solution core surrounded by a hydrophobic membrane, in the form of a lipid bilayer; hydrophilic solutes dissolved in the core cannot readily pass through the bilayer. Hydrophobic chemicals associate with the bilayer. A liposome can be hence loaded with hydrophobic and/or hydrophilic molecules. To deliver the molecules to a site of action, the lipid bilayer can fuse with other bilayers such as the cell membrane, thus delivering the liposome contents; this is a complex and non-spontaneous event, however. By preparing liposomes in a solution of DNA or drugs (which would normally be unable to diffuse through the membrane) they can be (indiscriminately) delivered past the lipid bilayer, but are then typically distributed non-homogeneously.

Liposomes are used as models for artificial cells. Liposomes can also be designed to deliver drugs in other ways. Liposomes that contain low (or high) pH can be constructed such that dissolved aqueous drugs will be charged in solution (i.e., the pH is outside the drug's pI range). As the pH naturally neutralizes within the liposome (protons can pass through some membranes), the drug will also be neutralized, allowing it to freely pass through a membrane. These liposomes work to deliver drug by diffusion rather than by direct cell fusion.

A similar approach can be exploited in the biodetoxification of drugs by injecting empty liposomes with a transmembrane pH gradient. In this case the vesicles act as sinks to scavenge the drug in the blood circulation and prevent its toxic effect. Another strategy for liposome drug delivery is to target endocytosis events. Liposomes can be made in a particular size range that makes them viable targets for natural macrophage phagocytosis. These liposomes may be digested while in the macrophage's phagosome, thus releasing its drug. Liposomes can also be decorated with opsonins and ligands to activate endocytosis in other cell types.

The use of liposomes for transformation or transfection of DNA into a host cell is known as lipofection.

In addition to gene and drug delivery applications, liposomes can be used as carriers for the delivery of dyes to textiles, pesticides to plants, enzymes and nutritional supplements to foods, and cosmetics to the skin.

Liposomes are also used as outer shells of some microbubble contrast agents used in contrast-enhanced ultrasound.

Lipid Signaling

Lipid signaling, broadly defined, refers to any biological signaling event involving a lipid messenger that binds a protein target, such as a receptor, kinase or phosphatase, which in turn mediate the effects of these lipids on specific cellular responses. Lipid signaling is thought to be qualitatively different from other classical signaling paradigms (such as monoamine neurotransmission) because lipids can freely diffuse through membranes. One consequence of this is that lipid messengers cannot be stored in vesicles prior to release and so are often biosynthesized "on demand" at their intended site of action. As such, many lipid signaling molecules cannot circulate freely in solution but, rather, exist bound to special carrier proteins in serum.

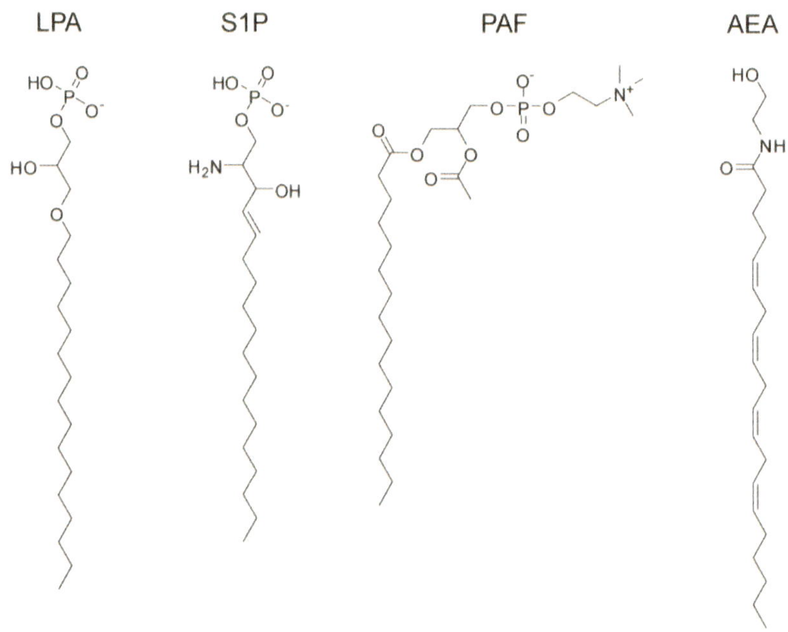

Common lipid signaling molecules:
lysophosphatidic acid (LPA)
sphingosine-1-phosphate (S1P)
platelet activating factor (PAF)
anandamide or arachidonoyl ethanolamine (AEA)

Sphingolipid Second Messengers

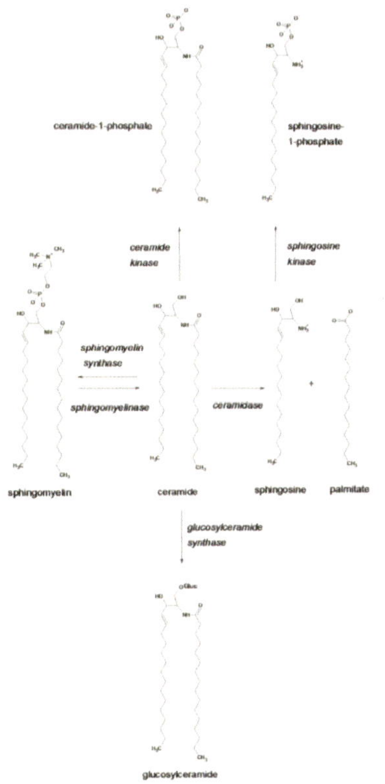

Sphingolipid second messengers. Ceramide is at the metabolic hub, leading to the formation of other sphingolipids.

Ceramide

Ceramide (Cer) can be generated by the breakdown of sphingomyelin (SM) by sphingomyelinases (SMases), which are enzymes that hydrolyze the phosphocholine group from the sphingosine backbone. Alternatively, this sphingosine-derived lipid (sphingolipid) can be synthesized from scratch (*de novo*) by the enzymes serine palmitoyl transferase (SPT) and ceramide synthase in organelles such as the endoplasmic reticulum (ER) and possibly, in the mitochondria-associated membranes (MAMs) and the perinuclear membranes. Being located in the metabolic hub, ceramide leads to the formation of other sphingolipids, with the C1 hydroxyl (-OH) group as the major site of modification. A sugar can be attached to ceramide (glycosylation) through the action of the enzymes, glucosyl or galactosyl ceramide synthases. Ceramide can also be broken down by enzymes called ceramidases, leading to the formation of sphingosine, Moreover, a phosphate group can be attached to ceramide (phosphorylation) by the enzyme, ceramide kinase. It is also possible to regenerate sphingomyelin from ceramide by accepting a phosphocholine headgroup from phosphatidylcholine (PC) by the action of an enzyme called sphingomyelin synthase. The latter process results in the formation of diacylglycerol (DAG) from PC.

Ceramide contains two hydrophobic ("water-fearing") chains and a neutral headgroup. Consequently, it has limited solubility in water and is restricted within the organelle where it was formed. Also, because of its hydrophobic nature, ceramide readily flip-flops across membranes as supported by studies in membrane models and membranes from red blood cells (erythrocytes). However, ceramide can possibly interact with other lipids to form bigger regions called microdomains which restrict its flip-flopping abilities. This could have immense effects on the signaling functions of ceramide because it is known that ceramide generated by acidic SMase enzymes in the outer leaflet of an organelle membrane may have different roles compared to ceramide that is formed in the inner leaflet by the action of neutral SMase enzymes.

Ceramide mediates many cell-stress responses, including the regulation of programmed cell death (apoptosis) and cell aging (senescence). Numerous research works have focused interest on defining the direct protein targets of action of ceramide. These include enzymes called ceramide-activated Ser-Thr phosphatases (CAPPs), such as protein phosphatase 1 and 2A (PP1 and PP2A), which were found to interact with ceramide in studies done in a controlled environment outside of a living organism (*in vitro*). On the other hand, studies in cells have shown that ceramide-inducing agents such as tumor necrosis factor-alpha α (TNFα) and palmitate induce the ceramide-dependent removal of a phosphate group (dephosphorylation) of the retinoblastoma gene product RB and the enzymes, protein kinases B (AKT protein family) and C α (PKB and PKCα). Moreover, there is also sufficient evidence which implicates ceramide to the activation of the kinase suppressor of Ras (KSR), PKCζ, and cathepsin D. Interestingly, cathepsin D has been proposed as the main target for ceramide formed in organelles called lysosomes, making lysosomal acidic SMase enzymes one of the key players in the mitochondrial pathway of apoptosis. Ceramide was also shown to activate PKCζ, implicating it to the inhibition of AKT, regulation of the voltage difference between the interior and exterior of the cell (membrane potential) and signaling functions that favor apoptosis. Chemotherapeutic agents such as daunorubicin and etoposide enhance the *de novo* synthesis of ceramide in studies done on mammalian cells. The same results were found for certain inducers of apoptosis particularly stimulators of receptors in a class of lymphocytes (a type of white blood cell) called B-cells. Regulation of the *de novo* synthesis of ceramide by palmitate may have a key role in diabetes and the metabolic syndrome. Experimental evidence shows that there is substantial increase of ceramide levels upon adding palmitate. Ceramide accumulation activates PP2A and the subsequent dephosphorylation and inactivation of AKT, a crucial mediator in metabolic control and insulin signaling. This results in a substantial decrease in insulin responsiveness (i.e. to glucose) and in the death of insulin-producing cells in the pancreas called islets of Langerhans. Inhibition of ceramide synthesis in mice via drug treatments or gene-knockout techniques prevented insulin resistance induced by fatty acids, glucocorticoids or obesity.

An increase in *in vitro* activity of acid SMase has been observed after applying multiple stress stimuli such as ultraviolet (UV) and ionizing radiation, binding of death receptors and chemotherapeutic agents such as platinum, histone deacetylase inhibitors and paclitaxel. In some studies, SMase activation results to its transport to the plasma membrane and the simultaneous formation of ceramide.

Ceramide transfer protein (CERT) transports ceramide from ER to the Golgi for the synthesis of SM. CERT is known to bind phosphatidylinositol phosphates, hinting its potential regulation via phosphorylation, a step of the ceramide metabolism that can be enzymatically regulated by protein kinases and phosphatases, and by inositol lipid metabolic pathways. Up to date, there are at least 26 distinct enzymes with varied subcellular localizations, that act on ceramide as either a substrate or product. Regulation of ceramide levels can therefore be performed by one of these enzymes in distinct organelles by particular mechanisms at various times.

Sphingosine

Sphingosine (Sph) is formed by the action of ceramidase (CDase) enzymes on ceramide in the lysosome. Sph can also be formed in the extracellular (outer leaflet) side of the plasma membrane by the action of neutral CDase enzyme. Sph then is either recycled back to ceramide or phosphorylated by one of the sphingosine kinase enzymes, SK1 and SK2. The product sphingosine-1-phosphate (S1P) can be dephosphorylated in the ER to regenerate sphingosine by certain S1P phosphatase enzymes within cells, where the salvaged Sph is recycled to ceramide. Sphingosine is a single-chain lipid (usually 18 carbons in length), rendering it to have sufficient solubility in water. This explains its ability to move between membranes and to flip-flop across a membrane. Estimates conducted at physiological pH show that approximately 70% of sphingosine remains in membranes while the remaining 30% is water-soluble. Sph that is formed has sufficient solubility in the liquid found inside cells (cytosol). Thus, Sph may come out of the lysosome and move to the ER without the need for transport via proteins or membrane-enclosed sacs called vesicles. However, its positive charge favors partitioning in lysosomes. It is proposed that the role of SK1 located near or in the lysosome is to 'trap' Sph via phosphorylation.

It is important to note that since sphingosine exerts surfactant activity, it is one of the sphingolipids found at lowest cellular levels. The low levels of Sph and their increase in response to stimulation of cells, primarily by activation of ceramidase by growth-inducing proteins such as platelet-derived growth factor and insulin-like growth factor, is consistent with its function as a second messenger. It was found that immediate hydrolysis of only 3 to 10% of newly generated ceramide may double the levels of Sph. Treatment of HL60 cells (a type of leukemia cell line) by a plant-derived organic compound called phorbol ester increased Sph levels threefold, whereby the cells differentiated into white blood cells called macrophages. Treatment of the same cells by exogenous Sph caused apoptosis. A specific protein kinase phosphorylates 14-3-3, otherwise known as sphingosine-dependent protein kinase 1 (SDK1), only in the presence of Sph.

Sph is also known to interact with protein targets such as the protein kinase H homologue (PKH) and the yeast protein kinase (YPK). These targets in turn mediate the effects of Sph and its related sphingoid bases, with known roles in regulating the actin cytoskeleton, endocytosis, the cell cycle and apoptosis. It is important to note however that the second messenger function of Sph is not yet established unambiguously.

Sphingosine-1-Phosphate

Sphingosine-1-phosphate (S1P), like Sph, is composed of a single hydrophobic chain and has sufficient solubility to move between membranes. S1P is formed by phosphorylation of sphingosine by sphingosine kinase (SK). The phosphate group of the product can be detached (dephosphorylated) to regenerate sphingosine via S1P phosphatase enzymes or S1P can be broken down by S1P lyase enzymes to ethanolamine phosphate and hexadecenal. Similar to Sph, its second messenger function is not yet clear. However, there is substantial evidence that implicates S1P to cell survival, cell migration, and inflammation. Certain growth-inducing proteins such as platelet-derived growth factor (PDGF), insulin-like growth factor (IGF) and vascular endothelial growth factor (VEGF) promote the formation of SK enzymes, leading to increased levels of S1P. Other factors that induce SK include cellular communication molecules called cytokines, such as tumor necrosis factor α (TNFα) and interleukin-1 (IL-1), hypoxia or lack of oxygen supply in cells, oxidized low-density lipoproteins (oxLDL) and several immune complexes.

S1P is probably formed at the inner leaflet of the plasma membrane in response to TNFα and other receptor activity-altering compounds called agonists. S1P, being present in low nanomolar concentrations in the cell, has to interact with high-affinity receptors that are capable of sensing their low levels. So far, the only identified receptors for S1P are the high-affinity G protein-coupled receptors (GPCRs), also known as S1P receptors (S1PRs). S1P is required to reach the extracellular side (outer leaflet) of the plasma membrane to interact with S1PRs and launch typical GPCR signaling pathways. However, the zwitterionic headgroup of S1P makes it unlikely to flip-flop spontaneously. To overcome this difficulty, the ATP-binding cassette (ABC) transporter C1 (ABCC1) serves as the "exit door" for S1P. On the other hand, the cystic fibrosis transmembrane regulator (CFTR) serves as the means of entry for S1P into the cell. In contrast to its low intracellular concentration, S1P is found in high nanomolar concentrations in serum where it is bound to albumin and lipoproteins. Inside the cell, S1P can induce calcium release independent of the S1PRs—the mechanism of which remains unknown. To date, the intracellular molecular targets for S1P are still unidentified.

The SK1-S1P pathway has been extensively studied in relation to cytokine action, with multiple functions connected to effects of TNFα and IL-1 favoring inflammation. Studies show that knockdown of key enzymes such as S1P lyase and S1P phosphatase increased prostaglandin production, parallel to increase of S1P levels. This strongly suggests that S1P is the mediator of SK1 action and not subsequent compounds. Research done on endothelial and smooth muscle cells is consistent to the hypothesis that S1P has a crucial role in regulating endothelial cell growth, and movement. Recent work on a sphingosine analogue, FTY270, demonstrates its ability to act as a potent compound that alters the activity of S1P receptors (agonist). FTY270 was further verified in clinical tests to have roles in immune modulation, such as that on multiple sclerosis. This highlights the importance of S1P in the regulation of lymphocyte function and immunity. Most of the studies on S1P are used to further understand diseases such as cancer, arthritis and inflammation, diabetes, immune function and neurodegenerative disorders.

Glucosylceramide

Glucosylceramides (GluCer) are the most widely distributed glycosphingolipids in cells serving as precursors for the formation of over 200 known glycosphingolipids. GluCer is formed by the glycosylation of ceramide in an organelle called Golgi via enzymes called glucosylceramide synthase (GCS) or by the breakdown of complex glycosphingolipids (GSLs) through the action of specific hydrolase enzymes. In turn, certain β-glucosidases hydrolyze these lipids to regenerate ceramide. GluCer appears to be synthesized in the inner leaflet of the Golgi. Studies show that GluCer has to flip to the inside of the Golgi or transfer to the site of GSL synthesis to initiate the synthesis of complex GSLs. Transferring to the GSL synthesis site is done with the help of a transport protein known as four phosphate adaptor protein 2 (FAPP2) while the flipping to the inside of the Golgi is made possible by the ABC transporter P-glycoprotein, also known as the multi-drug resistance 1 transporter (MDR1). GluCer is implicated in post-Golgi trafficking and drug resistance particularly to chemotherapeutic agents. For instance, a study demonstrated a correlation between cellular drug resistance and modifications in GluCer metabolism.

In addition to their role as building blocks of biological membranes, glycosphingolipids have long attracted attention because of their supposed involvement in cell growth, differentiation, and formation of tumors. The production of GluCer from Cer was found to be important in the growth of neurons or brain cells. On the other hand, pharmacological inhibition of GluCer synthase is being considered a technique to avoid insulin resistance.

Ceramide-1-Phosphate

Ceramide-1-phosphate (C1P) is formed by the action of ceramide kinase (CK) enzymes on Cer. C1P carry ionic charge at neutral pH and contain two hydrophobic chains making it relatively insoluble in aqueous environment. Thus, C1P reside in the organelle where it was formed and is unlikely to spontaneously flip-flop across membrane bilayers.

C1P activate phospholipase A2 and is found, along with CK, to be a mediator of arachidonic acid released in cells in response to a protein called interleukin-1β (IL-1β) and a lipid-soluble molecule that transports calcium ions (Ca^{2+}) across the bilayer, also known as calcium ionophore. C1P was also previously reported to encourage cell division (mitogenic) in fibroblasts, block apoptosis by inhibiting acid SMase in white blood cells within tissues (macrophages) and increase intracellular free calcium concentrations in thyroid cells. C1P also has known roles in vesicular trafficking, cell survival, phagocytosis ("cell eating") and macrophage degranulation.

Phosphatidylinositol bisphosphate (PIP2) Lipid Agonist

PIP_2 binds directly to ion channels and modulates their activity. PIP_2 was shown to directly agonizes Inward rectifying potassium channels(K_{ir}). In this regard intact PIP_2 signals as a bona fide neurotransmitter-like ligand. PIP_2's interaction with many ion channels suggest that the intact form of PIP_2 has an important signaling role independent of second messenger signaling.

Second Messengers from Phosphatidylinositol

Phosphatidylinositol Bisphosphate (PIP2) Second Messenger Systems

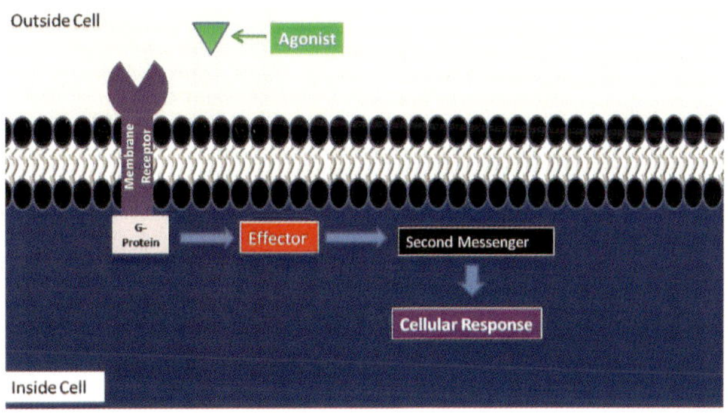

Cartoon of second messenger systems. Figure adapted From Barbraham Institute Mike Berridge.

A general second messenger system mechanism can be broken down into four steps. First, the agonist activates a membrane-bound receptor. Second, the activated G-protein produces a primary effector. Third, the primary effect stimulates the second messenger synthesis. Fourth, the second messenger activates a certain cellular process.

The G-protein coupled receptors for the PIP_2 messenger system produces two effectors, phospholipase C (PLC) and phosphoinositide 3-kinase (PI3K). PLC as an effector produces two different second messengers, inositol triphosphate (IP_3) and Diacylglycerol (DAG).

IP_3 is soluble and diffuses freely into the cytoplasm. As a second messenger, it is recognized by the inositol triphosphate receptor (IP3R), a Ca^{2+} channel in the endoplasmic reticulum (ER) membrane, which stores intracellular Ca^{2+}. The binding of IP_3 to IP3R releases Ca^{2+} from the ER into the normally Ca^{2+}-poor cytoplasm, which then triggers various events of Ca^{2+} signaling. Specifically in blood vessels, the increase in Ca^{2+} concentration from IP_3 releases nitric oxide, which then diffuses into the smooth muscle tissue and causes relaxation.

DAG remains bound to the membrane by its fatty acid "tails" where it recruits and activates both conventional and novel members of the protein kinase C family. Thus, both IP_3 and DAG contribute to activation of PKCs.

Phosphoinositide 3-kinase (PI3K) as an effector phosphorylates phosphatidylinositol bisphosphate (PIP_2) to produce phosphatidylinositol (3,4,5)-trisphosphate (PIP_3). PIP_3 has been shown to activate protein kinase B, increase binding to extracellular proteins and ultimately enhance cell survival.

Activators of G-protein Coupled Receptors

Lysophosphatidic Acid (LPA)

LPA is the result of phospholipase A2 action on phosphatidic acid. The SN-1 position can contain either an ester bond or an ether bond, with ether LPA being found at elevated levels in certain cancers. LPA binds the high-affinity G-protein coupled receptors LPA1, LPA2, and LPA3 (also known as EDG2, EDG4, and EDG7, respectively).

Sphingosine-1-phosphate (S1P)

S1P is present at high concentrations in plasma and secreted locally at elevated concentrations at sites of inflammation. It is formed by the regulated phosphorylation of sphingosine. It acts through five dedicated high-affinity G-protein coupled receptors, S1P1 - S1P5. Interestingly, targeted deletion of S1P1 results in lethality in mice and deletion of S1P2 results in seizures and deafness. Additionally, a mere 3- to 5-fold elevation in serum S1P concentrations induces sudden cardiac death by an S1P3-receptor specific mechanism.

Platelet Activating Factor (PAF)

PAF is a potent activator of platelet aggregation, inflammation, and anaphylaxis. It is similar to the ubiquitous membrane phospholipid phosphatidylcholine except that it contains an acetyl-group in the SN-2 position and the SN-1 position contains an ether-linkage. PAF signals through a dedicated G-protein coupled receptor, PAFR and is inactivated by PAF acetylhydrolase.

Endocannabinoids

The endogenous cannabinoids, or endocannabinoids, are endogenous lipids that activate cannabinoid receptors. The first such lipid to be isolated was anandamide which is the arachidonoyl amide of ethanolamine. Anandamide is formed via enzymatic release from N-arachidonoyl phosphatidylethanolamine by the N-acyl phosphatidylethanolamine phospholipase D (NAPE-PLD). Anandamide activates both the CB1 receptor, found primarily in the central nervous system, and the CB2 receptor which is found primarily in lymphocytes and the periphery. It is found at very low levels (nM) in most tissues and is inactivated by the fatty acid amide hydrolase. Subsequently, another endocannabinoid was isolated, 2-arachidonoylglycerol, which is produced when phospholipase C releases diacylglycerol which is then converted to 2-AG by diacylglycerol lipase. 2-AG can also activate both cannabinoid receptors and is inactivated by monoacylglycerol lipase. It is present at approximately 100-times the concentration of anandamide in most tissues. Elevations in either of these lipids causes analgesia and anti-inflammation and tissue protection during states of ischemia, but the precise roles played by these various endocannabinoids are still not totally known and intensive research into their function, metabolism, and regulation is ongoing. One saturated lipid from this class, often called an endocannabinoid, but with no relevant affinity for the CB1 and CB 2 receptor is palmitoylethanolamide. This signaling lipid has great affinity for the GRP55 receptor and the PPAR alpha receptor. It has been identified as an anti-inflammatory compound already in 1957, and as an analgesic compound in 1975. It was Rita Levi-Montalcini. who first identified one of its biological mechanisms of action, the inhibition of activated mast cells. Palmitoylethanolamide is the only endocannabinoid available on the market for treatment, as a food supplement.

Prostaglandins

Prostaglandins are formed through oxidation of arachidonic acid by cyclooxygenases and other prostaglandin synthases. There are currently nine known G-protein coupled receptors (eicosanoid receptors) that largely mediate prostaglandin physiology.

FAHFA

FAHFAs (fatty acid esters of hydroxy fatty acids) are formed in adipose tissue, improve glucose tolerance and also reduce adipose tissue inflammation. Palmitic acid esters of hydroxy-stearic acids (PAHSAs) are among the most bioactive members able to activate G-protein coupled receptors 120. Docosahexaenoic acid ester of hydroxy-linoleic acid (DHAHLA) exert anti-inflammatory and pro-resolving properties.

Retinol Derivatives

Retinaldehyde is a retinol (vitamin A) derivative responsible for vision. It binds rhodopsin, a well-characterized GPCR that binds all-cis retinal in its inactive state. Upon photoisomerization by a photon the cis-retinal is converted to trans-retinal causing activation of rhodopsin which ultimately leads to depolarization of the neuron thereby enabling visual perception.

Activators of Nuclear Receptors

Steroid Hormones

This large and diverse class of steroids are biosynthesized from isoprenoids and structurally resemble cholesterol. Mammalian steroid hormones can be grouped into five groups by the receptors to which they bind: glucocorticoids, mineralocorticoids, androgens, estrogens, and progestogens.

Retinoic Acid

Retinol (vitamin A) can be metabolized to retinoic acid which activates nuclear receptors such as the RAR to control differentiation and proliferation of many types of cells during development.

Prostaglandins

The majority of prostaglandin signaling occurs via GPCRs although certain prostaglandins activate nuclear receptors in the PPAR family.

Lipid Metabolism

Lipid metabolism is the synthesis and degradation of lipids in cells. Lipid metabolism is the break down or storage of fats for energy; these fats are obtained from consuming food and absorbing them or they are synthesized by an animal's liver. Lipogenesis is the process of synthesizing these fats. The majority of lipids found in the human body from ingesting food are Triglycerides. Since lipids are fats, lipid metabolism is often considered the digestion and absorption process of dietary fats. Lipid metabolism often begins with hydrolysis, which occurs when a chemical breaks down as a reaction to coming in contact with water. Lipid metabolism does exist in plants, though the processes differ in some ways when compared to animals.

Lipid Digestion

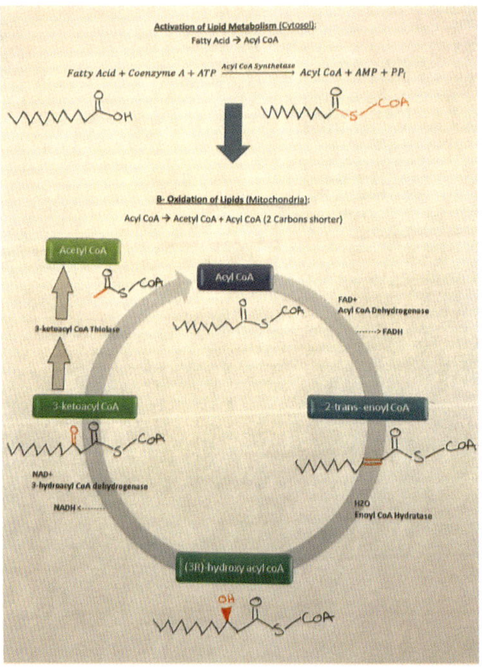

Digestion is the first step to lipid metabolism, and is the process of breaking the triglycerides down into smaller monoglyceride units with the help of lipase enzymes. Digestion of fats begin in the mouth through chemical digestion by lingual lipase. Lipids then continue to the stomach where chemical digestion continues by gastric lipase and mechanical digestion begins (Peristalsis). The majority of lipid digestion and absorption, however, occurs once the fats reach the small intestines. Chemicals from the pancreas (pancreatic lipase family and Bile salt-dependent lipase) are transported to the small intestines to help breakdown the triglycerols, along with further mechanical digestion, until they are individual fatty acid units able to be absorbed into the small intestines. It is the pancreatic lipase that is responsible for signaling for the hydrolysis of the triglycerides into separate free fatty acids and glycerol units.

Lipid Absorption

The second step in lipid metabolism is the absorption of fats. Absorption of fats occurs only in the small intestines, and only after the fats are broken down into fatty acids. The diagram to the left shows how fatty acids are absorbed into the lining of the small intestines and further broken down into smaller components to be used to create energy to fuel the body. Fatty acid metabolism begins in the cytoplasm of epithelial cells as Acyl-CoA synthetase and hydrolysis of ATP cleaves a phosphate group and catalyzes the addition of Coenzyme A to the fatty acid. The resulting Acyl-CoA activates the process of beta oxidation to further break down into Acetyl-CoA (which is used in the Citric acid cycle to produce energy). The process of beta oxidation occurs in the mitochondria of the small intestine epithelium, and require the following enzymes: Acyl CoA dehydrogenase, Enoyl-CoA hydratase, 3-hydroxyacyl-CoA dehydrogenase, and 3-ketoacyl-CoA thiolase.

Transporting Lipids

Since lipids (fats) are hydrophobic, hydrolysis in lipid metabolism occurs in the cytoplasm which ends up creating glycerol and fatty acids. Due to the hydrophobic nature of lipids they require special transport proteins known as lipoproteins, which are hydrophilic. Lipoproteins are categorized by their density levels. The varying densities between the types of lipoproteins are characteristic to what type of fats they transport. A number of these lipoproteins are synthesized in the liver, but not all of them originate from this organ.

Lipid Catabolism

Once the chilomicrons (or other lipoproteins) travel through the tissues, these particles will be broken down by lipoprotein lipase in the luminal surface of endothelial cells in capillaries to release tryglycerides.Tryglycerides will get broken down into fatty acids and glycerol before entering cells and remaining cholesterol will again travel through the blood to the liver.

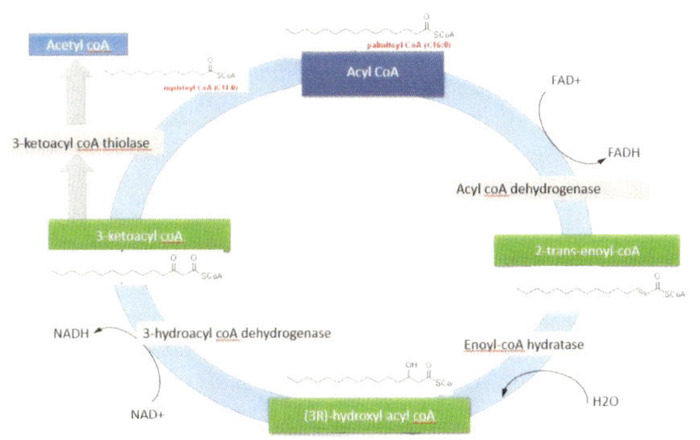

In the cytosol of the cell (for example a muscle cell), the glycerol will be converted to glyceraldehyde 3-phosphate, which is an intermediate in the glycolysis, to get further oxidized and produce energy. However, the main steps of fatty acids catabolism occur in the mitochondria. Long chain fatty acids (more than 14 carbon) need to be converted to Fatty acyl-CoA in order to pass across the mitochondria membrane. Fatty acid catabolism begins in the cytoplasm of cells as Acyl-CoA synthetase uses the energy from cleavage of an ATP to catalyze the addition of Coenzyme A to the fatty acid. The resulting Acyl-CoA cross the mitochondria membrane and enter the process of beta oxidation. The main products of the beta oxidation pathway are Acetyl-CoA (which is used in the Citric acid cycle to produce energy), NADH and FADH. The process of beta oxidation requires the following enzymes: Acyl CoA dehydrogenase, Enoyl-CoA hydratase, 3-hydroxyacyl-CoA dehydrogenase, and 3-ketoacyl-CoA thiolase. The diagram to the left shows how fatty acids are converted into Acetyl-coA. The overall net reaction, using palmitoyl CoA (16:0) as a model substrate is:

$$7\,FAD + 7\,NAD^{+} + 7\,CoASH + 7\,H_2O + H(CH_2CH_2)_7CH_2CO-SCoA \rightarrow 8\,CH_3CO-SCoA + 7\,FADH_2 + 7\,NADH + 7\,H^{+}$$

Lipid Biosynthesis

In addition to dietary fats, storage lipids stored in the adipose tissues are one of the main sources of energy for living organisms. Triacylglycerols, lipd membrane and cholesterol can be synthesized by the organisms through various pathways.

Lipid Metabolism Disorders

Lipid Metabolism Disorders are illnesses where trouble occurs in breaking down or synthesizing fats (or fat-like substances). A good deal of the time these disorders are hereditary, meaning it's a condition that is passed along from parent to child through their genes. Gaucher's Disease (Type I, Type II, and Type III), Neimann-Pick Disease, Tay-Sachs Disease, and Fabry's Disease are all diseases where those afflicted can have a disorder of their body's lipid metabolism. Rarer diseases concerning a disorder of the lipid metabolism are Sitosterolemia, Wolman's Disease, Refsum's Disease, and Cerebrotendinous Xanthomatosis.

Lipolysis

Lipolysis is the breakdown of lipids and involves hydrolysis of triglycerides into glycerol and free fatty acids. Predominantly occurring in adipose tissue, lipolysis is used to mobilize stored energy during fasting or exercise. Lipolysis is directly induced in adipocytes by glucagon, epinephrine, norepinephrine, growth hormone, atrial natriuretic peptide, brain natriuretic peptide, and cortisol.

Mechanisms

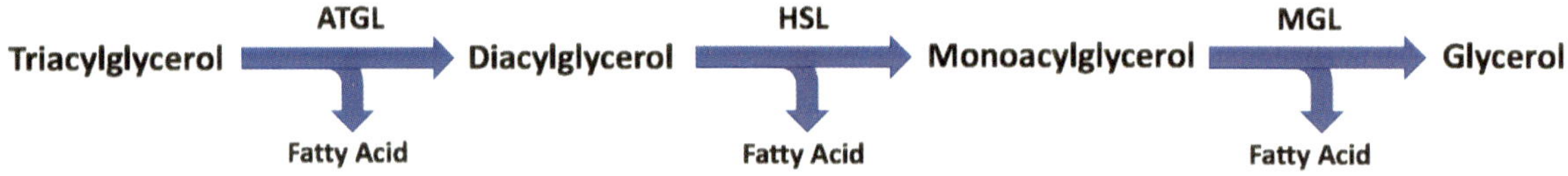

This image illustrates the three separate steps of hydrolysis involved in lipolysis.

In the first step, triacylglycerol is hydrolyzed to make diacylglycerol and this is catalyzed by adipose triglyceride lipase (ATGL). In the second step, diacylglycerol is hydrolyzed to make monoacylglycerol and this is catalyzed by hormone-sensitive lipase (HSL). In the last step, monoacylglycerol is hydrolyzed to make glycerol and this is catalyzed by monoacylglycerol lipase (MGL).

In adipose tissue, intracellular triglycerides are stored in cytoplasmic lipid droplets. When lipases are phosphorylated, they access lipid droplets and through multiple steps of hydrolysis, breakdown triglycerides into fatty acids and glycerol. Each step of hydrolysis leads to the removal of one fatty acid. The first step and the rate-limiting step of lipolysis is carried out by adipose triglyceride lipase (ATGL). This enzyme catalyzes the hydrolysis of triacylglycerol to diacylglycerol. Subsequently, hormone-sensitive lipase (HSL) catalyzes the hydrolysis of diacylglycerol to monoacylglycerol and monoacylglycerol lipase (MGL) catalyzes the hydrolysis of monoacylglycerol to glycerol. Perilipin 1A is a key protein regulator of lipolysis in adipose tissue. This lipid droplet-associated protein, when deactivated, will prevent the interaction of lipases with triglycerides in the lipid droplet and grasp the ATGL co-activator, comparative gene identification 58 (CGI-58) (a.k.a. ABHD5). When perilipin 1A is phosphorylated by PKA, it releases CGI-58 and it expedites the docking of phosphorylated lipases to the lipid droplet. CGI-58 can be further phosphorylated by PKA to assist in its dispersal to the cytoplasm. In the cytoplasm, CGI-58 can co-activate ATGL. ATGL activity is also impacted by the negative regulator of lipolysis, G0/G1 switch gene 2 (G0S2). When expressed, G0S2 acts as a competitive inhibitor in the binding of CGI-58. Fat-specific protein 27 (FSP-27) (a.k.a CIDEC) is also a negative regulator of lipolysis. FSP-27 expression is negatively correlated with ATGL mRNA levels.

Regulation

Lipolysis can be regulated through cAMP's binding and activation of protein kinase A (PKA). PKA can phosphorylate lipases, perilipin 1A, and CGI-58 to increase the rate of lipolysis. Catecholamines bind to 7TM receptors (G protein-coupled receptors) on the adipocyte cell membrane, which activate adenylate cyclase. This results in increased production of cAMP, which activates PKA and leads to an increased rate of lipolysis. Insulin counter-regulates this increase in lipolysis when it binds to insulin receptors on the adipocyte cell membrane. Insulin receptors activate insulin-like receptor substrates. These substrates activate phosphoinositide 3-kinases (PI-3K) which then phosphorylate protein kinase B (PKB) (a.k.a. Akt). PKB subsequently phosphorylates phosphodiesterase 3B (PD3B) which converts cAMP, produced by adenylate cyclase, into 5'AMP. Due to the reduced levels of cAMP, insulin decreases the rate of lipolysis. Insulin has additional actions in the mediobasal hypothalamus. It has been shown to suppress lipolysis due to lower sympathetic nervous outflow to white adipose tissue. The regulation of this process involves interactions between insulin receptors and gangliosides present in the neuronal cell membrane.

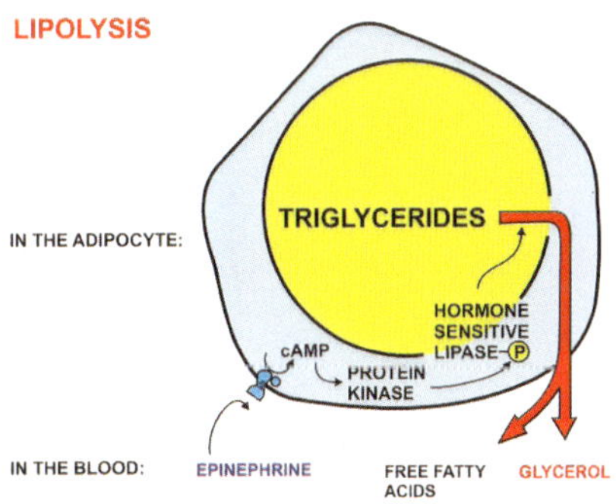

Illustration of the activation of lipolysis in an adipocyte

Induced by high epinephrine and low insulin levels in the blood. Epinephrine binds to beta-adrenergic receptors on the cell membrane of the adipocyte, which causes cAMP to be generated inside the cell. The cAMP activate protein kinases, which phosphorylate and thus, activate hormone-sensitive lipases in the adipocyte. These lipases cleave free fatty acids from their attachment to glycerol in the lipid droplet of the adipocyte. The free fatty acids and glycerol are then released into the blood. The activity of hormone sensitive lipase is regulated by the circulating hormones, insulin, glucagon, norepinephrine, and epinephrine.

In Blood

Triglycerides are transported through the blood to appropriate tissues (adipose, muscle, etc.) by lipoproteins such as Very-Low-Density-Lipoproteins (VLDL). Triglycerides present on the VLDL undergo lipolysis by the cellular lipases of target tissues, which yields glycerol and free fatty acids. Free fatty acids released into the blood are then available for cellular uptake. Free fatty acids not immediately taken up by cells may bind to albumin for transport to surrounding tissues that require energy. Serum albumin is the major carrier of free fatty acids in the blood. The glycerol also enters the bloodstream and is absorbed by the liver or kidney where it is converted to glycerol 3-phosphate by the enzyme glycerol kinase. Hepatic glycerol 3-phosphate is converted mostly into dihydroxyacetonephosphate (DHAP) and then glyceraldehyde 3-phosphate (GA3P) to rejoin the glycolysis and gluconeogenesis pathway.

Lipogenesis

Lipogenesis is the process by which acetyl-CoA is converted to fatty acids. The former is an intermediate stage in metabolism of simple sugars, such as glucose, a source of energy of living organisms. Through lipogenesis and subsequent triglyceride synthesis, the energy can be efficiently stored in the form of fats.

Lipogenesis encompasses both the process of fatty acid synthesis and triglyceride synthesis (where fatty acids are esterified to glycerol). The products are secreted from the liver in the form of very-low-density lipoproteins (VLDL). VLDL particles are secreted directly into blood, where they mature and function to deliver the endogenously derived lipids to peripheral tissues.

Fatty Acid Synthesis

Fatty acids synthesis starts with acetyl-CoA and builds up by the addition of two-carbon units. The synthesis occurs in the cytoplasm of the cell, in contrast to the degradation (oxidation), which occurs in the mitochondria. Many of the enzymes for the fatty acid synthesis are organized into a multienzyme complex called fatty acid synthase. The major sites of fatty acid synthesis are adipose tissue and the liver.

Control and Regulation

Hormonal Regulation

Insulin is a peptide hormone that is critical for managing the body's metabolism. Insulin is released by the pancreas when blood sugar levels rise, and it has many effects that broadly promote the absorption and storage of sugars, including lipogenesis.

Insulin stimulates lipogenesis primarily by activating two enzymatic pathways. Pyruvate dehydrogenase (PDH), converts pyruvate into acetyl-CoA. Acetyl-CoA carboxylase (ACC), converts acetyl-CoA produced by PDH into malonyl-CoA. Malonyl-CoA provides the two-carbon building blocks that are used to create larger fatty acids.

Insulin stimulation of lipogenesis also occurs through the promotion of glucose uptake by adipose tissue. The increase in the uptake of glucose can occur through the use of glucose transporters directed to the plasma membrane or through the activation of lipogenic and glycolytic enzymes via covalent modification. The hormone has also been found to have long term effects on lipogenic gene expression. It is hypothesized that this effect occurs through the transcription factor SREBP-1, where the association of insulin and SREBP-1 lead to the gene expression of glucokinase. The interaction of glucose and lipogenic gene expression is assumed to be managed by the increasing concentration of an unknown glucose metabolite through the activity of glucokinase.

Another hormone that may affects lipogenesis through the SREBP-1 pathway is leptin. It is involved in the process by limiting fat storage through inhibition of glucose intake and interfering with other adipose metabolic pathways. The inhibition of lipogenesis occurs through the down regulation of fatty acid and triglyceride gene expression. Through the promotion of fatty acid oxidation and lipogenesis inhibition, leptin was found to control the release of stored glucose from adipose tissues.

Other hormones that prevent the stimulation of lipogenesis in adipose cells are growth hormones (GH). Growth hormones result in loss of fat but stimulates muscle gain. One proposed mechanism for how the hormone works is that growth hormones affects insulin signaling thereby decreasing insulin sensitivity and in turn down regulating fatty acid synthase expression. Another proposed mechanism suggests that growth hormones may phosphorylate with STAT5A and STAT5B, transcription factors that are a part of the Signal Transducer And Activator Of Transcription (STAT) family.

There is also evidence suggesting that acylation stimulating protein (ASP) promotes the aggregation of triglycerides in adipose cells. This aggregation of triglycerides occurs through the increase in the synthesis of triglyceride production.

Transcriptional Regulation

SREBPs have been found to play a role with the nutritional or hormonal effects on the lipogenic gene expression. SREBP-2 has a well defined mode of action of the different members of this transcriptional family. At high levels of free cholesterol in the cell SREBP-2 is found bound to the endoplasmic reticulum as an immature precursor. when the level of cholesterol drops then the SREBP-2 is proteolytically cleaved releasing the mature fragment so it can move to the nucleus and bind to the sterol response element in the promoter region of target genes. These genes are then activated for transcription.

It has been indicated that SREBP-2 promote the expression of genes involved in cholesterol metabolism in liver cells. It has been indicated that SREBP-1 plays a role in the activation of genes connected with lipogenesis in liver. Studies have found that an over expression of SREBP-1a or SREBP-1c in mouse liver cells results in the build-up of hepatic triglycerides and higher expression levels of lipogenic genes.

Lipogenic gene expression in the liver via glucose and insulin is moderated by SREBP-1. The effect of glucose and insulin on the transcriptional factor can occur through various pathways. There is evidence suggesting that insulin promotes SREBP-1 mRNA expression in adipocytes and hepatocytes, it has also been suggested that the hormone increases transcriptional activation by SREBP-1 through MAP-kinase-dependent phosphorylation regardless of changes in the mRNA levels. Along with insulin glucose also have been shown to promote SREBP-1 activity and mRNA expression.

PDH Dephosphorylation

Insulin stimulates the activity of pyruvate dehydrogenase phosphatase. The phosphatase removes the phosphate from pyruvate dehydrogenase activating it and allowing for conversion of pyruvate to acetyl-CoA. This mechanism leads to the increased rate of catalysis of this enzyme, so increases the levels of acetyl-CoA. Increased levels of acetyl-CoA will increase the flux through not only the fat synthesis pathway but also the citric acid cycle.

Acetyl-CoA Carboxylase

Insulin affects ACC in a similar way to PDH. It leads to its dephosphorylation via activation of PP2A phosphatase whose activity results in the activation of the enzyme. Glucagon has an antagonistic effect and increases phosphorylation, deactivation, thereby inhibiting ACC and slowing fat synthesis.

Affecting ACC affects the rate of acetyl-CoA conversion to malonyl-CoA. Increased malonyl-CoA level pushes the equilibrium over to increase production of fatty acids through biosynthesis. Long chain fatty acids are negative allosteric regulators of ACC and so when the cell has sufficient long chain fatty acids, they will eventually inhibit ACC activity and stop fatty acid synthesis.

AMP and ATP concentrations of the cell act as a measure of the ATP needs of a cell. When ATP is depleted, there is a rise in 5'AMP. This rise activates AMP-activated protein kinase, which phosphorylates ACC and thereby inhibits fat synthesis. This is a useful way to ensure that glucose is not diverted down a storage pathway in times when energy levels are low.

ACC is also activated by citrate. When there is abundant acetyl-CoA in the cell cytoplasm for fat synthesis, it proceeds at an appropriate rate.

Note: Research now shows that glucose metabolism (exact metabolite to be determined), aside from insulin's influence on lipogenic enzyme genes, can induce the gene products for liver's pyruvate kinase, acetyl-CoA carboxylase, and fatty acid synthase. These genes are induced by the transcription factors ChREBP/Mlx via high blood glucose levels and presently unknown signaling events. Insulin induction of SREBP-1c is also involved in cholesterol metabolism.

Fatty Acid Esterification

Experiments were conducted to study in vivo the over-all fatty acid specificity of the mechanisms involved in chylomicron cholesterol ester and triglyceride formation during fat absorption in the rat. Mixtures containing similar amounts of two, three, or four C14-labeled fatty acids (palmitic, stearic, oleic, and linoleic acids), but with varying ratios of unlabeled fatty acids, were given by gastric intubation to rats with cannulated thoracic ducts. The chyle or chylomicron lipid so obtained was chromatographed on silicic acid columns to separate cholesterol esters and glycerides (the latter being 98.2% triglycerides). After assaying each lipid class for total radioactivity, gas-liquid chromatography was employed to measure the total mass and the distribution of mass and of radioactivity in the individual fatty acid components of each lipid fraction. The specific radioactivity of each fatty acid in each fraction could then be calculated. The data provided quantitative information on the relative specificity of incorporation of each fatty acid into each chylomicron lipid class and on the relative extent to which each fatty acid in each lipid fraction was diluted with endogenous fatty acid. With the exception of a slight discrimination against stearic acid, the processes of fatty acid absorption and chylomicron triglyceride formation displayed no specificity for one fatty acid relative to another. In contrast, chylomicron cholesterol ester formation showed marked specificity for oleic acid, relative to the other three fatty acids. This specificity was not significantly altered by varying the composition of the test meal, by including cholesterol in the test meal, or by feeding the animal a high-cholesterol diet for several weeks preceding the study. Considerable dilution of the dietary fatty acids with endogenous fatty acids was observed. In one experiment, 43% of the chylomicron triglyceride fatty acids was of endogenous origin. Relatively more (54%) of the cholesterol ester fatty acids was of endogenous origin.

Fatty Acid

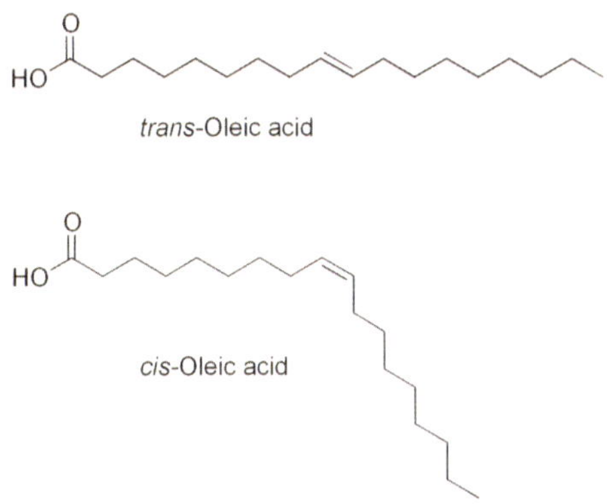

Comparison of the *trans* isomer Elaidic acid (bottom) and the *cis* isomer oleic acid (top).

In chemistry, particularly in biochemistry, a fatty acid is a carboxylic acid with a long aliphatic chain, which is either saturated or unsaturated. Most naturally occurring fatty acids have an unbranched chain of an even number of carbon atoms, from 4 to 28. Fatty acids are usually derived from triglycerides or phospholipids. Fatty acids are important dietary sources of fuel for animals because, when metabolized, they yield large quantities of ATP. Many cell types can use either glucose or fatty acids for this purpose. Long-chain fatty acids cannot cross the blood–brain barrier (BBB) and so cannot be used as fuel by the cells of the central nervous system; however, free short-chain fatty acids and medium-chain fatty acids can cross the BBB, in addition to glucose and ketone bodies.

Types of Fatty Acids

Fatty acids that have carbon–carbon double bonds are known as unsaturated. Fatty acids without double bonds are known as saturated. They differ in length as well.

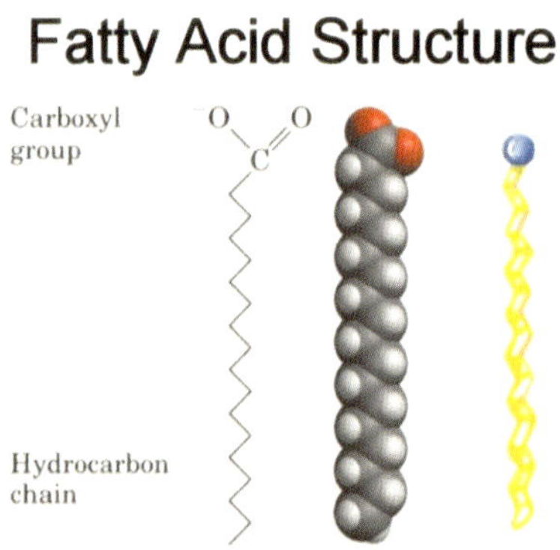

Three-dimensional representations of several fatty acids

Length of Free Fatty Acid Chains

Fatty acid chains differ by length, often categorized as short to very long.

- Short-chain fatty acids (SCFA) are fatty acids with aliphatic tails of five or fewer carbons (e.g. butyric acid).
- Medium-chain fatty acids (MCFA) are fatty acids with aliphatic tails of 6 to 12 carbons, which can form medium-chain triglycerides.
- Long-chain fatty acids (LCFA) are fatty acids with aliphatic tails of 13 to 21 carbons.
- Very long chain fatty acids (VLCFA) are fatty acids with aliphatic tails of 22 or more carbons.

Unsaturated Fatty Acids

Unsaturated fatty acids have one or more double bonds between carbon atoms. (Pairs of carbon atoms connected by double bonds can be saturated by adding hydrogen atoms to them, converting the double bonds to single bonds. Therefore, the double bonds are called unsaturated.)

The two carbon atoms in the chain that are bound next to either side of the double bond can occur in a *cis* or *trans* configuration.

cis

A *cis* configuration means that the two hydrogen atoms adjacent to the double bond stick out on the same side of the chain. The rigidity of the double bond freezes its conformation and, in the case of the *cis* isomer, causes the chain to bend and restricts the conformational freedom of the fatty acid. The more double bonds the chain has in the *cis* configuration, the less flexibility it has. When a chain has many *cis* bonds, it becomes quite curved in its most accessible conformations. For example, oleic acid, with one double bond, has a "kink" in it, whereas linoleic acid, with two double bonds, has a more pronounced bend. α-Linolenic acid, with three double bonds, favors a hooked shape. The effect of this is that, in restricted environments, such as when fatty acids are part of a phospholipid in a lipid bilayer, or triglycerides in lipid droplets, cis bonds limit the ability of fatty acids to be closely packed, and therefore can affect the melting temperature of the membrane or of the fat.

trans

A *trans* configuration, by contrast, means that the adjacent two hydrogen atoms lie on *opposite* sides of the chain. As a result, they do not cause the chain to bend much, and their shape is similar to straight saturated fatty acids.

In most naturally occurring unsaturated fatty acids, each double bond has three *n* carbon atoms after it, for some n, and all are cis bonds. Most fatty acids in the *trans* configuration (trans fats) are not found in nature and are the result of human processing (e.g., hydrogenation).

The differences in geometry between the various types of unsaturated fatty acids, as well as between saturated and unsaturated fatty acids, play an important role in biological processes, and in the construction of biological structures (such as cell membranes).

Examples of Unsaturated Fatty Acids				
Common name	Chemical structure	Δ^{x}	C:D	n−x
Myristoleic acid	$CH_3(CH_2)_3CH{=}CH(CH_2)_7COOH$	cis-Δ^{9}	14:1	n−5
Palmitoleic acid	$CH_3(CH_2)_5CH{=}CH(CH_2)_7COOH$	cis-Δ^{9}	16:1	n−7
Sapienic acid	$CH_3(CH_2)_8CH{=}CH(CH_2)_4COOH$	cis-Δ^{6}	16:1	n−10
Oleic acid	$CH_3(CH_2)_7CH{=}CH(CH_2)_7COOH$	cis-Δ^{9}	18:1	n−9
Elaidic acid	$CH_3(CH_2)_7CH{=}CH(CH_2)_7COOH$	trans-Δ^{9}	18:1	n−9
Vaccenic acid	$CH_3(CH_2)_5CH{=}CH(CH_2)_9COOH$	trans-Δ^{11}	18:1	n−7
Linoleic acid	$CH_3(CH_2)_4CH{=}CHCH_2CH{=}CH(CH_2)_7COOH$	cis,cis-Δ^{9},Δ^{12}	18:2	n−6
Linoelaidic acid	$CH_3(CH_2)_4CH{=}CHCH_2CH{=}CH(CH_2)_7COOH$	trans,trans-Δ^{9},Δ^{12}	18:2	n−6
α-Linolenic acid	$CH_3CH_2CH{=}CHCH_2CH{=}CHCH_2CH{=}CH(CH_2)_7COOH$	cis,cis,cis-Δ^{9},Δ^{12},Δ^{15}	18:3	n−3
Arachidonic acid	$CH_3(CH_2)_4CH{=}CHCH_2CH{=}CHCH_2CH{=}CHCH$-$_2CH{=}CH(CH_2)_3COOH$[NIST]	cis,cis,cis,-cis-$\Delta^{5}\Delta^{8}$,Δ^{11},Δ^{14}	20:4	n−6
Eicosapentaenoic acid	$CH_3CH_2CH{=}CHCH_2CH{=}CHCH_2CH{=}CHCH_2CH{=}CH$-$CH_2CH{=}CH(CH_2)_3COOH$	cis,cis,cis,cis,-cis-Δ^{5},Δ^{8},Δ^{11},Δ^{14},Δ^{17}	20:5	n−3
Erucic acid	$CH_3(CH_2)_7CH{=}CH(CH_2)_{11}COOH$	cis-Δ^{13}	22:1	n−9
Docosahexaenoic acid	$CH_3CH_2CH{=}CHCH_2CH{=}CHCH_2CH{=}CHCH_2CH{=}CH$-$CH_2CH{=}CHCH_2CH{=}CH(CH_2)_2COOH$	cis,cis,cis,cis,cis,-cis-Δ^{4},Δ^{7},Δ^{10},Δ^{13},Δ^{16},Δ^{19}	22:6	n−3

Essential Fatty Acids

Fatty acids that are required by the human body but cannot be made in sufficient quantity from other substrates, and therefore must be obtained from food, are called essential fatty acids. There are two series of essential fatty acids: one has a double bond three carbon atoms away from the methyl end; the other has a double bond six carbon atoms away from the methyl end. Humans lack the ability to introduce double bonds in fatty acids beyond carbons 9 and 10, as counted from the carboxylic acid side. Two essential fatty acids are linoleic acid (LA) and alpha-linolenic acid (ALA). These fatty acids are widely distributed in plant oils. The human body has a limited ability to convert ALA into the longer-chain omega-3 fatty acids — eicosapentaenoic acid (EPA) and docosahexaenoic acid (DHA), which can also be obtained from fish. Omega-3 and omega-6 fatty acids are biosynthetic precursors to endocannabinoids with antinociceptive, anxiolytic, and neurogenic properties.

Saturated Fatty Acids

Saturated fatty acids have no double bonds. Thus, saturated fatty acids are saturated with hydrogen (since double bonds reduce the number of hydrogens on each carbon). Because saturated fatty acids have only single bonds, each carbon atom within the chain has 2 hydrogen atoms (except for the omega carbon at the end that has 3 hydrogens).

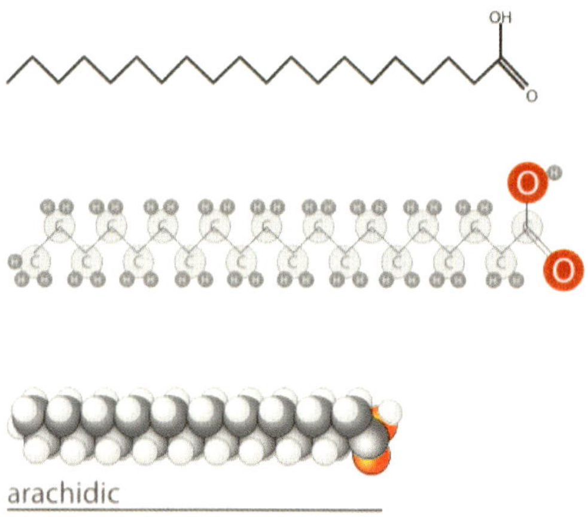

Arachidic acid, a saturated fatty acid.

Examples of Saturated Fatty Acids		
Common name	Chemical structure	C:D
Caprylic acid	$CH_3(CH_2)_6COOH$	8:0
Capric acid	$CH_3(CH_2)_8COOH$	10:0
Lauric acid	$CH_3(CH_2)_{10}COOH$	12:0
Myristic acid	$CH_3(CH_2)_{12}COOH$	14:0
Palmitic acid	$CH_3(CH_2)_{14}COOH$	16:0
Stearic acid	$CH_3(CH_2)_{16}COOH$	18:0
Arachidic acid	$CH_3(CH_2)_{18}COOH$	20:0
Behenic acid	$CH_3(CH_2)_{20}COOH$	22:0
Lignoceric acid	$CH_3(CH_2)_{22}COOH$	24:0
Cerotic acid	$CH_3(CH_2)_{24}COOH$	26:0

Nomenclature

Numbering of the Carbon Atoms in a Fatty Acid

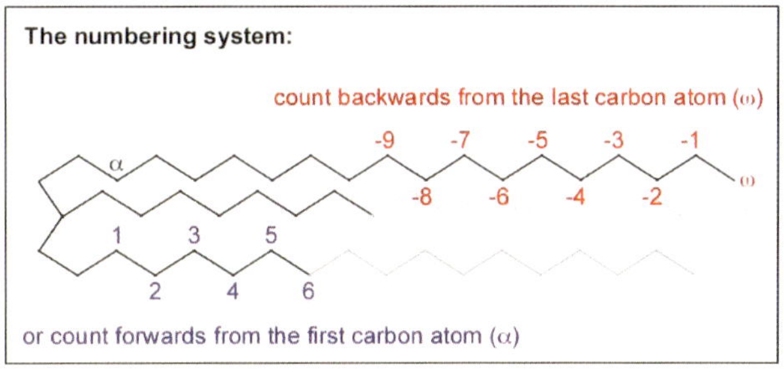

Numbering of carbon atoms

The position of the carbon atoms in a fatty acid can be indicated from the COOH- (or carboxy) end, or from the $-CH_3$ (or methyl) end. If indicated from the -COOH end, then the C-1, C-2, C-3, ….(etc.) notation is used (blue numerals in the diagram on the right, where C-1 is the –COOH carbon). If the position is counted from the other, $-CH_3$, end then the position is indicated by the ω-*n* notation (numerals in red, where ω-1 refers to the methyl carbon).

The positions of the double bonds in a fatty acid chain can, therefore, be indicated in two ways, using the C-*n* or the ω-*n* notation. Thus, in an 18 carbon fatty acid, a double bond between C-12 (or ω-7) and C-13 (or ω-6) is reported either as Δ^{12} if counted from the –COOH end (indicating only the "beginning" of the double bond), or as ω-6 (or omega-6) if counting from the $-CH_3$ end. The "Δ" is the letter "delta", which translates into "D" (for Double bond) in the Roman alphabet. Omega (ω) is the last letter in the alphabet, and is therefore used to indicate the "last" carbon atom in the fatty acid chain. Since the ω-*n* notation is used almost exclusively to indicate the positions of the double bonds close to the $-CH_3$ end in essential fatty acids, there is no necessity for an equivalent "Δ"-like notation - the use of the "ω-*n*" notation always refers to the position of a double bond.

Fatty acids with an odd number of carbon atoms are called odd-chain fatty acids, whereas the rest are even-chain fatty acids. The difference is relevant to gluconeogenesis.

Naming of Fatty Acids

The following table describes the most common systems of naming fatty acids.

System	Example	Explanation
Trivial nomenclature	Palmitoleic acid	Trivial names (or common names) are non-systematic historical names, which are the most frequent naming system used in literature. Most common fatty acids have trivial names in addition to their systematic names. These names frequently do not follow any pattern, but they are concise and often unambiguous.
Systematic nomenclature	(9Z)-octadecenoic acid	Systematic names (or IUPAC names) derive from the standard IUPAC Rules for the Nomenclature of Organic Chemistry, published in 1979, along with a recommendation published specifically for lipids in 1977. Counting begins from the carboxylic acid end. Double bonds are labelled with cis-/trans- notation or E-/Z- notation, where appropriate. This notation is generally more verbose than common nomenclature, but has the advantage of being more technically clear and descriptive.

Δ^x nomenclature	cis,cis-Δ^9,Δ^{12} octadecadienoic acid	In Δ^x (or delta-x) nomenclature, each double bond is indicated by Δ^x, where the double bond is located on the xth carbon–carbon bond, counting from the carboxylic acid end. Each double bond is preceded by a cis- or trans- prefix, indicating the configuration of the molecule around the bond. For example, linoleic acid is designated "cis-Δ^9, cis-Δ^{12} octadecadienoic acid". This nomenclature has the advantage of being less verbose than systematic nomenclature, but is no more technically clear or descriptive.
n−x nomenclature	n−3	n−x (n minus x; also ω−x or omega-x) nomenclature both provides names for individual compounds and classifies them by their likely biosynthetic properties in animals. A double bond is located on the xth carbon–carbon bond, counting from the terminal methyl carbon (designated as n or ω) toward the carbonyl carbon. For example, α-Linolenic acid is classified as a n−3 or omega-3 fatty acid, and so it is likely to share a biosynthetic pathway with other compounds of this type. The ω−x, omega-x, or "omega" notation is common in popular nutritional literature, but IUPAC has deprecated it in favor of n−x notation in technical documents. The most commonly researched fatty acid biosynthetic pathways are n−3 and n−6.
Lipid numbers	18:3 18:3ω6 18:3, cis,cis,-cis-Δ^9,Δ^{12},Δ^{15}	Lipid numbers take the form C:D, where C is the number of carbon atoms in the fatty acid and D is the number of double bonds in the fatty acid (if more than one, the double bonds are assumed to be interrupted by CH 2 units, i.e., at intervals of 3 carbon atoms along the chain). This notation can be ambiguous, as some different fatty acids can have the same numbers. Consequently, when ambiguity exists this notation is usually paired with either a Δ^x or n−x term.

Esterified, Free, Unsaturated, Conjugated

When fatty acids circulating in the plasma (plasma fatty acids) are not in their glycerol ester form (glycerides), they are known as non-esterified fatty acids (NEFAs) or free fatty acids (FFAs). The latter term can be viewed as a misnomer because they are transported complexed with a transport protein, such as albumin, as opposed to being unattached to any other molecule. But the term conveys the idea that they are circulating and available for metabolism.

Fatty acids can exist in various states of saturation. Unsaturated fatty acids include monounsaturated fatty acids (MUFAs) and polyunsaturated fatty acids (PUFAs). Conjugated fatty acids are a subset of PUFAs.

Production

Industrial

Fatty acids are usually produced industrially by the hydrolysis of triglycerides, with the removal of glycerol. Phospholipids represent another source. Some fatty acids are produced synthetically by hydrocarboxylation of alkenes.

By Animals

In animals, fatty acids are formed from carbohydrates predominantly in the liver, adipose tissue, and the mammary glands during lactation.

Carbohydrates are converted into pyruvate by glycolysis as the first important step in the conversion of carbohydrates into fatty acids. Pyruvate is then dehydrogenated to form acetyl-CoA in the mitochondrion. However, this acetyl CoA needs to be transported into cytosol where the synthesis of fatty acids occurs. This cannot occur directly. To obtain cytosolic acetyl-CoA, citrate (produced by the condensation of acetyl-CoA with oxaloacetate) is removed from the citric acid cycle and carried across the inner mitochondrial membrane into the cytosol. There it is cleaved by ATP citrate lyase into acetyl-CoA and oxaloacetate. The oxaloacetate is returned to the mitochondrion as malate. The cytosolic acetyl-CoA is carboxylated by acetyl CoA carboxylase into malonyl-CoA, the first committed step in the synthesis of fatty acids.

Malonyl-CoA is then involved in a repeating series of reactions that lengthens the growing fatty acid chain by two carbons at a time. Almost all natural fatty acids, therefore, have even numbers of carbon atoms. When synthesis is complete the free fatty acids are nearly always combined with glycerol (three fatty acids to one glycerol molecule) to form triglycerides, the main storage form of fatty acids, and thus of energy in animals. However, fatty acids are also important components of the phospholipids that form the phospholipid bilayers out of which all the membranes of the cell are constructed (the cell wall, and the membranes that enclose all the organelles within the cells, such as the nucleus, the mitochondria, endoplasmic reticulum, and the Golgi apparatus).

The "uncombined fatty acids" or "free fatty acids" found in the circulation of animals come from the breakdown (or lipolysis) of stored triglycerides. Because they are insoluble in water, these fatty acids are transported bound to plasma albumin. The levels of "free fatty acids" in the blood are limited by the availability of albumin binding sites. They can be taken up from the blood by all cells that have mitochondria (with the exception of the cells of the central nervous system). Fatty acids can only be broken down in mitochondria, by means of beta-oxidation followed by further combustion in the citric acid cycle to CO_2 and water. Cells in the central nervous system, which, although they possess mitochondria, cannot take free fatty acids up from the blood, as the blood-brain barrier is impervious to most free fatty acids, excluding short-chain fatty acids and medium-chain fatty acids. These cells have to manufacture their own fatty acids from carbohydrates, as described above, in order to produce and maintain the phospholipids of their cell membranes, and those of their organelles.

Fatty Acids in Dietary Fats

The following table gives the fatty acid, vitamin E and cholesterol composition of some common dietary fats.

	Saturated	Monounsaturated	Polyunsaturated	Cholesterol	Vitamin E
	g/100g	g/100g	g/100g	mg/100g	mg/100g
Animal fats					
Lard	40.8	43.8	9.6	93	0.60
Duck fat	33.2	49.3	12.9	100	2.70
Butter	54.0	19.8	2.6	230	2.00
Vegetable fats					
Coconut oil	85.2	6.6	1.7	0	.66
Cocoa butter	60.0	32.9	3.0	0	1.8
Palm kernel oil	81.5	11.4	1.6	0	3.80
Palm oil	45.3	41.6	8.3	0	33.12

Cottonseed oil	25.5	21.3	48.1	0	42.77
Wheat germ oil	18.8	15.9	60.7	0	136.65
Soybean oil	14.5	23.2	56.5	0	16.29
Olive oil	14.0	69.7	11.2	0	5.10
Corn oil	12.7	24.7	57.8	0	17.24
Sunflower oil	11.9	20.2	63.0	0	49.00
Safflower oil	10.2	12.6	72.1	0	40.68
Hemp oil	10	15	75	0	12.34
Canola/Rapeseed oil	5.3	64.3	24.8	0	22.21

Reactions of Fatty Acids

Fatty acids exhibit reactions like other carboxylic acids, i.e. they undergo esterification and acid-base reactions.

Acidity

Fatty acids do not show a great variation in their acidities, as indicated by their respective pK_a Nonanoic acid, for example, has a pK_a of 4.96, being only slightly weaker than acetic acid (4.76). As the chain length increases, the solubility of the fatty acids in water decreases very rapidly, so that the longer-chain fatty acids have minimal effect on the pH of an aqueous solution. Even those fatty acids that are insoluble in water will dissolve in warm ethanol, and can be titrated with sodium hydroxide solution using phenolphthalein as an indicator. This analysis is used to determine the free fatty acid content of fats; i.e., the proportion of the triglycerides that have been hydrolyzed.

Hydrogenation and Hardening

Hydrogenation of unsaturated fatty acids is widely practiced, typical conditions involving 2.0–3.0 MPa of H2 pressure, 150 °C, and nickel supported on silica. This treatment affords saturated fatty acids, as reflected in their iodine number. Hydrogenated fatty acids are less prone toward rancidification. Since the saturated fatty acids are higher melting than the unsaturated precursors, the process is called hardening. Related technology is used to convert vegetable oils into margarine. The hydrogenation of triglycerides is advantageous because the carboxylic acids degrade the nickel catalysts, affording nickel soaps. During partial hydrogenation, unsaturated fatty acids can be isomerized from *cis* to *trans* configuration.

More forcing hydrogenation, i.e. using higher pressures of H_2 and higher temperatures, converts fatty acids into fatty alcohols. Fatty alcohols are, however, more easily produced from fatty acid esters.

In the Varrentrapp reaction certain unsaturated fatty acids are cleaved in molten alkali, a reaction at one time of relevance to structure elucidation.

Auto-oxidation and Rancidity

Unsaturated fatty acids undergo a chemical change known as auto-oxidation. The process requires oxygen (air) and is accelerated by the presence of trace metals. Vegetable oils resist this process to a small degree because they contain antioxidants, such as tocopherol. Fats and oils often are treated with chelating agents such as citric acid to remove the metal catalysts.

Ozonolysis

Unsaturated fatty acids are susceptible to degradation by ozone. This reaction is practiced in the production of azelaic acid ($(CH_2)_7(CO_2H)_2$) from oleic acid.

Analysis

In chemical analysis, fatty acids are separated by gas chromatography of methyl esters; additionally, a separation of unsaturated isomers is possible by argentation thin-layer chromatography.

Circulation

Digestion and Intake

Short- and medium-chain fatty acids are absorbed directly into the blood via intestine capillaries and travel through the portal vein just as other absorbed nutrients do. However, long-chain fatty acids are not directly released into the intestinal capillaries. Instead they are absorbed into the fatty walls of the intestine villi and reassembled again into triglycerides. The triglycerides are coated with cholesterol and protein (protein coat) into a compound called a chylomicron.

From within the cell, the chylomicron is released into a lymphatic capillary called a lacteal, which merges into larger lymphatic vessels. It is transported via the lymphatic system and the thoracic duct up to a location near the heart (where the arteries and veins are larger). The thoracic duct empties the chylomicrons into the bloodstream via the left subclavian vein. At this point the chylomicrons can transport the triglycerides to tissues where they are stored or metabolized for energy.

Metabolism

Fatty acids (provided either by ingestion or by drawing on triglycerides stored in fatty tissues) are distributed to cells to serve as a fuel for muscular contraction and general metabolism. They are broken down to CO_2 and water by the intra-cellular mitochondria, releasing large amounts of energy, captured in the form of ATP through beta oxidation and the citric acid cycle.

Distribution

Blood fatty acids are in different forms in different stages in the blood circulation. They are taken in through the intestine in chylomicrons, but also exist in very low density lipoproteins (VLDL) and low density lipoproteins (LDL) after processing in the liver. In addition, when released from adipocytes, fatty acids exist in the blood as free fatty acids.

It is proposed that the blend of fatty acids exuded by mammalian skin, together with lactic acid and pyruvic acid, is distinctive and enables animals with a keen sense of smell to differentiate individuals.

Lipid Droplet

Lipid droplets, also referred to as lipid bodies, oil bodies or adiposomes, are lipid-rich cellular organelles that regulate the storage and hydrolysis of neutral lipids and are found largely in the adipose tissue. They also serve as a reservoir for cholesterol and acyl-glycerols for membrane formation and maintenance. Lipid droplets are found in all eukaryotic organisms and store a large portion of lipids in mammalian adipocytes. Initially, these lipid droplets were considered to merely serve as fat depots, but since the discovery in the 1990s of proteins in the lipid droplet coat that regulate lipid droplet dynamics and lipid metabolism, lipid droplets are seen as highly dynamic organelles that play a very important role in the regulation of intracellular lipid storage and lipid metabolism. The role of lipid droplets outside of lipid and cholesterol storage has recently begun to be elucidated and includes a close association to inflammatory responses through the synthesis and metabolism of eicosanoids and to metabolic disorders such as obesity, cancer, and atherosclerosis. In non-adipocytes, lipid droplets are known to play a role in protection from lipotoxicity by storage of fatty acids in the form of neutral triacylglycerol, which consists of three fatty acids bound to glycerol. Alternatively, fatty acids can be converted to lipid intermediates like diacylglycerol (DAG), ceramides and fatty acyl-CoAs. These lipid intermediates can impair insulin signaling, which is referred to as lipid-induced insulin resistance and lipotoxicity. Lipid droplets also serve as platforms for protein binding and degradation. Finally, lipid droplets are known to be exploited by pathogens such as the hepatitis C virus, the dengue virus and *chlamydia trachomatis* among others.

Structure

Lipid droplets are composed of a neutral lipid core consisting mainly of triacylglycerols (TAGs) and cholesteryl esters surrounded by a phospholipid monolayer. The surface of lipid droplets is decorated by a number of proteins which are involved in the regulation of lipid metabolism. The first and best-characterized family of lipid droplet coat proteins is the perilipin protein family, consisting of five proteins. These include perilipin 1 (PLIN1), perilipin 2 (PLIN2/ ADRP), perilipin 3 (PLIN3/ TIP47), perilipin 4 (PLIN4/ S3-12) and perilipin 5 (PLIN5/ OXPAT/ LSDP5/ MLDP). Proteomics studies have elucidated the association of many other families of proteins to the lipid surface including proteins involved in membrane trafficking, vesicle docking, endocytosis and exocytosis. Analysis of the lipid composition of lipid droplets has revealed the presence of a diverse set of phospholipid species; phosphatidylcholine and phosphatidylethanolamine are the most abundant, followed by phosphatidylinositol.

Lipid droplets vary greatly in size, ranging from 20-40 nm to 100 um. In adipocytes, lipid bodies tend to be larger and they may compose the majority of the cell, while in other cells they may only be induced under certain conditions and are considerably smaller in size.

Formation

Although the exact mechanism of formation of lipid droplets is still unknown, it is proposed that they bud off the membrane of the endoplasmic reticulum as TAGs are collected between the two layers of its phospholipid membrane. Lipid droplet growth can consequently happen through direct diffusion of fatty acids, endocytosis of sterols or by fusion of smaller lipid droplets through the aid of SNARE proteins. Lipid droplets have also been observed to be created by the fission of existing lipid droplets, though this is thought to be less common than *de novo* formation.

The formation of lipid droplets from the endoplasmic reticulum begins with the synthesis of the neutral lipids to be transported. The manufacture of TAGs from diacylglycerol (by the addition of a fatty acyl chain) is catalyzed by the DGAT proteins, though the extent to which these and other proteins are required depends on cell type. Neither DGAT1 nor DGAT2 is singularly essential for TAG synthesis or droplet formation, though mammalian cells lacking both cannot form lipid droplets and have severely stunted TAG synthesis. DGAT1, which seems to prefer exogenous fatty acid substrates, is not essential for life; DGAT2, which seems to prefer endogenously synthesized fatty acids, is.

In non-adipocytes, lipid storage, lipid droplet synthesis and lipid droplet growth can be induced by various stimuli including growth factors, long-chain unsaturated fatty acids (including oleic acid and arachidonic acid), oxidative stress and inflammatory stimuli such bacterial lipopolysaccharides, various microbial pathogens, platelet-activating factor, eicosanoids, and cytokines.

Lipid Peroxidation

Lipid peroxidation is the oxidative degradation of lipids. It is the process in which free radicals "steal" electrons from the lipids in cell membranes, resulting in cell damage. This process proceeds by a free radical chain reaction mechanism. It most often affects polyunsaturated fatty acids, because they contain multiple double bonds in between which lie methylene bridges (-CH_2-) that possess especially reactive hydrogen atoms. As with any radical reaction, the reaction consists of three major steps: initiation, propagation, and termination. The chemical products of this oxidation are known as lipid peroxides or lipid oxidation products (LOPs).

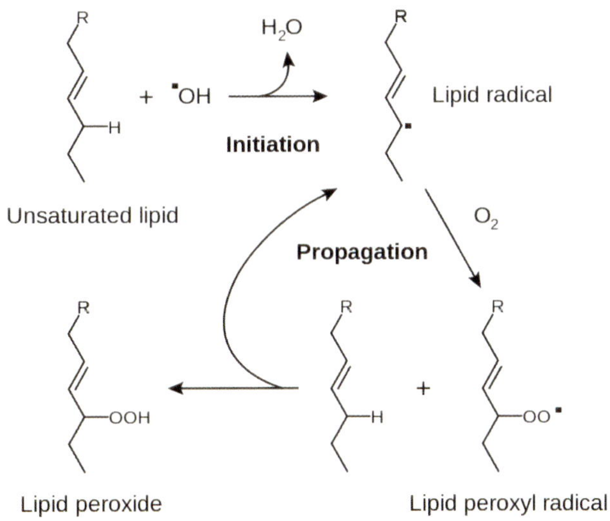

Mechanism of lipid peroxidation.

Initiation

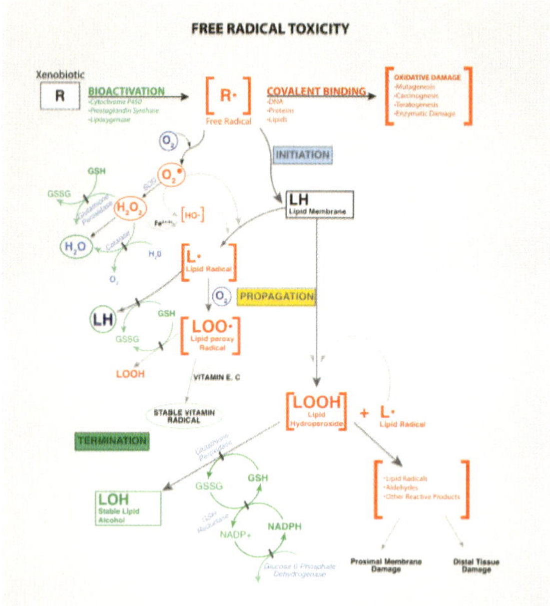

Free radical mechanisms in tissue injury. Lipid peroxidation induced by xenobiotics and the subsequent detoxification by cellular enzymes (termination).

Initiation is the step in which a fatty acid radical is produced. The most notable initiators in living cells are reactive oxygen species (ROS), such as OH· and HOO·, which combines with a hydrogen atom to make water and a fatty acid radical.

Propagation

The fatty acid radical is not a very stable molecule, so it reacts readily with molecular oxygen, thereby creating a peroxyl-fatty acid radical. This radical is also an unstable species that reacts with another free fatty acid, producing a different fatty acid radical and a lipid peroxide, or a cyclic peroxide if it had reacted with itself. This cycle continues, as the new fatty acid radical reacts in the same way.

Termination

When a radical reacts with a non-radical, it always produces another radical, which is why the process is called a "chain reaction mechanism". The radical reaction stops when two radicals react and produce a non-radical species. This happens only when the concentration of radical species is high enough for there to be a high probability of collision of two radicals. Living organisms have different molecules that speed up termination by neutralizing free radicals and, therefore, protecting the cell membrane. One important such antioxidant is vitamin E. Another important antioxidant is vitamin C. Other anti-oxidants made within the body include the enzymes superoxide dismutase, catalase, and peroxidase.

Final Products of Lipid Peroxidation

The end products of lipid peroxidation are reactive aldehydes, such as malondialdehyde (MDA) and 4-hydroxynonenal (HNE), the second one being known also as "second messenger of free radicals" and major bioactive marker of lipid peroxidation, due to its numerous biological activities resembling activities of reactive oxygen species.

Hazards

If not terminated fast enough, there will be damage to the cell membrane, which consists mainly of lipids. Phototherapy may cause hemolysis by rupturing red blood cell cell membranes in this way.

In addition, end-products of lipid peroxidation may be mutagenic and carcinogenic. For instance, the end-product malondialdehyde reacts with deoxyadenosine and deoxyguanosine in DNA, forming DNA adducts to them, primarily M_1G.

The toxicity of lipid hydroperoxides to animals is best illustrated by the lethal phenotype of glutathione peroxidase 4 (GPX4) knockout mice. These animals do not survive past embryonic day 8, indicating that the removal of lipid hydroperoxides is essential for mammalian life.

Inhibiting Lipid Peroxidation

The chain reaction format of the lipid peroxidation cycle and tight control of the levels of antioxidants "dissolved" in lipid bilayers (vitamin E is present in lipid membranes at levels not exceeding the ratio of 1 molecule per 2000 fatty acid residues) explain the inefficiency of antioxidant approaches at keeping lipid peroxidation in check. An alternative, non-antioxidant based method employs deuteration of polyunsaturated fatty acids (PUFA) at the methylene bridges (bis-allylic sites) between double bonds, which leads to the inhibition of the chain reaction courtesy of a kinetic isotope effect. Such D-PUFAs, for example, 11,11-D2-ethyl linoleate, suppress lipid peroxidation even at relatively low levels of incorporation into membranes.

Tests

Certain diagnostic tests are available for the quantification of the end-products of lipid peroxidation, to be specific, malondialdehyde (MDA). The most commonly used test is called a TBARS Assay (thiobarbituric acid reactive substances assay). Thiobarbituric acid reacts with malondialdehyde to yield a fluorescent product. However, there are other sources of malondialdehyde, so this test is not completely specific for lipid peroxidation.

In recent years, development of immunochemical detection of HNE-histidine adducts opened more advanced methodological possibilities for qualitative and quantitative detection of lipid peroxidation in various human and animal tissues as well as in body fluids, including human serum and plasma samples.

References

- McNaught, A. D.; Wilkinson, A., eds. (1997). "lipids". Compendium of Chemical Terminology (the "Gold Book") (2nd ed.). Oxford: Blackwell Scientific Publications. doi:10.1351/goldbook. ISBN 0-9678550-9-8
- Subramaniam S, Fahy E, Gupta S, Sud M, Byrnes RW, Cotter D, Dinasarapu AR, Maurya MR (October 2011). "Bioinformatics and systems biology of the lipidome". Chemical Reviews. 111 (10): 6452–90. doi:10.1021/cr200295k. PMC 3383319. PMID 21939287
- Geoff Watts (2010-06-12). "Alec Douglas Bangham". The Lancet. 375(9731): 2070. doi:10.1016/S0140-

6736(10)60950-6. Retrieved 2014-10-01

- Barani, H; Montazer, M (2008). "A review on applications of liposomes in textile processing". Journal of liposome research. 18 (3): 249–62. doi:10.1080/08982100802354665. PMID 18770074
- Michelle A, Hopkins J, McLaughlin CW, Johnson S, Warner MQ, LaHart D, Wright JD (1993). Human Biology and Health. Englewood Cliffs, New Jersey, USA: Prentice Hall. ISBN 978-0-13-981176-0
- Mashaghi S, Jadidi T, Koenderink G, Mashaghi A (February 2013). "Lipid nanotechnology". International Journal of Molecular Sciences. 14 (2): 4242–82. doi:10.3390/ijms14024242. PMC 3588097. PMID 23429269
- "Hydrolysis - Chemistry Encyclopedia - structure, reaction, water, proteins, examples, salt, molecule". www.chemistryexplained.com. Retrieved 2016-11-01
- Murray, I.; Sniderman, A. D.; Cianflone, K. (1999-09-01). "Enhanced triglyceride clearance with intraperitoneal human acylation stimulating protein in C57BL/6 mice". The American Journal of Physiology. 277 (3 Pt 1): E474–480. ISSN 0002-9513. PMID 10484359
- Vance JE, Vance DE (2002). Biochemistry of Lipids, Lipoproteins and Membranes. Amsterdam: Elsevier. ISBN 978-0-444-51139-3
- Hunter JE (November 2006). "Dietary trans fatty acids: review of recent human studies and food industry responses". Lipids. 41 (11): 967–92. doi:10.1007/s11745-006-5049-y. PMID 17263298
- IUPAC Compendium of Chemical Terminology (2nd ed.). International Union of Pure and Applied Chemistry. 1997. doi:10.1351/pac199567081307. ISBN 0-521-51150-X. Retrieved 2007-10-31
- Meure, LA; Knott, R; Foster, NR; Dehghani, F (2009). "The depressurization of an expanded solution into aqueous media for the bulk production of liposomes". Langmuir : the ACS journal of surfaces and colloids. 25 (1): 326–37. doi:10.1021/la802511a. PMID 19072018
- Hunt SM, Groff JL, Gropper SA (1995). Advanced Nutrition and Human Metabolism. Belmont, California: West Pub. Co. p. 98. ISBN 978-0-314-04467-9
- Coleman RA, Lee DP (March 2004). "Enzymes of triacylglycerol synthesis and their regulation". Progress in Lipid Research. 43 (2): 134–76. doi:10.1016/S0163-7827(03)00051-1. PMID 14654091
- Altar, Ted. "More Than You Wanted To Know About Fats/Oils". Sundance Natural Foods. Retrieved 2006-08-31
- Etherton, T. D. (2000-11-01). "The biology of somatotropin in adipose tissue growth and nutrient partitioning". The Journal of Nutrition. 130 (11): 2623–2625. doi:10.1093/jn/130.11.2623. ISSN 0022-3166. PMID 11053496
- Fezza F, De Simone C, Amadio D, Maccarrone M (2008). "Fatty acid amide hydrolase: a gate-keeper of the endocannabinoid system". Sub-Cellular Biochemistry. Subcellular Biochemistry. 49: 101–32. doi:10.1007/978-1-4020-8831-5_4. ISBN 978-1-4020-8830-8. PMID 18751909
- "USDA National Nutrient Database for Standard Reference". U.S. Department of Agriculture. Archived from the original on 2015-03-03. Retrieved 2010-02-17

Chapter 4

Proteins: Design, Structure and Metabolism

The scientific field of the study of proteins is known as proteomics. Proteins are macromolecules which perform an array of functions in living organisms. Some of their primary functions include biosynthesis and chemical synthesis, cell signaling, DNA replication, etc. The following chapter discusses the structure, design and metabolism of proteins in detail.

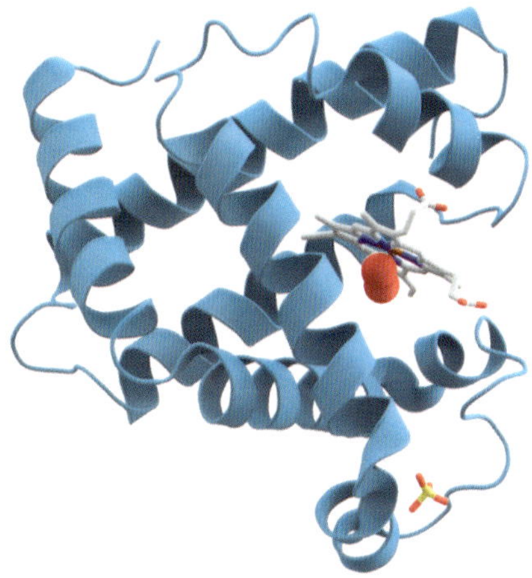

A representation of the 3D structure of the protein myoglobin showing turquoise α-helices. This protein was the first to have its structure solved by X-ray crystallography. Towards the right-center among the coils, a prosthetic group called a heme group (shown in gray) with a bound oxygen molecule (red).

Proteins are large biomolecules, or macromolecules, consisting of one or more long chains of amino acid residues. Proteins perform a vast array of functions within organisms, including catalysing metabolic reactions, DNA replication, responding to stimuli, and transporting molecules from one location to another. Proteins differ from one another primarily in their sequence of amino acids, which is dictated by the nucleotide sequence of their genes, and which usually results in protein folding into a specific three-dimensional structure that determines its activity.

A linear chain of amino acid residues is called a polypeptide. A protein contains at least one long polypeptide. Short polypeptides, containing less than 20–30 residues, are rarely considered to be proteins and are commonly called peptides, or sometimes oligopeptides. The individual amino acid residues are bonded together by peptide bonds and adjacent amino acid residues. The sequence of amino acid residues in a protein is defined by the sequence of a gene, which is encoded in the genetic code. In general, the genetic code specifies 20 standard amino acids; however, in certain organisms the genetic code can include selenocysteine and—in certain archaea—pyrrolysine. Shortly after or even during synthesis, the residues in a protein are often chemically modified by post-translational modification, which alters the physical and chemical properties, folding, stability, activity, and ultimately, the function of the proteins. Sometimes proteins have non-peptide groups attached, which can be called prosthetic groups or cofactors. Proteins can also work together to achieve a particular function, and they often associate to form stable protein complexes.

Once formed, proteins only exist for a certain period of time and are then degraded and recycled by the cell's machinery through the process of protein turnover. A protein's lifespan is measured in terms of its half-life and covers a wide range. They can exist for minutes or years with an average lifespan of 1–2 days in mammalian cells. Abnormal or misfolded proteins are degraded more rapidly either due to being targeted for destruction or due to being unstable.

Like other biological macromolecules such as polysaccharides and nucleic acids, proteins are essential parts of organisms and participate in virtually every process within cells. Many proteins are enzymes that catalyse biochemical reactions and are vital to metabolism. Proteins also have structural or mechanical functions, such as actin and myosin in muscle and the proteins in the cytoskeleton, which form a system of scaffolding that maintains cell shape. Other proteins are important in cell signaling, immune responses, cell adhesion, and the cell cycle. In animals, proteins are needed in the diet to provide the essential amino acids that cannot be synthesized. Digestion breaks the proteins down for use in the metabolism.

Proteins may be purified from other cellular components using a variety of techniques such as ultracentrifugation, precipitation, electrophoresis, and chromatography; the advent of genetic engineering has made possible a number of methods to facilitate purification. Methods commonly used to study protein structure and function include immunohistochemistry, site-directed mutagenesis, X-ray crystallography, nuclear magnetic resonance and mass spectrometry.

Protein Design

Protein design is the rational design of new protein molecules to fold to a target protein structure, with the end goal of designing novel function and/or behavior. Proteins can be designed from scratch (*de novo* design) or by making calculated variants of a known protein structure and its sequence (termed *protein redesign*). Rational protein design approaches make protein-sequence predictions that will fold to specific structures. These predicted sequences can then be validated experimentally through methods such as peptide synthesis, site-directed mutagenesis, or artificial gene synthesis.

Rational protein design dates back to the mid-1970s, although initial protein design approaches were based mostly on sequence composition and did not account for specific interactions between side-chains at the atomic level. Recently, however, improvements in molecular force fields, protein design algorithms, and structural bioinformatics, such as libraries of amino acid conformations, have enabled the development of advanced computational protein design tools. These computational tools can make complex calculations on protein energetics and flexibility, and perform searches over vast configuration spaces, which would be unfeasible to perform manually. Due to the development of computational protein design programs, and important successes in the field, rational protein design has become one of the most important tools in protein engineering.

The goal in rational protein design is to predict amino acid sequences that will fold to a specific protein structure. Although the number of possible protein sequences is vast, growing exponentially with the size of the protein chain, only a subset of them will fold reliably and quickly to one native state. Protein design involves identifying novel sequences within this subset. The native state of a protein is the conformational free energy minimum for the chain. Thus, protein design is the search for sequences that have the chosen structure as a free energy minimum. In a sense, it is the reverse of protein structure prediction. In design, a tertiary structure is specified, and a sequence that will fold to it is identified. Hence, it is also termed *inverse folding*. Protein design is then an optimization problem: using some scoring criteria, an optimized sequence that will fold to the desired structure is chosen.

When the first proteins were rationally designed during the 1970s and 1980s, the sequence for these was optimized manually based on analyses of other known proteins, the sequence composition, amino acid charges, and the geometry of the desired structure. The first designed proteins are attributed to Bernd Gutte, who designed a reduced version of a known catalyst, bovine ribonuclease, and tertiary structures consisting of beta-sheets and alpha-helices, including a binder of DDT. Urry and colleagues later designed elastin-like fibrous peptides based on rules on sequence composition. Richardson and coworkers designed a 79-residue protein with no sequence homology to a known protein. In the 1990s, the advent of powerful computers, libraries of amino acid conformations, and force fields developed mainly for molecular dynamics simulations enabled the development of structure-based computational protein design tools. Following the development of these computational tools, great success has been achieved over the last 30 years in protein design. The first protein successfully designed completely *de novo* was done by Stephen Mayo and coworkers in 1997, and, shortly after, in 1999 Peter S. Kim and coworkers designed dimers, trimers, and tetramers of unnatural right-handed coiled coils. In 2003, David Baker's laboratory designed a full protein to a fold never seen before in nature. Later, in 2008, Baker's group computationally designed enzymes for two different reactions. In 2010, one of the most powerful broadly neutralizing antibodies was isolated from patient serum using a computationally designed protein probe. Due to these and other successes, protein design has become one of the most important tools available for protein engineering. There is great hope that the design of new proteins, small and large, will have uses in biomedicine and bioengineering.

Underlying Models of Protein Structure and Function

Protein design programs use computer models of the molecular forces that drive proteins in *in vivo* environments. In order to make the problem tractable, these forces are simplified by protein design models. Although protein design programs vary greatly, they have to address four main modeling questions: What is the target structure of the design, what flexibility is allowed on the target structure, which sequences are included in the search, and which force field will be used to score sequences and structures.

Target Structure

Protein function is heavily dependent on protein structure, and rational protein design uses this relationship to design function by designing proteins that have a target structure or fold. Thus, by definition, in rational protein design the target structure or ensemble of structures must be known beforehand. This contrasts with other forms of protein engineering, such as directed evolution, where a variety of methods are used to find proteins that achieve a specific function, and with protein structure prediction where the sequence is known, but the structure is unknown.

The Top7 protein was one of the first proteins designed for a fold that had never been seen before in nature

Most often, the target structure is based on a known structure of another protein. However, novel folds not seen in nature have been made increasingly possible. Peter S. Kim and coworkers designed trimers and tetramers of unnatural coiled coils, which had not been seen before in nature. The protein Top7, developed in David Baker's lab, was designed completely using protein design algorithms, to a completely novel fold. More recently, Baker and coworkers developed a series of principles to design ideal globular-protein structures based on protein folding funnels that bridge between secondary structure prediction and tertiary structures. These principles, which build on both protein structure prediction and protein design, were used to design five different novel protein topologies.

Sequence Space

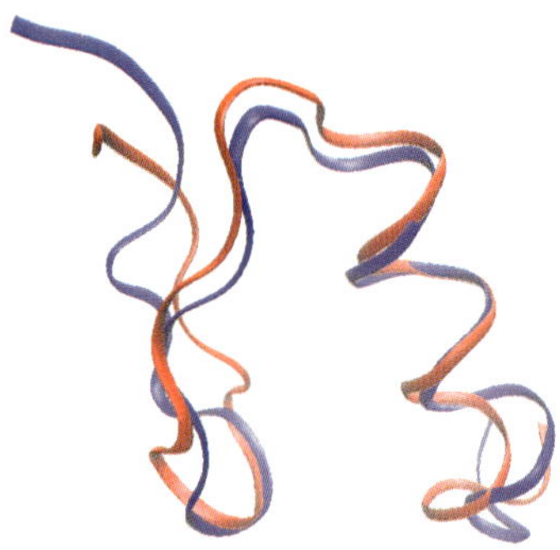

FSD-1 (shown in blue, PDB id: 1FSV) was the first *de novo* computational design of a full protein. The target fold was that of the zinc finger in residues 33–60 of the structure of protein Zif268 (shown in red, PDB id: 1ZAA). The designed sequence had very little sequence identity with any known protein sequence.

In rational protein design, proteins can be redesigned from the sequence and structure of a known protein, or completely from scratch in *de novo* protein design. In protein redesign, most of the residues in the sequence are maintained as their wild-type amino-acid while a few are allowed to mutate. In *de novo* design, the entire sequence is designed anew, based on no prior sequence.

Both *de novo* designs and protein redesigns can establish rules on the sequence space: the specific amino acids that are allowed at each mutable residue position. For example, the composition of the surface of the RSC3 probe to select HIV-broadly neutralizing antibodies was restricted based on evolutionary data and charge balancing. Many of the earliest attempts on protein design were heavily based on empiric *rules* on the sequence space. Moreover, the design of fibrous proteins usually follows strict rules on the sequence space. Collagen-based designed proteins, for example, are often composed of Gly-Pro-X repeating patterns. The advent of computational techniques allows designing proteins with no human intervention in sequence selection.

Structural Flexibility

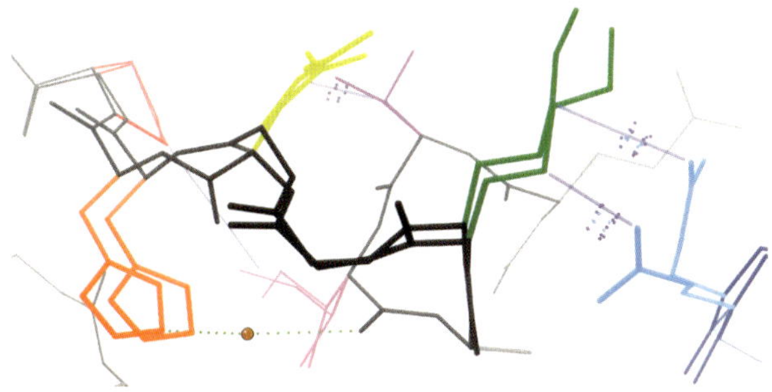

In protein design, the target structure (or structures) of the protein are known. However, a rational protein design approach must model some *flexibility* on the target structure in order to increase the number of sequences that can be designed for that structure and to minimize the chance of a sequence folding to a different structure. For example, in a protein redesign of one small amino acid (such as alanine) in the tightly packed core of a protein, very few mutants would be predicted by a rational design approach to fold to the target structure, if the surrounding side-chains are not allowed to be repacked.

Thus, an essential parameter of any design process is the amount of flexibility allowed for both the side-chains and the backbone. In the simplest models, the protein backbone is kept rigid while some of the protein side-chains are allowed to change conformations. However, side-chains can have many degrees of freedom in their bond lengths, bond angles, and χ dihedral angles. To simplify this space, protein design methods use rotamer libraries that assume ideal values for bond lengths and bond angles, while restricting χ dihedral angles to a few oft-observed low-energy conformations termed rotamers.

Rotamer libraries describe rotamers based on an analysis of many protein structures. Backbone-independent rotamer libraries describe all rotamers. Backbone-dependent rotamer libraries, in contrast, describe the rotamers as how likely they are to appear depending on the protein backbone arrangement around the side chain. The rotamers described by rotamer libraries are usually regions in space. Most protein design programs use one conformation (e.g., the modal value for rotamer dihedrals in space) or several points in the region described by the rotamer; the OSPREY protein design program, in contrast, models the entire continuous region.

Although rational protein design must preserve the general backbone fold a protein, allowing some backbone flexibility can significantly increase the number of sequences that fold to the structure while maintaining the general fold of the protein. Backbone flexibility is especially important in protein redesign because sequence mutations often result in small changes to the backbone structure. Moreover, backbone flexibility can be essential for more advanced applications of protein design, such as binding prediction and enzyme design. Some models of protein design backbone flexibility include small and continuous global backbone movements, discrete backbone samples around the target fold, backrub motions, and protein loop flexibility.

Energy Function

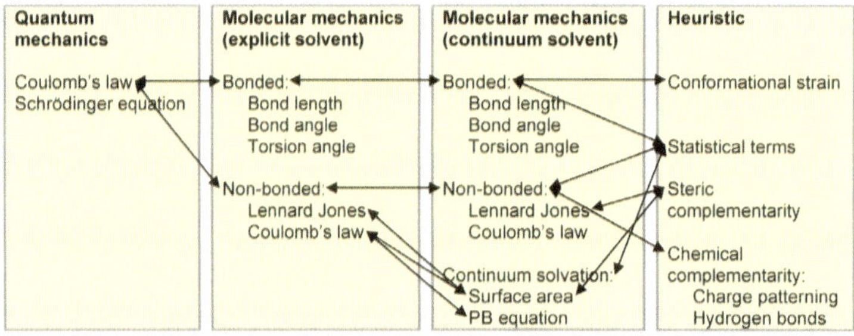

Comparison of various potential energy functions. The most accurate energy are those that use quantum mechanical calculations, but these are too slow for protein design. On the other extreme, heuristic energy functions, are based on statistical terms and are very fast. In the middle are molecular mechanics energy functions that are physically-based but are not as computationally expensive as quantum mechanical simulations.

Rational protein design techniques must be able to discriminate sequences that will be stable under the target fold from those that would prefer other low-energy competing states. Thus, protein design requires accurate energy functions that can rank and score sequences by how well they fold to the target structure. At the same time, however, these energy functions must consider the computational challenges behind protein design. One of the most challenging requirements for successful design is an energy function that is both accurate and simple for computational calculations.

The most accurate energy functions are those based on quantum mechanical simulations. However, such simulations are too slow and typically impractical for protein design. Instead, many protein design algorithms use either physics-based energy functions adapted from molecular mechanics simulation programs, knowledge based energy-functions, or a hybrid mix of both. The trend has been toward using more physics-based potential energy functions.

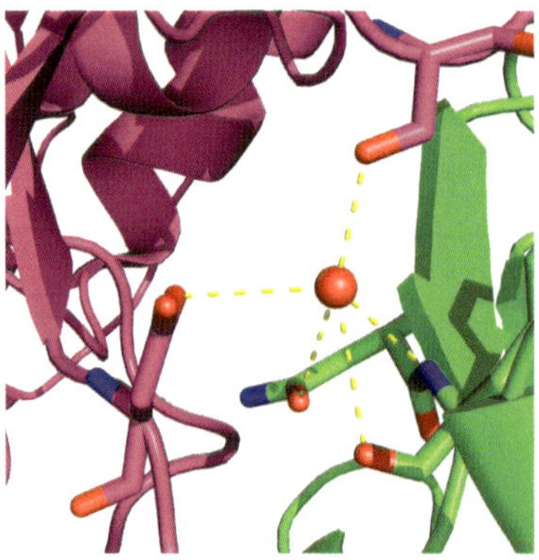

Water-mediated hydrogen bonds play a key role in protein–protein binding. One such interaction is shown between residues D457, S365 in the heavy chain of the HIV-broadly-neutralizing antibody VRC01 (green) and residues N58 and Y59 in the HIV envelope protein GP120 (purple).

Physics-based energy functions, such as AMBER and CHARMM, are typically derived from quantum mechanical simulations, and experimental data from thermodynamics, crystallography, and spectroscopy. These energy functions typically simplify physical energy function and make them pairwise decomposable, meaning that the total energy of a protein conformation can be calculated by adding the pairwise energy between each atom pair, which makes them attractive for optimization algorithms. Physics-based energy functions typically model an attractive-repulsive Lennard-Jones term between atoms and a pairwise electrostatics coulombic term between non-bonded atoms.

Statistical potentials, in contrast to physics-based potentials, have the advantage of being fast to compute, of accounting implicitly of complex effects and being less sensitive to small changes in the protein structure. These energy functions are based on deriving energy values from frequency of appearance on a structural database.

Protein design, however, has requirements that can sometimes be limited in molecular mechanics force-fields. Molecular mechanics force-fields, which have been used mostly in molecular dynamics simulations, are optimized for the simulation of single sequences, but protein design searches through many conformations of many sequences. Thus, molecular mechanics force-fields must be tailored for protein design. In practice, protein design energy functions often incorporate both statistical terms and physics-based terms. For example, the Rosetta energy function, one of the most-used energy functions, incorporates physics-based energy terms originating in the CHARMM energy function, and statistical energy terms, such as rotamer probability and knowledge-based electrostatics. Typically, energy functions are highly customized between laboratories, and specifically tailored for every design.

Challenges for Effective Design Energy Functions

Water makes up most of the molecules surrounding proteins and is the main driver of protein structure. Thus, modeling the interaction between water and protein is vital in protein design. The number of water molecules that interact with a protein at any given time is huge and each one has a large number of degrees of freedom and interaction partners. Instead, protein design programs model most of such water molecules as a continuum, modeling both the hydrophobic effect and solvation polarization.

Individual water molecules can sometimes have a crucial structural role in the core of proteins, and in protein–protein or protein–ligand interactions. Failing to model such waters can result in mispredictions of the optimal sequence of a protein–protein interface. As an alternative, water molecules can be added to rotamers.

As an Optimization Problem

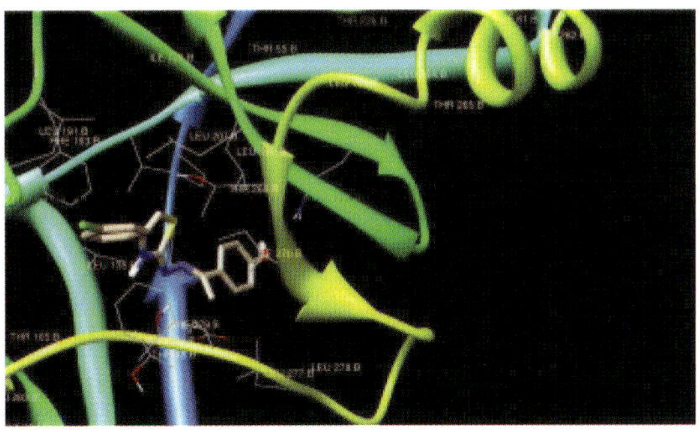

The goal of protein design is to find a protein sequence that will fold to a target structure. A protein design algorithm must, thus, search all the conformations of each sequence, with respect to the target fold, and rank sequences according to the lowest-energy conformation of each one, as determined by the protein design energy function. Thus, a typical input to the protein design algorithm is the target fold, the sequence space, the structural flexibility, and the energy function, while the output is one or more sequences that are predicted to fold stably to the target structure.

The number of candidate protein sequences, however, grows exponentially with the number of protein residues; for example, there are 20^{100} protein sequences of length 100. Furthermore, even if amino acid side-chain conformations are limited to a few rotamers, this results in an exponential number of conformations for each sequence. Thus, in our 100 residue protein, and assuming that each amino acid has exactly 10 rotamers, a search algorithm that searches this space will have to search over 200^{100} protein conformations.

The most common energy functions can be decomposed into pairwise terms between rotamers and amino acid types, which casts the problem as a combinatorial one, and powerful optimization algorithms can be used to solve it. In those cases, the total energy of each conformation belonging to each sequence can be formulated as a sum of individual and pairwise terms between residue positions. If a designer is interested only in the best sequence, the protein design algorithm only requires the lowest-energy conformation of the lowest-energy sequence. In these cases, the amino acid identity of each rotamer can be ignored and all rotamers belonging to different amino acids can be treated the same. Let r_i be a rotamer at residue position i in the protein chain, and $E(r_i)$ the potential energy between the internal atoms of the rotamer. Let $E(r_i, r_j)$ be the potential energy between r_i and rotamer r_j at residue position j. Then, we define the optimization problem as one of finding the conformation of minimum energy (E_T):

$$\min E_T = \sum_i \left[E_i(r_i) + \sum_{i \neq j} E_{ij}(r_i, r_j) \right]$$

The problem of minimizing E_T is an NP-hard problem. Even though the class of problems is NP-hard, in practice many instances of protein design can be solved exactly or optimized satisfactorily through heuristic methods.

Algorithms

Several algorithms have been developed specifically for the protein design problem. These algorithms can be divided into two broad classes: exact algorithms, such as dead-end elimination, that lack runtime guarantees but guarantee the quality of the solution; and heuristic algorithms, such as Monte Carlo, that are faster than exact algorithms but have no guarantees on the optimality of the results. Exact algorithms guarantee that the optimization process produced the optimal according to the protein design model. Thus, if the predictions of exact algorithms fail when these are experimentally validated, then the source of error can be attributed to the energy function, the allowed flexibility, the sequence space or the target structure (e.g., if it cannot be designed for)

Some protein design algorithms are listed below. Although these algorithms address only the most basic formulation of the protein design problem, Equation above, when the optimization goal changes because designers introduce improvements and extensions to the protein design model, such as improvements to the structural flexibility allowed (e.g., protein backbone flexibility) or including sophisticated energy terms, many of the extensions on protein design that improve modeling are built atop these algorithms. For example, Rosetta Design incorporates sophisticated energy terms, and backbone flexibility using Monte Carlo as the underlying optimizing algorithm. OSPREY's algorithms build on the dead-end elimination algorithm and A* to incorporate continuous backbone and side-chain movements. Thus, these algorithms provide a good perspective on the different kinds of algorithms available for protein design.

With Mathematical Guarantees

Dead-end Elimination

The dead-end elimination (DEE) algorithm reduces the search space of the problem iteratively by removing rotamers that can be provably shown to be not part of the global lowest energy conformation (GMEC). On each iteration, the dead-end elimination algorithm compares all possible pairs of rotamers at each residue position, and removes each rotamer r'_i that can be shown to always be of higher energy than another rotamer r_i and is thus not part of the GMEC:

$$E(r_i') + \sum_{j \neq i} \min_{r_j} E(r_i', r_j) > E(r_i) + \sum_{j \neq i} \max_{r_j} E(r_i, r_j)$$

Other powerful extensions to the dead-end elimination algorithm include the pairs elimination criterion, and the generalized dead-end elimination criterion. This algorithm has also been extended to handle continuous rotamers with provable guarantees.

Although the Dead-end elimination algorithm runs in polynomial time on each iteration, it cannot guarantee convergence. If, after a certain number of iterations, the dead-end elimination algorithm does not prune any more rotamers, then either rotamers have to be merged or another search algorithm must be used to search the remaining search space. In such cases, the dead-end elimination acts as a pre-filtering algorithm to reduce the search space, while other algorithms, such as A*, Monte Carlo, Linear Programming, or FASTER are used to search the remaining search space.

Branch and Bound

The protein design conformational space can be represented as a tree, where the protein residues are ordered in an arbitrary way, and the tree branches at each of the rotamers in a residue. Branch and bound algorithms use this representation to efficiently explore the conformation tree: At each *branching*, branch and bound algorithms *bound* the conformation space and explore only the promising branches.

A popular search algorithm for protein design is the A* search algorithm. A* computes a lower-bound score on each partial tree path that lower bounds (with guarantees) the energy of each of the expanded rotamers. Each partial conformation is added to a priority queue and at each iteration the partial path with the lowest lower bound is popped from the queue and expanded. The algorithm stops once a full conformation has been enumerated and guarantees that the conformation is the optimal.

The A* score f in protein design consists of two parts, f=g+h. g is the exact energy of the rotamers that have already been assigned in the partial conformation. h is a lower bound on the energy of the rotamers that have not yet been assigned. Each is designed as follows, where d is the index of the last assigned residue in the partial conformation.

$$g = \sum_{i=1}^{d} (E(r_i) + \sum_{j=i+1}^{d} E(r_i, r_j))$$

$$h = \sum_{j=d+1}^{n} \left[\min_{r_j}(E(r_j) + \sum_{i=1}^{d} E\left(r_i, r_j\right) + \sum_{k=j+1}^{n} \min_{r_k} E(r_j, r_k)) \right]$$

Integer Linear Programming

The problem of optimizing E_T (Equation) can be easily formulated as an integer linear program (ILP). One of the most powerful formulations uses binary variables to represent the presence of a rotamer and edges in the final solution, and constraints the solution to have exactly one rotamer for each residue and one pairwise interaction for each pair of residues:

$$\min \sum_{i} \sum_{r_i} E_i(r_i) q_i(r_i) + \sum_{j \neq i} \sum_{r_j} E_{ij}(r_i, r_j) q_{ij}(r_i, r_j)$$

s.t.

$$\sum_{r_i} q_i(r_i) = 1, \forall i$$

$$\sum_{r_j} q_{ij}(r_i, r_j) = q_i(r_i), \forall i, r_i, j$$

$$q_i, q_{ij} \in \{0,1\}$$

ILP solvers, such as CPLEX, can compute the exact optimal solution for large instances of protein design problems. These solvers use a linear programming relaxation of the problem, where q_i and q_{ij} are allowed to take continuous values, in combination with a branch and cut algorithm to search only a small portion of the conformation space for the optimal solution. ILP solvers have been shown to solve many instances of the side-chain placement problem.

Message-passing based Approximations to the Linear Programming Dual

ILP solvers depend on linear programming (LP) algorithms, such as the Simplex or barrier-based methods to perform the LP relaxation at each branch. These LP algorithms were developed as general-purpose optimization methods and are not optimized for the protein design problem (Equation (1)). In consequence, the LP relaxation becomes the bottleneck of ILP solvers when the problem size is large. Recently, several alternatives based on message-passing algorithms have been designed specifically for the optimization of the LP relaxation of the protein design problem. These algorithms can approximate both the dual or the primal instances of the integer programming, but in order to maintain guarantees on optimality, they are most useful when used to approximate the dual of the protein design problem, because approximating the dual guarantees that no solutions are missed. Message-passing based approximations include the *tree reweighted max-product message passing* algorithm, and the *message passing linear programming* algorithm.

Optimization Algorithms without Guarantees

Monte Carlo and Simulated Annealing

Monte Carlo is one of the most widely used algorithms for protein design. In its simplest form, a Monte Carlo algorithm selects a residue at random, and in that residue a randomly chosen rotamer (of any amino acid) is evaluated. The new energy of the protein, E_{new} is compared against the old energy E_{old} and the new rotamer is *accepted* with a probability of:

$$p = e^{-\beta(E_{\text{new}} - E_{\text{old}}))},$$

where β is the Boltzmann constant and the temperature T can be chosen such that in the initial rounds it is high and it is slowly annealed to overcome local minima.

FASTER

The FASTER algorithm uses a combination of deterministic and stochastic criteria to optimize amino acid sequences. FASTER first uses DEE to eliminate rotamers that are not part of the optimal solution. Then, a series of iterative steps optimize the rotamer assignment.

Belief Propagation

In belief propagation for protein design, the algorithm exchanges messages that describe the *belief* that each residue has about the probability of each rotamer in neighboring residues. The algorithm updates messages on every iteration and iterates until convergence or until a fixed number of iterations. Convergence is not guaranteed in protein design. The message $m_{i\to j}(r_j$ that a residue i sends to every rotamer $(r_j$ at neighboring residue j is defined as:

$$m_{i\to j}(r_j) = \max_{r_i}\left(e^{\frac{-E_i(r_i)-E_{ij}(r_i,r_j)}{T}}\right)\prod_{k\in N(i)\setminus j} m_{k\to i(r_i)}$$

Both max-product and sum-product belief propagation have been used to optimize protein design.

Applications and Examples of Designed Proteins

Enzyme Design

The design of new enzymes is a use of protein design with huge bioengineering and biomedical applications. In general, designing a protein structure can be different from designing an enzyme, because the design of enzymes must consider many states involved in the catalytic mechanism. However protein design is a prerequisite of *de novo* enzyme design because, at the very least, the design of catalysts requires a scaffold in which the catalytic mechanism can be inserted.

Great progress in *de novo* enzyme design, and redesign, was made in the first decade of the 21st century. In three major studies, David Baker and coworkers *de novo* designed enzymes for the retro-aldol reaction, a Kemp-elimination reaction, and for the Diels-Alder reaction. Furthermore, Stephen Mayo and coworkers developed an iterative method to design the most efficient known enzyme for the Kemp-elimination reaction. Also, in the laboratory of Bruce Donald, computational protein design was used to switch the specificity of one of the protein domains of the nonribosomal peptide synthetase that produces Gramicidin S, from its natural substrate phenylalanine to other noncognate substrates including charged amino acids; the redesigned enzymes had activities close to those of the wild-type.

Design for Affinity

Protein–protein interactions are involved in most biotic processes. Many of the hardest-to-treat diseases, such as Alzheimer's, many forms of cancer (e.g., TP53), and human immunodeficiency virus (HIV) infection involve protein–protein interactions. Thus, to treat such diseases, it is desirable to design protein or protein-like therapeutics that bind one of the partners of the interaction and, thus, disrupt the disease-causing interaction. This requires designing protein-therapeutics for *affinity* toward its partner.

Protein–protein interactions can be designed using protein design algorithms because the principles that rule protein stability also rule protein–protein binding. Protein–protein interaction design, however, presents challenges not commonly present in protein design. One of the most important challenges is that, in general, the interfaces between proteins are more polar than protein cores, and binding involves a tradeoff between desolvation and hydrogen bond formation. To overcome this challenge, Bruce Tidor and coworkers developed a method to improve the affinity of antibodies by focusing on electrostatic contributions. They found that, for the antibodies designed in the study, reducing the desolvation costs of the residues in the interface increased the affinity of the binding pair.

Scoring Binding Predictions

Protein design energy functions must be adapted to score binding predictions because binding involves a trade-off between the lowest-energy conformations of the free proteins (E_P and E_L) and the lowest-energy conformation of the bound complex (E_{PL}):

$$\Delta_G = E_{PL} - E_P - E_L.$$

The K* algorithm approximates the binding constant of the algorithm by including conformational entropy into the free energy calculation. The K* algorithm considers only the lowest-energy conformations of the free and bound complexes (denoted by the sets P, L, and PL) to approximate the partition functions of each complex:

$$K^* = \frac{\sum_{x \in PL} e^{-E(x)/RT}}{\sum_{x \in P} e^{-E(x)/RT} \sum_{x \in L} e^{-E(x)/RT}}$$

Design for Specificity

The design of protein–protein interactions must be highly specific because proteins can interact with a large number of proteins; successful design requires selective binders. Thus, protein design algorithms must be able to distinguish between on-target (or *positive design*) and off-target binding (or *negative design*). One of the most prominent examples of design for specificity is the design of specific bZIP-binding peptides by Amy Keating and coworkers for 19 out of the 20 bZIP families; 8 of these peptides were specific for their intended partner over competing peptides. Further, positive and negative design was also used by Anderson and coworkers to predict mutations in the active site of a drug target that conferred resistance to a new drug; positive design was used to maintain wild-type activity, while negative design was used to disrupt binding of the drug. Recent computational redesign by Costas Maranas and coworkers was also capable of experimentally switching the cofactor specificity of *Candida boidinii* xylose reductase from NADPH to NADH.

Protein Resurfacing

Protein resurfacing consists of designing a protein's surface while preserving the overall fold, core, and boundary regions of the protein intact. Protein resurfacing is especially useful to alter the binding of a protein to other proteins. One of the most important applications of protein resurfacing was the design of the RSC3 probe to select broadly neutralizing HIV antibodies at the NIH Vaccine Research Center. First, residues outside of the binding interface between the gp120 HIV envelope protein and the formerly discovered b12-antibody were selected to be designed. Then, the sequence spaced was selected based on evolutionary information, solubility, similarity with the wild-type, and other considerations. Then the RosettaDesign software was used to find optimal sequences in the selected sequence space. RSC3 was later used to discover the broadly neutralizing antibody VRC01 in the serum of a long-term HIV-infected non-progressor individual.

Design of Globular Proteins

Globular proteins are proteins that contain a hydrophobic core and a hydrophilic surface. Globular proteins often assume a stable structure, unlike fibrous proteins, which have multiple conformations. The three-dimensional structure of globular proteins is typically easier to determine through X-ray crystallography and nuclear magnetic resonance than both fibrous proteins and membrane proteins, which makes globular proteins more attractive for protein design than the other types of proteins. Most successful protein designs have involved globular proteins. Both RSD-1, and Top7 were *de novo* designs of globular proteins. Five more protein structures were designed, synthesized, and verified in 2012 by the Baker group. These new proteins serve no biotic function, but the structures are intended to act as building-blocks that can be expanded to incorporate functional active sites. The structures were found computationally by using new heuristics based on analyzing the connecting loops between parts of the sequence that specify secondary structures.

Design of Transmembrane Proteins

Membrane proteins are inherently hard to design, in large part because it is hard to validate the designs experimentally for several reasons. Membrane proteins are hard to purify. Their structure is hard to characterize because they adopt their native conformation only in the presence of a membrane. Crystallization is inherently hard, and NMR studies of membrane proteins can fail because of their size.

Design of Fibrous Proteins

Fibrous proteins, such as elastin or collagen, typically have no one structure. However, such proteins likely lack random folds, and instead have folds defined within an ensemble of structures. This ensemble defines their behavior. Thus, in theory it is possible to rationally design fibrous proteins by selecting a sequence that will populate a specific ensemble.

Other Applications

One of the most desirable uses for protein design is for biosensors, proteins that will sense the presence of specific compounds. Some attempts in the design of biosensors include sensors for unnatural molecules including TNT. More recently, Kuhlman and coworkers designed a biosensor of the PAK1.

Protein Structure

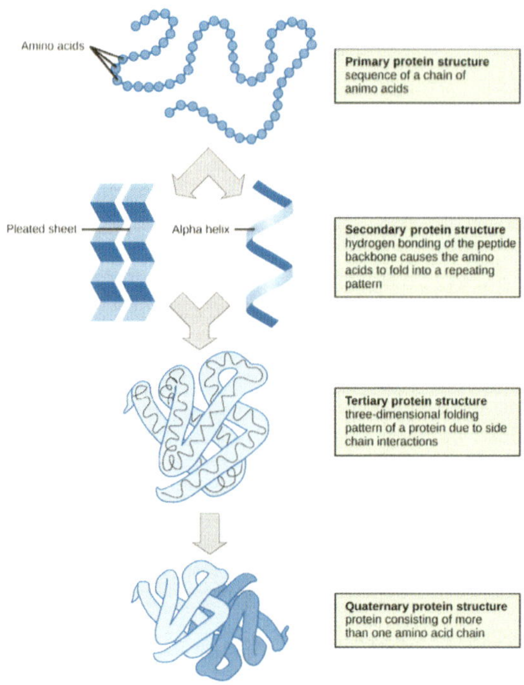

Interactive diagram of protein structure.

Protein structure is the three-dimensional arrangement of atoms in an amino acid-chain molecule. Proteins are polymers – specifically polypeptides – formed from sequences of amino acids, the monomers of the polymer. A single amino acid monomer may also be called a residue indicating a repeating unit of a polymer. Proteins form by amino acids undergoing condensation reactions, in which the amino acids lose one water molecule per reaction in order to attach to one another with a peptide bond. By convention, a chain under 30 amino acids is often identified as a peptide, rather than a protein. To be able to perform their biological function, proteins fold into one or more specific spatial conformations driven by a number of non-covalent interactions such as hydrogen bonding, ionic interactions, Van der Waals forces, and hydrophobic packing. To understand the functions of proteins at a molecular level, it is often necessary to determine their three-dimensional structure. This is the topic of the scientific field of structural biology, which employs techniques such as X-ray crystallography, NMR spectroscopy, and dual polarisation interferometry to determine the structure of proteins.

Protein structures range in size from tens to several thousand amino acids. By physical size, proteins are classified as nanoparticles, between 1–100 nm. Very large aggregates can be formed from protein subunits. For example, many thousands of actin molecules assemble into a microfilament.

A protein may undergo reversible structural changes in performing its biological function. The alternative structures of the same protein are referred to as different conformational isomers, or simply, conformations, and transitions between them are called conformational changes.

Levels of Protein Structure

There are four distinct levels of protein structure.

Primary Structure

The primary structure of a protein refers to the sequence of amino acids in the polypeptide chain. The primary structure is held together by peptide bonds that are made during the process of protein biosynthesis. The two ends of the polypeptide chain are referred to as the carboxyl terminus (C-terminus) and the amino terminus (N-terminus) based on the nature of the free group on each extremity. Counting of residues always starts at the N-terminal end (NH_2-group), which is the end where the amino group is not involved in a peptide bond. The primary structure of a protein is determined by the gene corresponding to the protein. A specific sequence of nucleotides in DNA is transcribed into mRNA, which is read by the ribosome in a process called translation. The sequence of amino acids in insulin was discovered by Frederick Sanger, establishing that proteins have defining amino acid sequences. The sequence of a protein is unique to that protein, and defines the structure and function of the protein. The sequence of a protein can be determined by methods such as Edman degradation or tandem mass spectrometry. Often, however, it is read directly from the sequence of the gene using the genetic code. It is strictly recommended to use the words "amino acid residues" when discussing proteins because when a peptide bond is formed, a water molecule is lost, and therefore proteins are made up of amino acid residues. Post-translational modification such as disulfide bond formation, phosphorylations and glycosylations are usually also considered a part of the primary structure, and cannot be read from the gene. For example, insulin is composed of 51 amino acids in 2 chains. One chain has 31 amino acids, and the other has 20 amino acids.

Secondary Structure

Secondary structure refers to highly regular local sub-structures on the actual polypeptide backbone chain. Two main types of secondary structure, the α-helix and the β-strand or β-sheets, were suggested in 1951 by Linus Pauling and coworkers. These secondary structures are defined by patterns of hydrogen bonds between the main-chain peptide groups. They have a regular geometry, being constrained to specific values of the dihedral angles ψ and φ on the Ramachandran plot. Both the α-helix and the β-sheet represent a way of saturating all the hydrogen bond donors and acceptors in the peptide backbone. Some parts of the protein are ordered but do not form any regular structures. They should not be confused with random coil, an unfolded polypeptide chain lacking any fixed three-dimensional structure. Several sequential secondary structures may form a "supersecondary unit".

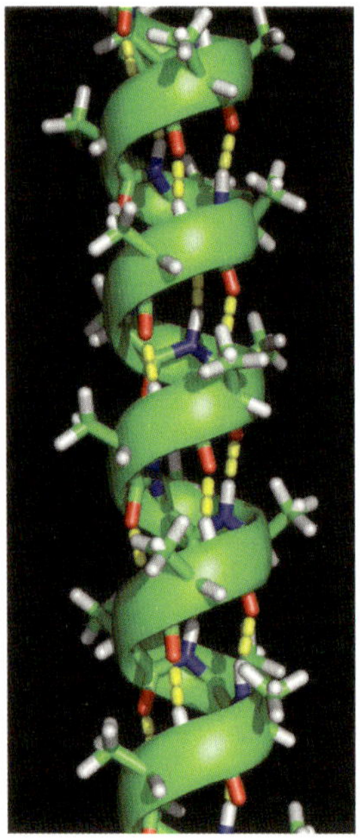

An α-helix with hydrogen bonds (yellow dots)

Tertiary Structure

Tertiary structure refers to the three-dimensional structure of monomeric and multimeric protein molecules. The α-helixes and β-pleated-sheets are folded into a compact globular structure. The folding is driven by the *non-specific* hydrophobic interactions, the burial of hydrophobic residues from water, but the structure is stable only when the parts of a protein domain are locked into place by *specific* tertiary interactions, such as salt bridges, hydrogen bonds, and the tight packing of side chains and disulfide bonds. The disulfide bonds are extremely rare in cytosolic proteins, since the cytosol (intracellular fluid) is generally a reducing environment.

Quaternary Structure

Quaternary structure is the three-dimensional structure consisting of the aggregation of two or more individual polypeptide chains (subunits) that operate as a single functional unit. In this context, the quaternary structure is stabilized by the same non-covalent interactions and disulfide bonds as the tertiary structure. Complexes of two or more polypeptides (i.e. multiple subunits) are called multimers. Specifically it would be called a dimer if it contains two subunits, a trimer if it contains three subunits, a tetramer if it contains four subunits, and a pentamer if it contains five subunits. The subunits are frequently related to one another by symmetry operations, such as a 2-fold axis in a dimer. Multimers made up of identical subunits are referred to with a prefix of "homo-" (e.g. a homotetramer) and those made up of different subunits are referred to with a prefix of "hetero-", for example, a heterotetramer, such as the two alpha and two beta chains of hemoglobin.

Protein Structure and Stability

Protein stability is depended upon a few factors such as 1) Non-covalent electrostatic interactions 2) Hydrophobic interactions These interaction energies are from the order of 20-40 kJ/mol. Proteins are very sensitive to changing temperatures and a change in temperature may result in unfolding or denaturation. Protein denaturation may result in loss of function, and loss of native state.

X-ray Crystallography, and calorimetry indicates that there is no general mechanism that describes the effect of temperature change on the functions and structure of proteins. This is due to the fact that proteins do not represent a uniform class of chemical entities from an energetic point of view. The structure and stability of an individual protein depends on the ratio of its polar and non-polar residues. They contribute to the conformational and the net enthalpies of local and non-local interactions.

Taking the weak intermolecular interactions responsible for structural integrity into consideration, it is hard to predict the effects of temperature because there are too many unknown factors contributing to the hypothetical free energy balance and its temperature dependence. Internal salt linkages produce thermal stability, and whether cold temperature results in the destabilization of these linkages is unknown.

In principle, the free energy of stabilization of soluble globular proteins does not exceed 50-100 kJ/mol. The stabilization is based on the equivalent of few hydrogen bonds, ion pairs, or hydrophobic interactions, even though numerous intramolecular interactions results in stabilization. Taking into consideration the large number of hydrogen bonds that take place for the stabilization of secondary structures, and the stabilization of the inner core through hydrophobic interactions, the free energy of stabilization emerges as small difference between large numbers. Therefore, the structure of a native protein is not optimized for the maximum stability.

Domains, Motifs, and Folds in Protein Structure

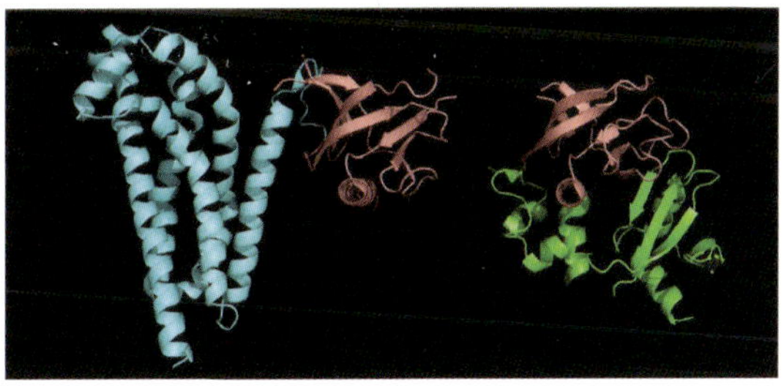

Protein domains. The two shown protein structures share a common domain (maroon), the PH domain, which is involved in phosphatidylinositol (3,4,5)-trisphosphate binding

Proteins are frequently described as consisting of several structural units. These units include domains, motifs, and folds. Despite the fact that there are about 100,000 different proteins expressed in eukaryotic systems, there are many fewer different domains, structural motifs and folds.

Structural Domain

A structural domain is an element of the protein's overall structure that is self-stabilizing and often folds independently of the rest of the protein chain. Many domains are not unique to the protein products of one gene or one gene family but instead appear in a variety of proteins. Domains often are named and singled out because they figure prominently in the biological function of the protein they belong to; for example, the "calcium-binding domain of calmodulin". Because they are independently stable, domains can be "swapped" by genetic engineering between one protein and another to make chimera proteins.

Structural and Sequence Motif

The structural and sequence motifs refer to short segments of protein three-dimensional structure or amino acid sequence that were found in a large number of different proteins.

Supersecondary Structure

The supersecondary structure refers to a specific combination of secondary structure elements, such as β-α-β units or a helix-turn-helix motif. Some of them may be also referred to as structural motifs.

Protein Fold

A protein fold refers to the general protein architecture, like a helix bundle, β-barrel, Rossman fold or different "folds" provided in the Structural Classification of Proteins database. A related concept is protein topology that refers to the arrangement of contacts within the protein.

Superdomain

A superdomain consists of two or more nominally unrelated structural domains that are inherited as a single unit and occur in different proteins. An example is provided by the protein tyrosine phosphatase domain and C2 domain pair in PTEN, several tensin proteins, auxilin and proteins in plants and fungi. The PTP-C2 superdomain evidently came into existence prior to the divergence of fungi, plants and animals is therefore likely to be about 1.5 billion years old.

Protein Folding

Once translated by a ribosome, each polypeptide folds into its characteristic three-dimensional structure from a random coil. Since the fold is maintained by a network of interactions between amino acids in the polypeptide, the native state of the protein chain is determined by the amino acid sequence (Anfinsen's dogma).

Protein Structure Determination

Around 90% of the protein structures available in the Protein Data Bank have been determined by X-ray crystallography. This method allows one to measure the three-dimensional (3-D) density distribution of electrons in the protein, in the crystallized state, and thereby infer the 3-D coordinates of all the atoms to be determined to a certain resolution. Roughly 9% of the known protein structures have been obtained by nuclear magnetic resonance (NMR) techniques. For larger protein complexes, cryo-electron microscopy can determine protein structures. The resolution is typically lower than that of X-ray crystallography, or NMR, but the maximum resolution is steadily increasing. This technique is still a particularly valuable for very large protein complexes such as virus coat proteins and amyloid fibers.

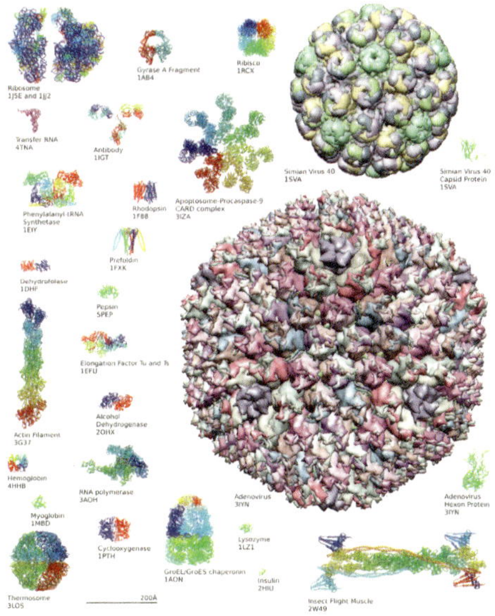

Examples of protein structures from the PDB

General secondary structure composition can be determined via circular dichroism. Vibrational spectroscopy can also be used to characterize the conformation of peptides, polypeptides, and proteins. Two-dimensional infrared spectroscopy has become a valuable method to investigate the structures of flexible peptides and proteins that cannot be studied with other methods. A more qualitative picture of protein structure is often obtained by proteolysis, which is also useful to screen for more crystallizable protein samples. Novel implementations of this approach, including fast parallel proteolysis (FASTpp), can probe the structured fraction and its stability without the need for purification. Once a protein's structure has been experimentally determined, further detailed studies can be done computationally, using molecular dynamic simulations of that structure.

Protein Sequence Analysis: Ensembles

Proteins are often thought of as relatively stable structures that have a set tertiary structure and experience conformational changes as a result of being modified by other proteins or as part of enzymatic activity. However proteins have varying degrees of stability and some of the less stable variants are intrinsically disordered proteins. These proteins exist and function in a relatively 'disordered' state lacking a stable tertiary structure. As a result, they are difficult to describe in a standard protein structure model that was designed for proteins with a fixed tertiary structure. Conformational ensembles have been devised as a way to provide a more accurate and 'dynamic' representation of the conformational state of intrinsically disordered proteins. Conformational ensembles function by attempting to represent the various conformations of intrinsically disordered proteins within an ensemble file (the type found at the Protein Ensemble Database).

Protein ensemble files are a representation of a protein that can be considered to have a flexible structure. Creating these files requires determining which of the various theoretically possible protein conformations actually exist. One approach is to apply computational algorithms to the protein data in order to try to determine the most likely set of conformations for an ensemble file.

There are multiple methods for preparing data for the Protein Ensemble Database that fall into two general methodologies – pool and molecular dynamics (MD) approaches (diagrammed in the figure). The pool based approach uses the protein's amino acid sequence to create a massive pool of random conformations. This pool is then subjected to more computational processing that creates a set of theoretical parameters for each conformation based on the structure. Conformational subsets from this pool whose average theoretical parameters closely match known experimental data for this protein are selected.

The molecular dynamics approach takes multiple random conformations at a time and subjects all of them to experimental data. Here the experimental data is serving as limitations to be placed on the conformations (e.g. known distances between atoms). Only conformations that manage to remain within the limits set by the experimental data are accepted. This approach often applies large amounts of experimental data to the conformations which is a very computationally demanding task.

Protein	Data Type	Protocol	PED ID
Sic1/Cdc4	NMR and SAXS	Pool-based	PED9AAA
p15 PAF	NMR and SAXS	Pool-based	PED6AAA
MKK7	NMR	Pool-based	PED5AAB
Beta-synuclein	NMR	MD-based	PED1AAD
P27 KID	NMR	MD-based	PED2AAA

(adapted from image in "Computational approaches for inferring the functions of intrinsically disordered proteins")

Protein Structure Databases

Protein structure database is a database that is modeled around the various experimentally determined protein structures. The aim of most protein structure databases is to organize and annotate the protein structures, providing the biological community access to the experimental data in a useful way. Data included in protein structure databases often includes 3D coordinates as well as experimental information, such as unit cell dimensions and angles for x-ray crystallography determined structures. Though most instances, in this case either proteins or a specific structure determinations of a protein, also contain sequence information and some databases even provide means for performing sequence based queries, the primary attribute of a structure database is structural information, whereas sequence databases focus on sequence information, and contain no structural information for the majority of entries. Protein structure databases are critical for many efforts in computational biology such as structure based drug design, both in developing the computational methods used and in providing a large experimental dataset used by some methods to provide insights about the function of a protein.

Structure Classification

Protein structures can be grouped based on their similarity or a common evolutionary origin. The Structural Classification of Proteins database and CATH database provide two different structural classifications of proteins. Shared structure between proteins is considered evidence of evolutionary relatedness between proteins and is used group proteins together into protein superfamilies.

Computational Prediction of Protein Structure

The generation of a protein sequence is much easier than the determination of a protein structure. However, the structure of a protein gives much more insight in the function of the protein than its sequence. Therefore, a number of methods for the computational prediction of protein structure from its sequence have been developed. *Ab initio* prediction methods use just the sequence of the protein. Threading and homology modeling methods can build a 3-D model for a protein of unknown structure from experimental structures of evolutionarily-related proteins, called a protein family.

Protein Primary Structure

Protein primary structure is the linear sequence of amino acids in a peptide or protein. By convention, the primary structure of a protein is reported starting from the amino-terminal (N) end to the carboxyl-terminal (C) end. Protein biosynthesis is most commonly performed by ribosomes in cells. Peptides can also be synthesized in the laboratory. Protein primary structures can be directly sequenced, or inferred from DNA sequences.

Interactive diagram of protein structure, using PCNA as an example. (PDB: 1AXC)

Formation

Biological

Amino acids are polymerised via peptide bonds to form a long backbone, with the different amino acid side chains protruding along it. In biological systems, proteins are produced during translation by a cell's ribosomes. Some organisms can also make short peptides by non-ribosomal peptide synthesis, which often use amino acids other than the standard 20, and may be cyclised, modified and cross-linked.

Chemical

Peptides can be synthesised chemically via a range of laboratory methods. Chemical methods typically synthesise peptides in the opposite order to biological protein synthesis (starting at the C-terminus).

Notation

Protein sequence is typically notated as a string of letters, listing the amino acids starting at the amino-terminal end through to the carboxyl-terminal end. Either a three letter code or single letter code can be used to represent the 20 naturally occurring amino acids, as well as mixtures or ambiguous amino acids (similar to nucleic acid notation).

Peptides can be directly sequenced, or inferred from DNA sequences. Large sequence databases now exist that collate known protein sequences.

20 natural amino acid notation		
Amino Acid	3-Letter	1-Letter
Alanine	Ala	A
Arginine	Arg	R
Asparagine	Asn	N
Aspartic acid	Asp	D
Cysteine	Cys	C
Glutamic acid	Glu	E
Glutamine	Gln	Q
Glycine	Gly	G
Histidine	His	H
Isoleucine	Ile	I
Leucine	Leu	L
Lysine	Lys	K
Methionine	Met	M
Phenylalanine	Phe	F
Proline	Pro	P
Serine	Ser	S
Threonine	Thr	T
Tryptophan	Trp	W
Tyrosine	Tyr	Y
Valine	Val	V

Ambiguous amino acid notation	
Symbol	Description
X	Any amino acid, or unknown
B	Aspartate or Asparagine

Z	Glutamate or Glutamine
J	Leucine or Isoleucine
Φ	Hydrophobic
Ω	Aromatic
Ψ	Aliphatic
π	Small
ζ	Hydrophilic
+	Positively charged
-	Negatively charged

Modification

In general, polypeptides are unbranched polymers, so their primary structure can often be specified by the sequence of amino acids along their backbone. However, proteins can become cross-linked, most commonly by disulfide bonds, and the primary structure also requires specifying the cross-linking atoms, e.g., specifying the cysteines involved in the protein's disulfide bonds. Other crosslinks include desmosine.

Isomerisation

The chiral centers of a polypeptide chain can undergo racemization. Although it does not change the sequence, it does affect the chemical properties of the sequence. In particular, the L-amino acids normally found in proteins can spontaneously isomerize at the C^{α} atom to form D-amino acids, which cannot be cleaved by most proteases. Additionally, proline can form stable trans-isomers at the peptide bond.

Posttranslational Modification

Finally, the protein can undergo a variety of posttranslational modifications, which are briefly summarized here.

The N-terminal amino group of a polypeptide can be modified covalently, e.g.,

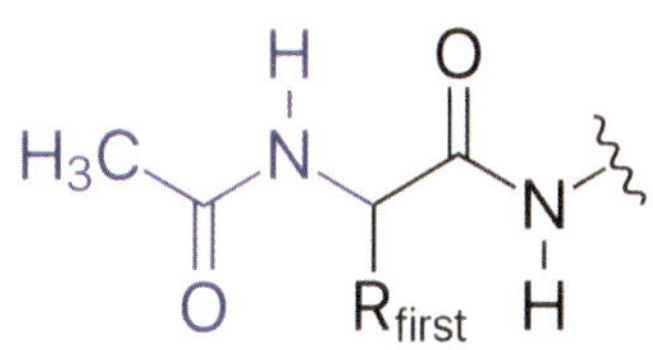

N-terminal acetylation

- acetylation -C(=O)-CH_3

  The positive charge on the N-terminal amino group may be eliminated by changing it to an acetyl group (N-terminal blocking).

- formylation -C(=O)H

  The N-terminal methionine usually found after translation has an N-terminus blocked with a formyl group. This formyl group (and sometimes the methionine residue itself, if followed by Gly or Ser) is removed by the enzyme deformylase.

- pyroglutamate

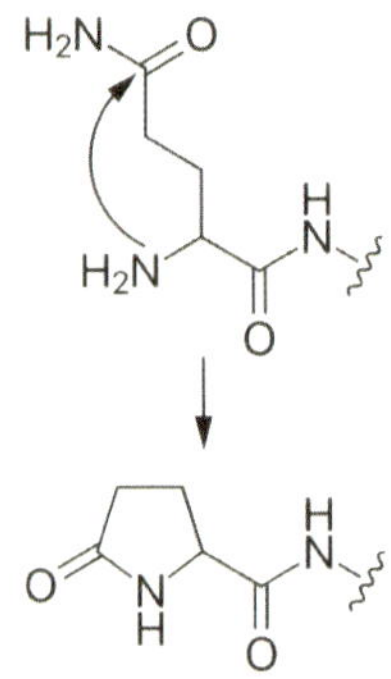

Formation of pyroglutamate from an N-terminal glutamine

  An N-terminal glutamine can attack itself, forming a cyclic pyroglutamate group.

- myristoylation -C(=O)-$(CH_2)_{12}$-CH_3

  Similar to acetylation. Instead of a simple methyl group, the myristoyl group has a tail of 14 hydrophobic carbons, which make it ideal for anchoring proteins to cellular membranes.

The C-terminal carboxylate group of a polypeptide can also be modified, e.g.,

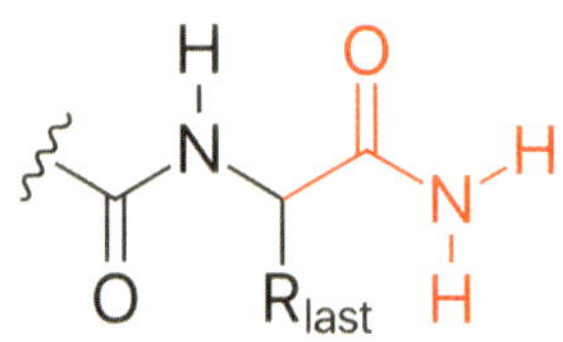

C-terminal amidation

- amidation

  The C-terminus can also be blocked (thus, neutralizing its negative charge) by amidation.

- glycosyl phosphatidylinositol (GPI) attachment

  Glycosyl phosphatidylinositol is a large, hydrophobic phospholipid prosthetic group that achors proteins to cellular membranes. It is attached to the polypeptide C-terminus through an amide linkage that then connects to ethanolamine, thence to sundry sugars and finally to the phosphatidylinositol lipid moiety.

Finally, the peptide side chains can also be modified covalently, e.g.,

- phosphorylation

  Aside from cleavage, phosphorylation is perhaps the most important chemical modification of proteins. A phosphate group can be attached to the sidechain hydroxyl group of serine, threonine and tyrosine residues, adding a negative charge at that site and producing an unnatural amino acid. Such reactions are catalyzed by kinases and the reverse reaction is catalyzed by phosphatases. The phosphorylated tyrosines are often used as "handles" by which proteins can bind to one another, whereas phosphorylation of Ser/Thr often induces conformational changes, presumably because of the introduced negative charge. The effects of phosphorylating Ser/Thr can sometimes be simulated by mutating the Ser/Thr residue to glutamate.

- glycosylation

  A catch-all name for a set of very common and very heterogeneous chemical modifications. Sugar moieties can be attached to the sidechain hydroxyl groups of Ser/Thr or to the sidechain amide groups of Asn. Such attachments can serve many functions, ranging from increasing solubility to complex recognition. All glycosylation can be blocked with certain inhibitors, such as tunicamycin.

- deamidation (succinimide formation)

  In this modification, an asparagine or aspartate side chain attacks the following peptide bond, forming a symmetrical succinimide intermediate. Hydrolysis of the intermediate produces either asparate or the β-amino acid, iso(Asp). For asparagine, either product results in the loss of the amide group, hence "deamidation".

- hydroxylation

  Proline residues may be hydroxylates at either of two atoms, as can lysine (at one atom). Hydroxyproline is a critical component of collagen, which becomes unstable upon its loss. The hydroxylation reaction is catalyzed by an enzyme that requires ascorbic acid (vitamin C), deficiencies in which lead to many connective-tissue diseases such as scurvy.

- methylation

  Several protein residues can be methylated, most notably the positive groups of lysine and arginine. Arginine residues interact with the nucleic acid phosphate backbone and commonly form hydrogen bonds with the base residues, particularly guanine, in protein–DNA complexes. Lysine residues can be singly, doubly and even triply methylated. Methylation does *not* alter the positive charge on the side chain, however.

- acetylation

  Acetylation of the lysine amino groups is chemically analogous to the acetylation of the N-terminus. Functionally, however, the acetylation of lysine residues is used to regulate the binding of proteins to nucleic acids. The cancellation of the positive charge on the lysine weakens the electrostatic attraction for the (negatively charged) nucleic acids.

- sulfation

  Tyrosines may become sulfated on their O^{η} atom. Somewhat unusually, this modification occurs in the Golgi apparatus, not in the endoplasmic reticulum. Similar to phosphorylated tyrosines, sulfated tyrosines are used for specific recognition, e.g., in chemokine receptors on the cell surface. As with phosphorylation, sulfation adds a negative charge to a previously neutral site.

- prenylation and palmitoylation -C(=O)-$(CH_2)_{14}$-CH_3

  The hydrophobic isoprene (e.g., farnesyl, geranyl, and geranylgeranyl groups) and palmitoyl groups may be added to the S^{γ} atom of cysteine residues to anchor proteins to cellular membranes. Unlike the GPI and myritoyl anchors, these groups are not necessarily added at the termini.

- carboxylation

  A relatively rare modification that adds an extra carboxylate group (and, hence, a double negative charge) to a glutamate side chain, producing a Gla residue. This is used to strengthen the binding to "hard" metal ions such as calcium.

- ADP-ribosylation

  The large ADP-ribosyl group can be transferred to several types of side chains within proteins, with heterogeneous effects. This modification is a target for the powerful toxins of disparate bacteria, e.g., *Vibrio cholerae*, *Corynebacterium diphtheriae* and *Bordetella pertussis*.

- ubiquitination and SUMOylation

  Various full-length, folded proteins can be attached at their C-termini to the sidechain ammonium groups of lysines of other proteins. Ubiquitin is the most common of these, and usually signals that the ubiquitin-tagged protein should be degraded.

Most of the polypeptide modifications listed above occur *post-translationally*, i.e., after the protein has been synthesized on the ribosome, typically occurring in the endoplasmic reticulum, a subcellular organelle of the eukaryotic cell.

Many other chemical reactions (e.g., cyanylation) have been applied to proteins by chemists, although they are not found in biological systems.

Cleavage and Ligation

In addition to those listed above, the most important modification of primary structure is peptide cleavage (by chemical hydrolysis or by proteases). Proteins are often synthesized in an inactive precursor form; typically, an N-terminal or C-terminal segment blocks the active site of the protein, inhibiting its function. The protein is activated by cleaving off the inhibitory peptide.

Some proteins even have the power to cleave themselves. Typically, the hydroxyl group of a serine (rarely, threonine) or the thiol group of a cysteine residue will attack the carbonyl carbon of the preceding peptide bond, forming a tetrahedrally bonded intermediate [classified as a hydroxyoxazolidine (Ser/Thr) or hydroxythiazolidine (Cys) intermediate]. This intermediate tends to revert to the amide form, expelling the attacking group, since the amide form is usually favored by free energy, (presumably due to the strong resonance stabilization of the peptide group). However, additional molecular interactions may render the amide form less stable; the amino group is expelled instead, resulting in an ester (Ser/Thr) or thioester (Cys) bond in place of the peptide bond. This chemical reaction is called an N-O acyl shift.

The ester/thioester bond can be resolved in several ways:

- Simple hydrolysis will split the polypeptide chain, where the displaced amino group becomes the new N-terminus. This is seen in the maturation of glycosylasparaginase.
- A β-elimination reaction also splits the chain, but results in a pyruvoyl group at the new N-terminus. This pyruvoyl group may be used as a covalently attached catalytic cofactor in some enzymes, especially decarboxylases such as S-adenosylmethionine decarboxylase (SAMDC) that exploit the electron-withdrawing power of the pyruvoyl group.
- Intramolecular transesterification, resulting in a *branched* polypeptide. In inteins, the new ester bond is broken by an intramolecular attack by the soon-to-be C-terminal asparagine.
- Intermolecular transesterification can transfer a whole segment from one polypeptide to another, as is seen in the Hedgehog protein autoprocessing.

Primary Structure in Other Molecules

Any linear-chain heteropolymer can be said to have a "primary structure" by analogy to the usage of the term for proteins, but this usage is rare compared to the extremely common usage in reference to proteins. In RNA, which also has extensive secondary structure, the linear chain of bases is generally just referred to as the "sequence" as it is in DNA (which usually forms a linear double helix with little secondary structure). Other biological polymers such as polysaccharides can also be considered to have a primary structure, although the usage is not standard.

Relation to Secondary and Tertiary Structure

The primary structure of a biological polymer to a large extent determines the three-dimensional shape (tertiary structure). Protein sequence can be used to predict local features, such as segments of secondary structure, or trans-membrane regions. However, the complexity of protein folding currently prohibits predicting the tertiary structure of a protein from its sequence alone. Knowing the structure of a similar homologous sequence (for example a member of the same protein family) allows highly accurate prediction of the tertiary structure by homology modeling. If the full-length protein sequence is available, it is possible to estimate its general biophysical properties, such as its isoelectric point.

Sequence families are often determined by sequence clustering, and structural genomics projects aim to produce a set of representative structures to cover the sequence space of possible non-redundant sequences.

Protein Secondary Structure

Protein secondary structure is the three dimensional form of *local segments* of proteins. The two most common secondary structural elements are alpha helices and beta sheets, though beta turns and omega loops occur as well. Secondary structure elements typically spontaneously form as an intermediate before the protein folds into its three dimensional tertiary structure.

Interactive diagram of protein structure, using PCNA as an example. (PDB: 1AXC)

Secondary structure is formally defined by the pattern of hydrogen bonds between the amino hydrogen and carboxyl oxygen atoms in the peptide backbone. Secondary structure may alternatively be defined based on the regular pattern of backbone dihedral angles in a particular region of the Ramachandran plot regardless of whether it has the correct hydrogen bonds.

The concept of secondary structure was first introduced by Kaj Ulrik Linderstrøm-Lang at Stanford in 1952. Other types of biopolymers such as nucleic acids also possess characteristic secondary structures.

Types

Structural features of the three major forms of protein helices			
Geometry attribute	α-helix	3_{10} helix	π-helix
Residues per turn	3.6	3.0	4.4
Translation per residue	1.5 Å (0.15 nm)	2.0 Å (0.20 nm)	1.1 Å (0.11 nm)
Radius of helix	2.3 Å (0.23 nm)	1.9 Å (0.19 nm)	2.8 Å (0.28 nm)
Pitch	5.4 Å (0.54 nm)	6.0 Å (0.60 nm)	4.8 Å (0.48 nm)

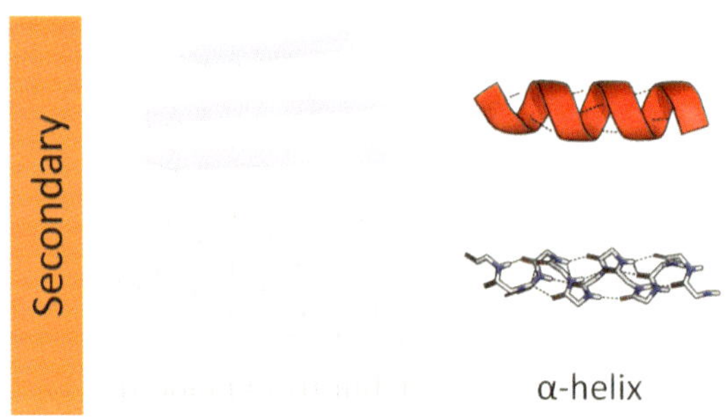

Interactive diagram of hydrogen bonds in protein secondary structure. Cartoon above, atoms below with nitrogen in blue, oxygen in red (PDB: 1AXC)

The most common secondary structures are alpha helices and beta sheets. Other helices, such as the 3_{10} helix and π helix, are calculated to have energetically favorable hydrogen-bonding patterns but are rarely observed in natural proteins except at the ends of α helices due to unfavorable backbone packing in the center of the helix. Other extended structures such as the polyproline helix and alpha sheet are rare in native state proteins but are often hypothesized as important protein folding intermediates. Tight turns and loose, flexible loops link the more "regular" secondary structure elements. The random coil is not a true secondary structure, but is the class of conformations that indicate an absence of regular secondary structure.

Amino acids vary in their ability to form the various secondary structure elements. Proline and glycine arc sometimes known as "helix breakers" because they disrupt the regularity of the α helical backbone conformation; however, both have unusual conformational abilities and are commonly found in turns. Amino acids that prefer to adopt helical conformations in proteins include methionine, alanine, leucine, glutamate and lysine ("MALEK" in amino-acid 1-letter codes); by contrast, the large aromatic residues (tryptophan, tyrosine and phenylalanine) and C^{β}-branched amino acids (isoleucine, valine, and threonine) prefer to adopt β-strand conformations. However, these preferences are not strong enough to produce a reliable method of predicting secondary structure from sequence alone.

Low frequency collective vibrations are thought to be sensitive to local rigidity within proteins, revealing beta structures to be generically more rigid than alpha or disordered proteins. Neutron scattering measurements have directly connected the spectral feature at ~1 THz to collective motions of the secondary structure of beta-barrel protein GFP.

Hydrogen bonding patterns in secondary structures may be significantly distorted, which makes automatic determination of secondary structure difficult. There are several methods for formally defining protein secondary structure (e.g., DSSP, DEFINE, STRIDE, ScrewFit, SST).

DSSP Classification

The Dictionary of Protein Secondary Structure, in short DSSP, is commonly used to describe the protein secondary structure with single letter codes. The secondary structure is assigned based on hydrogen bonding patterns as those initially proposed by Pauling et al. in 1951 (before any protein structure had ever been experimentally determined). There are eight types of secondary structure that DSSP defines:

- G = 3-turn helix (3_{10} helix). Min length 3 residues.
- H = 4-turn helix (α helix). Minimum length 4 residues.
- I = 5-turn helix (π helix). Minimum length 5 residues.
- T = hydrogen bonded turn (3, 4 or 5 turn).
- E = extended strand in parallel and/or anti-parallel β-sheet conformation. Min length 2 residues.
- B = residue in isolated β-bridge (single pair β-sheet hydrogen bond formation).
- S = bend (the only non-hydrogen-bond based assignment).

- C = coil (residues which are not in any of the above conformations).

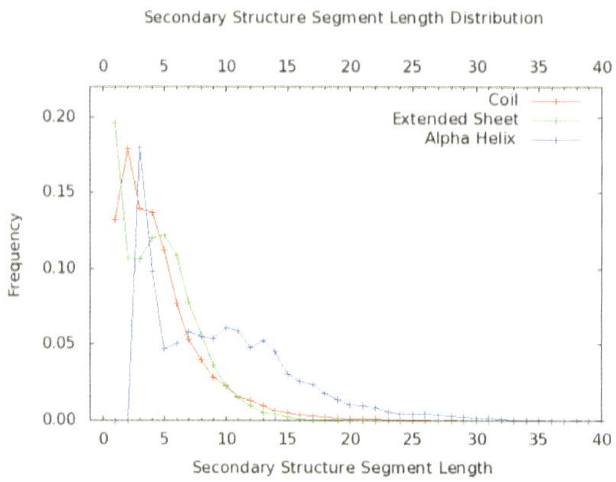

Distribution obtained from non-redundant pdb_select dataset; Secondary structure assigned by DSSP; 8 conformational states reduced to 3 states: H=HGI, E=EB, C=STC. Visible are mixtures of (gaussian) distributions, resulting also from the reduction of DSSP states.

'Coil' is often codified as ' ' (space), C (coil) or '–' (dash). The helices (G, H and I) and sheet conformations are all required to have a reasonable length. This means that 2 adjacent residues in the primary structure must form the same hydrogen bonding pattern. If the helix or sheet hydrogen bonding pattern is too short they are designated as T or B, respectively. Other protein secondary structure assignment categories exist (sharp turns, Omega loops, etc.), but they are less frequently used.

Secondary structure is defined by hydrogen bonding, so the exact definition of a hydrogen bond is critical. The standard hydrogen-bond definition for secondary structure is that of DSSP, which is a purely electrostatic model. It assigns charges of $\pm q_1 \approx 0.42e$ to the carbonyl carbon and oxygen, respectively, and charges of $\pm q_2 \approx 0.20e$ to the amide hydrogen and nitrogen, respectively. The electrostatic energy is

$$E = q_1 q_2 \left(\frac{1}{r_{ON}} + \frac{1}{r_{CH}} - \frac{1}{r_{OH}} - \frac{1}{r_{CN}} \right) \cdot 332 \text{ kcal/mol.}$$

According to DSSP, a hydrogen-bond exists if and only if E is less than −0.5 kcal/mol (−2.1 kJ/mol). Although the DSSP formula is a relatively crude approximation of the *physical* hydrogen-bond energy, it is generally accepted as a tool for defining secondary structure.

SST Classification

SST is a Bayesian method to assign secondary structure to protein coordinate data using the Shannon information criterion of Minimum Message Length (MML) inference. SST treats any assignment of secondary structure as a potential hypothesis that attempts to explain (compress) given protein coordinate data. The core idea is that the best secondary structural assignment is the one that can explain (compress) the coordinates of a given protein coordinates in the most economical way, thus linking the inference of secondary structure to lossless data compression. SST accurately delineates any protein chain into regions associated with the following assignment types:

- E = (Extended) strand of a β-pleated sheet
- G = Right-handed 310 helix
- H = Right-handed α-helix
- I = Right-handed π-helix
- g = Left-handed 3_{10} helix
- h = Left-handed α-helix
- i = Left-handed π-hclix
- 3 = 3_{10}-like Turn
- 4 = α-like Turn
- 5 = π-like Turn
- T = Unspecified Turn
- C = Coil
- - = Unassigned residue

SST detects π and 3_{10} helical caps to standard α-helices, and automatically assembles the various extended strands into consistent β-pleated sheets. It provides a readable output of dissected secondary structural elements, and a corresponding PyMol-loadable script to visualize the assigned secondary structural elements individually.

Experimental Determination

The rough secondary-structure content of a biopolymer (e.g., "this protein is 40% α-helix and 20% β-sheet.") can be estimated spectroscopically. For proteins, a common method is far-ultraviolet (far-UV, 170–250 nm) circular dichroism. A pronounced double minimum at 208 and 222 nm indicate α-helical structure, whereas a single minimum at 204 nm or 217 nm reflects random-coil or β-sheet structure, respectively. A less common method is infrared spectroscopy, which detects differences in the bond oscillations of amide groups due to hydrogen-bonding. Finally, secondary-structure contents may be estimated accurately using the chemical shifts of an initially unassigned NMR spectrum.

Prediction

Predicting protein tertiary structure from only its amino acid sequence is a very challenging problem, but using the simpler secondary structure definitions is more tractable.

Early methods of secondary-structure prediction were restricted to predicting the three predominate states: helix, sheet, or random coil. These methods were based on the helix- or sheet-forming propensities of individual amino acids, sometimes coupled with rules for estimating the free energy of forming secondary structure elements. Such methods were typically ~60% accurate in predicting which of the three states (helix/sheet/coil) a residue adopts. The first widely used technique to predict protein secondary structure from the amino acid sequence was the Chou–Fasman method.

A significant increase in accuracy (to nearly ~80%) was made by exploiting multiple sequence alignment; knowing the full distribution of amino acids that occur at a position (and in its vicinity, typically ~7 residues on either side) throughout evolution provides a much better picture of the structural tendencies near that position. For illustration, a given protein might have a glycine at a given position, which by itself might suggest a random coil there. However, multiple sequence alignment might reveal that helix-favoring amino acids occur at that position (and nearby positions) in 95% of homologous proteins spanning nearly a billion years of evolution. Moreover, by examining the average hydrophobicity at that and nearby positions, the same alignment might also suggest a pattern of residue solvent accessibility consistent with an α-helix. Taken together, these factors would suggest that the glycine of the original protein adopts α-helical structure, rather than random coil. Several types of methods are used to combine all the available data to form a 3-state prediction, including neural networks, hidden Markov models and support vector machines. Modern prediction methods also provide a confidence score for their predictions at every position.

Secondary-structure prediction methods was continuously benchmarked, e.g., EVA (benchmark). Based on these tests, the most accurate methods were Psipred, SAM, PORTER, PROF, and SABLE. The chief area for improvement appears to be the prediction of β-strands; residues confidently predicted as β-strand are likely to be so, but the methods are apt to overlook some β-strand segments (false negatives). There is likely an upper limit of ~90% prediction accuracy overall, due to the idiosyncrasies of the standard method (DSSP) for assigning secondary-structure classes (helix/strand/coil) to PDB structures, against which the predictions are benchmarked.

Accurate secondary-structure prediction is a key element in the prediction of tertiary structure, in all but the simplest (homology modeling) cases. For example, a confidently predicted pattern of six secondary structure elements βαββαβ is the signature of a ferredoxin fold.

Applications

Both protein and nucleic acid secondary structures can be used to aid in multiple sequence alignment. These alignments can be made more accurate by the inclusion of secondary structure information in addition to simple sequence information. This is sometimes less useful in RNA because base pairing is much more highly conserved than sequence. Distant relationships between proteins whose primary structures are unalignable can sometimes be found by secondary structure.

It has been shown that α-helices are more stable, robust to mutations and designable than β-strands in natural proteins, thus designing functional all-α proteins is likely to be easier that designing proteins with both helices and strands; this has been recently confirmed experimentally.

Protein Tertiary Structure

Protein tertiary structure is the three dimensional shape of a protein. The tertiary structure will have a single polypeptide chain "backbone" with one or more protein secondary structures, the protein domains. Amino acid side chains may interact and bond in a number of ways. The interactions and bonds of side chains within a particular protein determine its tertiary structure. The protein tertiary structure is defined by its atomic coordinates. These coordinates may refer either to a protein domain or to the entire tertiary structure. A number of tertiary structures may fold into a quaternary structure.

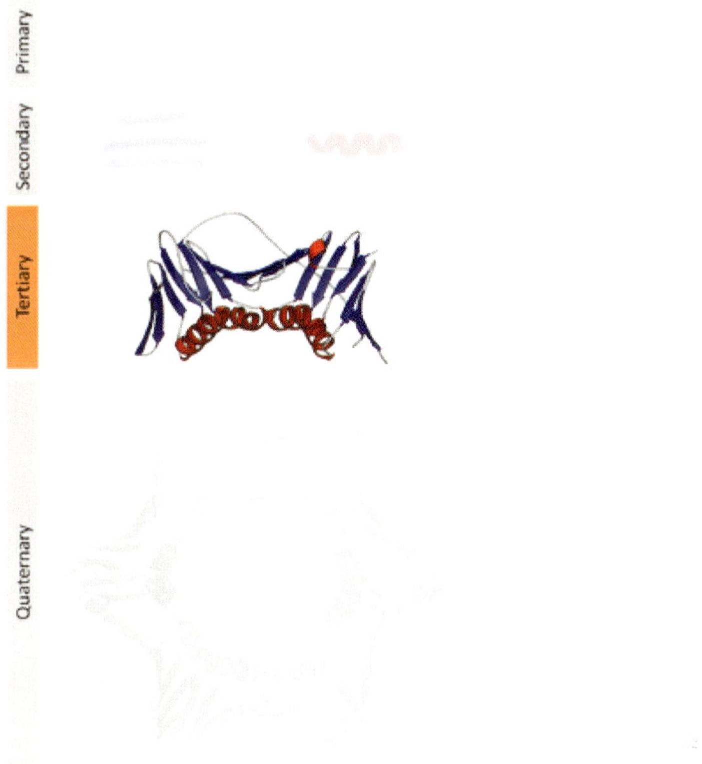

Interactive diagram of protein structure, using PCNA as an example. (PDB: 1AXC)

Determinants

Globular proteins have a core of hydrophobic amino acid residues and a surface region of water-exposed, charged, hydrophilic residues. This arrangement may stabilise interactions within the tertiary structure. For example, in secreted proteins, which are not bathed in cytoplasm, disulfide bonds between cysteine residues help to maintain the tertiary structure. There is a commonality of stable tertiary structures seen in proteins of diverse function and diverse evolution. For example, the TIM barrel, named for the enzyme triosephosphateisomerase, is a common tertiary structure as is the highly stable, dimeric, coiled coil structure. Hence, proteins may be classified by the structures they hold. Databases of proteins which use such a classification include *SCOP* and *CATH*.

Stability of Native States

The native state or native conformation of a protein is its most typical conformation in a cellular environment.

Chaperone Proteins

It is commonly assumed that the native state of a protein is also the most thermodynamically stable and that a protein will reach its native state, given its chemical kinetics, before it is translated. Protein chaperones within the cytoplasm of a cell assist a newly synthesised polypeptide to attain its native state. Some chaperone proteins are highly specific in their function, for example, protein disulfide isomerase; others are general in their function and may assist most globular proteins, for example, the prokaryotic GroEL/GroES system of proteins and the homologous eukaryotic heat shock proteins (the Hsp60/Hsp10 system).

Kinetic Traps

Folding kinetics may trap a protein in a high-energy conformation. The high-energy conformation may contribute to the function of the protein. For example, the influenza hemagglutinin protein is a single polypeptide chain which when activated, is proteolytically cleaved to form two polypeptide chains. The two chains are held in a high-energy conformation. When the local pH drops, the protein undergoes an energetically favorable conformational rearrangement that enables it to penetrate the host cell membrane.

Metastability

Some tertiary protein structures may exist in long-lived states that are not the expected most stable state. For example, many serpiginouss (serine protease inhibitors) show this metastability. They undergo a conformational change when a loop of the protein is cut by a protease.

Cytoplasmic Environment

Prediction of protein tertiary structure relies on knowing the protein's primary structure and comparing the possible predicted tertiary structure with known tertiary structures in protein data banks. This only takes into account the cytoplasmic environment present at the time of protein synthesis to the extent that a similar cytoplasmic environment may also have influenced the structure of the proteins recorded in the protein data bank.

Ligand Binding

The structure of a protein, for example an enzyme, may change upon binding of its natural ligands, for example a cofactor. In this case, the structure of the protein bound to the ligand is known as holo structure, of the unbound protein as apo structure.

Determination

The knowledge of the tertiary structure of soluble globular proteins is more advanced than that of membrane proteins because the former are easier to study with available technology.

X-ray Crystallography

X-ray crystallography is the most common tool used to determine protein structure. It provides high resolution of the structure but it does not give information about protein's conformational flexibility.

NMR

Protein NMR gives comparatively lower resolution of protein structure. It is limited to smaller proteins. However, it can provide information about conformational changes of a protein in solution.

Dual Polarisation Interferometry

Dual polarisation interferometry provides complimentary information about surface captured proteins. It assists in determining structure and conformation changes over time.

Projects

Prediction Algorithm

The Folding@home project at Stanford University is a distributed computing research effort which uses approximately 5 petaFLOPS (≈10 x86 petaFLOPS) of available computing. It aims to find an algorithm which will consistently predict protein tertiary and quaternary structures given the protein's amino acid sequence and its cellular conditions.

A list of software for protein tertiary structure prediction can be found at List of protein structure prediction software.

Protein Aggregation Diseases

Protein aggregation diseases such as Alzheimer's disease and Huntington's disease and prion diseases such as bovine spongiform encephalopathy can be better understood by constructing (and reconstructing) disease models. This is done by causing the disease in laboratory animals, for example, by administering a toxin, such as MPTP to cause Parkinson's disease, or through genetic manipulation. Protein structure prediction is a new way to create disease models, which may avoid the use of animals.

Protein Tertiary Structure Retrieval Project (CoMOGrad)

Matching patterns in tertiary structure of a given protein to huge number of known protein tertiary structures and retrieve most similar ones in ranked order is in the heart of many research areas like function prediction of novel proteins, study of evolution, disease diagnosis, drug discovery, antibody design etc. The CoMOGrad project at BUET is a research effort to device an extremely fast and much precise method for protein tertiary structure retrieval and develop online tool based on research outcome.

Protein Quaternary Structure

Protein quaternary structure is the number and arrangement of multiple folded protein subunits in a multi-subunit complex. It includes organisations from simple dimers to large homooligomers and complexes with defined or variable numbers of subunits. It can also refer to biomolecular complexes of proteins with nucleic acids and other cofactors.

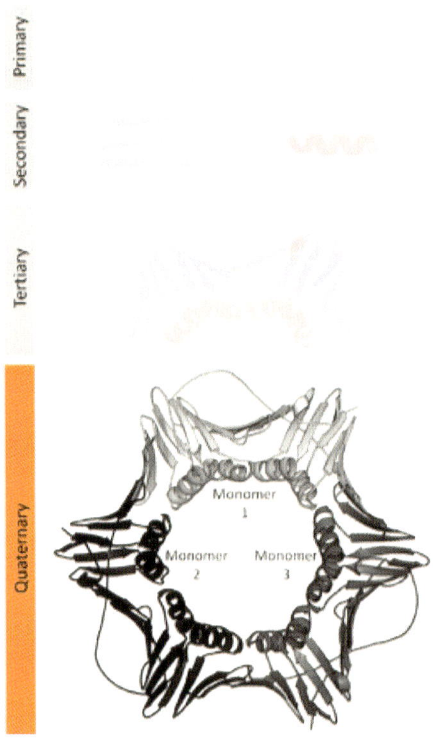

Interactive diagram of protein structure, using PCNA as an example. (PDB: 1AXC)

Description and Examples

Many proteins are actually assemblies of multiple polypeptide chains. The quaternary structure refers to the number and arrangement of the protein subunits with respect to one another. Examples of proteins with quaternary structure include hemoglobin, DNA polymerase, and ion channels.

Enzymes composed of subunits with diverse functions are sometimes called holoenzymes, in which some parts may be known as regulatory subunits and the functional core is known as the catalytic subunit. Other assemblies referred to instead as multiprotein complexes also possess quaternary structure. Examples include nucleosomes and microtubules. Changes in quaternary structure can occur through conformational changes within individual subunits or through reorientation of the subunits relative to each other. It is through such changes, which underlie cooperativity and allostery in "multimeric" enzymes, that many proteins undergo regulation and perform their physiological function.

The above definition follows a classical approach to biochemistry, established at times when the distinction between a protein and a functional, proteinaceous unit was difficult to elucidate. More recently, people refer to protein–protein interaction when discussing quaternary structure of proteins and consider all assemblies of proteins as protein complexes.

Nomenclature

The number of subunits in an oligomeric complex is described using names that end in -mer (for "part, subunit"). Formal and Greco-Latinate names are generally used for the first ten types and can be used for up to twenty subunits, whereas higher order complexes are usually described by the number of subunits, followed by -meric.

• 1 = monomer/subunit	• 7 = heptamer	• 13 = tridecamer	• 19 = nonadecamer
• 2 = dimer	• 8 = octamer	• 14 = tetradecamer	• 20 = eicosamer
• 3 = trimer	• 9 = nonamer	• 15 = pentadecamer*	• 21-mer
• 4 = tetramer	• 10 = decamer	• 16 = hexadecamer	• 22-mer
• 5 = pentamer	• 11 = undecamer	• 17 = heptadecamer*	• 23-mer*
• 6 = hexamer	• 12 = dodecamer	• 18 = octadecamer	• etc.

Although complexes higher than octamers are rarely observed for most proteins, there are some important exceptions. Viral capsids are often composed of multiples of 60 proteins. Several molecular machines are also found in the cell, such as the proteasome (four heptameric rings = 28 subunits), the transcription complex and the spliceosome. The ribosome is probably the largest molecular machine, and is composed of many RNA and protein molecules.

In some cases, proteins form complexes that then assemble into even larger complexes. In such cases, one uses the nomenclature, e.g., "dimer of dimers" or "trimer of dimers", to suggest that the complex might dissociate into smaller sub-complexes before dissociating into monomers.

Determination

Protein quaternary structure can be determined using a variety of experimental techniques that require a sample of protein in a variety of experimental conditions. The experiments often provide an estimate of the mass of the native protein and, together with knowledge of the masses and/or stoichiometry of the subunits, allow the quaternary structure to be predicted with a given accuracy. It is not always possible to obtain a precise determination of the subunit composition for a variety of reasons.

The number of subunits in a protein complex can often be determined by measuring the hydrodynamic molecular volume or mass of the intact complex, which requires native solution conditions. For *folded* proteins, the mass can be inferred from its volume using the partial specific volume of 0.73 ml/g. However, volume measurements are less certain than mass measurements, since *unfolded* proteins appear to have a much larger volume than folded proteins; additional experiments are required to determine whether a protein is unfolded or has formed an oligomer.

Prediction

Some bioinformatics methods were developed for predicting the quaternary structural attributes of proteins based on their sequence information by using various modes of pseudo amino acid composition.

Direct Mass Measurement of Intact Complexes

- sedimentation-equilibrium analytical ultracentrifugation
- electrospray mass spectrometry
- Mass Spectrometric Immunoassay MSIA

Direct Size Measurement of Intact Complexes

- static light scattering
- size exclusion chromatography (requires calibration)
- Dual polarisation interferometry

Indirect Size Measurement of Intact Complexes

- sedimentation-velocity analytical ultracentrifugation (measures the translational diffusion constant.)
- dynamic light scattering (measures the translational diffusion constant.)
- pulsed-gradient protein nuclear magnetic resonance (measures the translational diffusion constant.)
- fluorescence polarization (measures the rotational diffusion constant.)
- dielectric relaxation (measures the rotational diffusion constant.)
- Dual polarisation interferometry (measures the size and the density of the complex.)

Methods that measure the mass or volume under unfolding conditions (such as MALDI-TOF mass spectrometry and SDS-PAGE) are generally not useful, since non-native conditions usually cause the complex to dissociate into monomers. However, these may sometimes be applicable; for example, the experimenter may apply SDS-PAGE after first treating the intact complex with chemical cross-link reagents.

Protein–protein Interactions

Proteins are capable of forming very tight complexes. For example, ribonuclease inhibitor binds to ribonuclease A with a roughly 20 fM dissociation constant. Other proteins have evolved to bind specifically to unusual moieties on another protein, e.g., biotin groups (avidin), phosphorylated tyrosines (SH2 domains) or proline-rich segments (SH3 domains).

Protein Complex

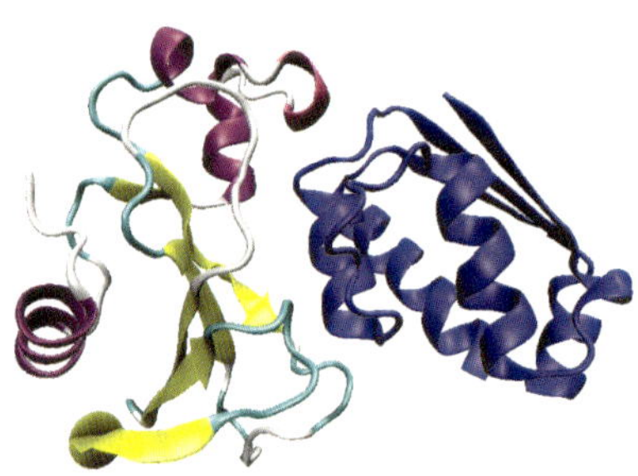

The *Bacillus amyloliquefaciens* ribonuclease barnase (colored) and its inhibitor barstar (blue) in a complex

A protein complex or multiprotein complex is a group of two or more associated polypeptide chains. Different polypeptide chains may have different functions. This is distinct from a multienzyme complex, in which multiple catalytic domains are found in a single polypeptide chain.

Protein complexes are a form of quaternary structure. Proteins in a protein complex are linked by non-covalent protein–protein interactions, and different protein complexes have different degrees of stability over time. These complexes are a cornerstone of many (if not most) biological processes and together they form various types of molecular machinery that perform a vast array of biological functions. The cell is seen to be composed of modular supramolecular complexes, each of which performs an independent, discrete biological function.

Through proximity, the speed and selectivity of binding interactions between enzymatic complex and substrates can be vastly improved, leading to higher cellular efficiency. Unfortunately, many of the techniques used to break open cells and isolate proteins are inherently disruptive to such large complexes, so it is often difficult to determine the components of a complex. Examples of protein complexes include the proteasome for molecular degradation and most RNA polymerases. In stable complexes, large hydrophobic interfaces between proteins typically bury surface areas larger than 2500 square Ås.

Function

Protein complex formation sometimes serves to activate or inhibit one or more of the complex members and in this way, protein complex formation can be similar to phosphorylation. Individual proteins can participate in the formation of a variety of different protein complexes. Different complexes perform different functions, and the same complex can perform very different functions that depend on a variety of factors. Some of these factors are:

- Which cellular compartment the complex exists in when it is contained
- Which stage in the cell cycle the complexes are present
- The nutritional status of the cell
- Others

Many protein complexes are well understood, particularly in the model organism *Saccharomyces cerevisiae* (a strain of yeast). For this relatively simple organism, the study of protein complexes is now being performed genome wide and the elucidation of most protein complexes of the yeast is ongoing.

Types of Protein Complexes

Obligate vs non-obligate Protein Complex

If a protein can form a stable well-folded structure on its own (without any other associated protein) *in vivo*, then the complexes formed by such proteins are termed "non-obligate protein complexes". However, some proteins can't be found to create a stable well-folded structure alone, but can be found as a part of a protein complex which stabilizes the constituent proteins. Such protein complexes are called "obligate protein complexes".

Transient vs Permanent/Stable Protein Complex

Transient protein complexes form and break down transiently *in vivo*, whereas permanent complexes have a relatively long half-life. Typically, the obligate interactions (protein–protein interactions in an obligate complex) are permanent, whereas non-obligate interactions have been found to be either permanent or transient. Note that there is no clear distinction between obligate and non-obligate interaction, rather there exist a continuum between them which depends on various conditions e.g. pH, protein concentration etc. However, there are important distinctions between the properties of transient and permanent/stable interactions: stable interactions are highly conserved but transient interactions are far less conserved, interacting proteins on the two sides of a stable interaction have more tendency of being co-expressed than those of a transient interaction (in fact, co-expression probability between two transiently interacting proteins is not higher than two random proteins), and transient interactions are much less co-localized than stable interactions. Though, transient by nature, transient interactions are very important for cell biology: human interactome is enriched in such interactions, these interactions are the dominating players of gene regulation and signal transduction, and proteins with *intrinsically disordered regions* (IDR: regions in protein that show dynamic inter-converting structures in the native state) are found to be enriched in transient regulatory and signaling interactions.

Fuzzy Complex

Fuzzy protein complexes have more than one structural form or dynamic structural disorder in the bound state. This means that proteins may not fold completely in either transient or permanent complexes. Consequently, specific complexes can have ambiguous interactions, which vary according to the environmental signals. Hence different ensemble of structures result in different (even opposite) biological functions. Post-translational modifications, protein interactions or alternative splicing modulate the conformational ensembles of fuzzy complexes, to fine-tune affinity or specificity of interactions. These mechanisms are often used for regulation within the eukaryotic transcription machinery.

Essential Proteins in Protein Complexes

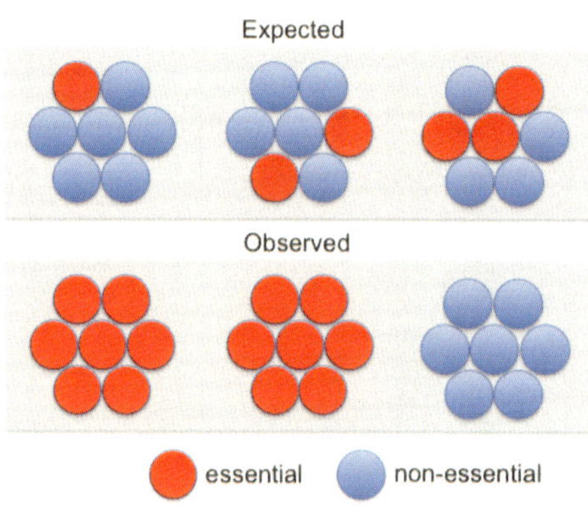

Essential proteins in yeast complexes occur much less randomly than expected by chance.

Although some early studies suggested a strong correlation between essentiality and protein interaction degree (the "centrality-lethality" rule) subsequent analyses have shown that this correlation is weak for binary or transient interactions (e.g., yeast two-hybrid). However, the correlation is robust for networks of stable co-complex interactions. In fact, a disproportionate number of essential genes belong to protein complexes. This led to the conclusion that essentiality is a property of molecular machines (i.e. complexes) rather than individual components. Wang et al. (2009) noted that larger protein complexes are more likely to be essential, explaining why essential genes are more likely to have high co-complex interaction degree. Ryan et al. (2013) referred to the observation that entire complexes appear essential as "modular essentiality". These authors also showed that complexes tend to be composed of either essential or non-essential proteins rather than showing a random distribution. However, this not an all or nothing phenomenon: only about 26% (105/401) of yeast complexes consist of solely essential or solely nonessential subunits.

In humans, genes whose protein products belong to the same complex are more likely to result in the same disease phenotype.

Homomultimeric and Heteromultimeric Proteins

The subunits of a multimeric protein may be identical as in a homomultimeric (homooligomeric) protein or different as in a heteromultimeric protein. Many soluble and membrane proteins form homomultimeric complexes in a cell, majority of proteins in the Protein Data Bank are homomultimeric. Homooligomers are responsible for the diversity and specificity of many pathways, may mediate and regulate gene expression, activity of enzymes, ion channels, receptors, and cell adhesion processes.

The voltage-gated potassium channels in the plasma membrane of a neuron are heteromultimeric proteins composed of four of forty known alpha subunits. Subunits must be of the same subfamily to form the multimeric protein channel. The tertiary structure of the channel allows ions to flow through the hydrophobic plasma membrane. Connexons are an example of a homomultimeric protein composed of six identical connexins. A cluster of connexons forms the gap-junction in two neurons that transmit signals through an electrical synapse.

Structure Determination

The molecular structure of protein complexes can be determined by experimental techniques such as X-ray crystallography, Single particle analysis or nuclear magnetic resonance. Increasingly the theoretical option of protein–protein docking is also becoming available. One method that is commonly used for identifying the meomplexes immunoprecipitation. Recently, Raicu and coworkers developed a method to determine the quaternary structure of protein complexes in living cells. This method is based on the determination of pixel-level Förster resonance energy transfer (FRET) efficiency in conjunction with spectrally resolved two-photon microscope. The distribution of FRET efficiencies are simulated against different models to get the geometry and stoichiometry of the complexes.

Assembly

Proper assembly of multiprotein complexes is important, since misassembly can lead to disastrous consequences. In order to study pathway assembly, researchers look at intermediate steps in the pathway. One such technique that allows one to do that is electrospray mass spectrometry, which can identify different intermediate states simultaneously. This has led to the discovery that most complexes follow an ordered assembly pathway. In the cases where disordered assembly is possible, the change from an ordered to a disordered state leads to a transition from function to dysfunction of the complex, since disordered assembly leads to aggregation.

The structure of proteins play a role in how the multiprotein complex assembles. The interfaces between proteins can be used to predict assembly pathways. The intrinsic flexibility of proteins also plays a role: more flexible proteins allow for a greater surface area available for interaction.

While assembly is a different process from disassembly, the two are reversible in both homomeric and heteromeric complexes. Thus, the overall process can be referred to as (dis)assembly.

Evolutionary Significance of Multiprotein Complex Assembly

In homomultimeric complexes, the homomeric proteins assemble in a way that mimics evolution. That is, an intermediate in the assembly process is present in the complex's evolutionary history. The opposite phenomenon is observed in heteromultimeric complexes, where gene fusion occurs in a manner that preserves the original assembly pathway.

Proteomics

Proteomics is the large-scale study of proteins. Proteins are vital parts of living organisms, with many functions. The term *proteomics* was coined in 1997 in analogy with genomics, the study of the genome. The word *proteome* is a portmanteau of *prot*ein and gen*ome*, and was coined by Marc Wilkins in 1994 while he was a PhD student at Macquarie University. Macquarie University also founded the first dedicated proteomics laboratory in 1995 (the Australian Proteome Analysis Facility – APAF).

Robotic preparation of MALDI mass spectrometry samples on a sample carrier.

The proteome is the entire set of proteins that are produced or modified by an organism or system. This varies with time and distinct requirements, or stresses, that a cell or organism undergoes. Proteomics is an interdisciplinary domain that has benefitted greatly from the genetic information of the Human Genome Project; it also covers emerging scientific research and the exploration of proteomes from the overall level of intracellular protein composition, structure, and its own unique activity patterns. It is an important component of functional genomics.

While *proteomics* generally refers to the large-scale experimental analysis of proteins, it is often specifically used for protein purification and mass spectrometry.

Complexity of the Problem

After genomics and transcriptomics, proteomics is the next step in the study of biological systems. It is more complicated than genomics because an organism's genome is more or less constant, whereas the proteome differs from cell to cell and from time to time. Distinct genes are expressed in different cell types, which means that even the basic set of proteins that are produced in a cell needs to be identified.

In the past this phenomenon was done by RNA analysis, but it was found not to correlate with protein content. It is now known that mRNA is not always translated into protein, and the amount of protein produced for a given amount of mRNA depends on the gene it is transcribed from and on the current physiological state of the cell. Proteomics confirms the presence of the protein and provides a direct measure of the quantity present.

Post-translational Modifications

Not only does the translation from mRNA cause differences, but many proteins are also subjected to a wide variety of chemical modifications after translation. Many of these post-translational modifications are critical to the protein's function.

Phosphorylation

One such modification is phosphorylation, which happens to many enzymes and structural proteins in the process of cell signaling. The addition of a phosphate to particular amino acids—most commonly serine and threonine mediated by serine/threonine kinases, or more rarely tyrosine mediated by tyrosine kinases—causes a protein to become a target for binding or interacting with a distinct set of other proteins that recognize the phosphorylated domain.

Because protein phosphorylation is one of the most-studied protein modifications, many "proteomic" efforts are geared to determining the set of phosphorylated proteins in a particular cell or tissue-type under particular circumstances. This alerts the scientist to the signaling pathways that may be active in that instance.

Ubiquitination

Ubiquitin is a small protein that can be affixed to certain protein substrates by enzymes called E3 ubiquitin ligases. Determining which proteins are poly-ubiquitinated helps understand how protein pathways are regulated. This is, therefore, an additional legitimate "proteomic" study. Similarly, once a researcher determines which substrates are ubiquitinated by each ligase, determining the set of ligases expressed in a particular cell type is helpful.

Additional Modifications

In addition to phosphorylation and ubiquitination, proteins can be subjected to (among others) methylation, acetylation, glycosylation, oxidation and nitrosylation. Some proteins undergo all these modifications, often in time-dependent combinations. This illustrates the potential complexity of studying protein structure and function.

Distinct Proteins are Made Under Distinct Settings

A cell may make different sets of proteins at different times or under different conditions, for example during development, cellular differentiation, cell cycle, or carcinogenesis. Further increasing proteome complexity, as mentioned, most proteins can undergo a wide range of post-translational modifications.

Therefore, a "proteomics" study can quickly become complex, even if the topic of study is restricted. In more ambitious settings, such as when a biomarker for a specific cancer subtype is sought, the proteomics scientist might elect to study multiple blood serum samples from multiple cancer patients to minimise confounding factors and account for experimental noise. Thus, complicated experimental designs are sometimes necessary to account for the dynamic complexity of the proteome.

Limitations of Genomics and Proteomics Studies

Proteomics gives a different level of understanding than genomics for many reasons:

- the level of transcription of a gene gives only a rough estimate of its *level of translation* into a protein. An mRNA produced in abundance may be degraded rapidly or translated inefficiently, resulting in a small amount of protein.
- as mentioned above many proteins experience *post-translational modifications* that profoundly affect their activities; for example some proteins are not active until they become phosphorylated. Methods such as phosphoproteomics and glycoproteomics are used to study post-translational modifications.
- many transcripts give rise to more than one protein, through alternative splicing or alternative post-translational modifications.
- many proteins form complexes with other proteins or RNA molecules, and only function in the presence of these other molecules.
- protein degradation rate plays an important role in protein content.

Reproducibility. One major factor affecting reproducibility in proteomics experiments is the simultaneous elution of many more peptides than can be measured by mass spectrometers. This causes stochastic differences between experiments due to data-dependent acquisition of tryptic peptides. Although early large-scale shotgun proteomics analyses showed considerable variability between laboratories, presumably due in part to technical and experimental differences between labs, reproducibility has been improved in more recent mass spectrometry analysis, particularly on the protein level and using Orbitrap mass spectrometers. Notably, targeted proteomics shows increased reproducibility and repeatability compared with shotgun methods, although at the expense of data density and effectiveness.

Methods of Studying Proteins

In proteomics, there are multiple methods to study proteins. Generally, proteins can either be detected using antibodies (immunoassays) or using mass spectrometry. If a complex biological sample is analyzed, either a very specific antibody needs to be used in quantitative dot blot analysis (qdb), or then biochemical separation needs to be used before the detection step as there are too many analytes in the sample to perform accurate detection and quantification.

Protein Detection with Antibodies (Immunoassays)

Antibodies to particular proteins or to their modified forms have been used in biochemistry and cell biology studies. These are among the most common tools used by molecular biologists today. There are several specific techniques and protocols that use antibodies for protein detection. The enzyme-linked immunosorbent assay (ELISA) has been used for decades to detect and quantitatively measure proteins in samples. The Western blot can be used for detection and quantification of individual proteins, where in an initial step a complex protein mixture is separated using SDS-PAGE and then the protein of interest is identified using an antibody.

Modified proteins can be studied by developing an antibody specific to that modification. For example, there are antibodies that only recognize certain proteins when they are tyrosine-phosphorylated, known as phospho-specific antibodies. Also, there are antibodies specific to other modifications. These can be used to determine the set of proteins that have undergone the modification of interest.

Disease detection at the molecular level is driving the emerging revolution of early diagnosis and treatment. A challenge facing the field is that protein biomarkers for early diagnosis can be present in very low abundance. The lower limit of detection with conventional immunoassay technology is the upper femtomolar range (10(-13) M). Digital immunoassay technology has improved detection sensitivity three logs, to the attomolar range (10(-16) M). This capability has the potential to open new advances in diagnostics and therapeutics, but such technologies have been relegated to manual procedures that are not well suited for efficient routine use.

Antibody-free Protein Detection

While protein detection with antibodies are still very common in molecular biology, also other methods have been developed that do not rely on an antibody. These methods offer various advantages, for instance they are often able to determine the sequence of a protein or peptide, they may have higher throughput than antibody-based and they sometimes can identify and quantify proteins for which no antibody exists.

Detection Methods

One of the earliest method for protein analysis has been Edman degradation (introduced in 1967) where a single peptide is subjected to multiple steps of chemical degradation to resolve its sequence. These methods have mostly been supplanted by technologies that offer higher throughput.

More recent methods use mass spectrometry-based techniques, a development that was made possible by the discovery of "soft ionization" methods such as matrix-assisted laser desorption/ionization (MALDI) and electrospray ionization (ESI) developed in the 1980s. These methods gave rise to the top-down and the bottom-up proteomics workflows where often additional separation is performed before analysis.

Separation Methods

For the analysis of complex biological samples, a reduction of sample complexity is required. This can be performed off-line by one-dimensional or two dimensional separation. More recently, on-line methods have been developed where individual peptides (in bottom-up proteomics approaches) are separated using Reversed-phase chromatography and then directly ionized using ESI; the direct coupling of separation and analysis explains the term "on-line" analysis.

Hybrid Technologies

There are several hybrid technologies that use antibody-based purification of individual analytes and then perform mass spectrometric analysis for identification and quantification. Examples of these methods are the MSIA (mass spectrometric immunoassay) developed by Randall Nelson in 1995 and the SISCAPA (Stable Isotope Standard Capture with Anti-Peptide Antibodies) method, introduced by Leigh Anderson in 2004.

Current Research Methodologies

Fluorescence two-dimensional differential gel electrophoresis (2-D DIGE) can be used to quantify variation in the 2-D DIGE process and establish statistically valid thresholds for assigning quantitative changes between samples.

Comparative proteomic analysis can reveal the role of proteins in complex biological systems, including reproduction. For example, treatment with the insecticide triazophos causes an increase in the content of brown planthopper (*Nilaparvata lugens* (Stål)) male accessory gland proteins (Acps) that can be transferred to females via mating, causing an increase in fecundity (i.e. birth rate) of females. To identify changes in the types of accessory gland proteins (Acps) and reproductive proteins that mated female planthoppers received from male planthoppers, researchers conducted a comparative proteomic analysis of mated *N. lugens* females. The results indicated that these proteins participate in the reproductive process of *N. lugens* adult females and males.

Proteome analysis of *Arabidopsis peroxisomes* has been established as the major unbiased approach for identifying new peroxisomal proteins on a large scale.

There are many approaches to characterizing the human proteome, which is estimated to contain between 20,000 and 25,000 non-redundant proteins. The number of unique protein species will likely increase by between 50,000 and 500,000 due to RNA splicing and proteolysis events, and when post-translational modification are also considered, the total number of unique human proteins is estimated to range in the low millions.

In addition, the first promising attempts to decipher the proteome of animal tumors have recently been reported. This method used as a functional method in *Macrobrachium rosenbergii* protein profiling.

High-throughput Proteomic Technologies

Proteomics has steadily gained momentum over the past decade with the evolution of several approaches. Few of these are new and others build on traditional methods. Mass spectrometry-based methods and micro arrays are the most common technologies for large-scale study of proteins.

Mass Spectrometry and Protein Profiling

There are two mass spectrometry-based methods currently used for protein profiling. The more established and widespread method uses high resolution, two-dimensional electrophoresis to separate proteins from different samples in parallel, followed by selection and staining of differentially expressed proteins to be identified by mass spectrometry. Despite the advances in 2DE and its maturity, it has its limits as well. The central concern is the inability to resolve all the proteins within a sample, given their dramatic range in expression level and differing properties.

The second quantitative approach uses stable isotopc tags to differentially label proteins from two different complex mixtures. Here, the proteins within a complex mixture are labeled first isotopically, and then digested to yield labeled peptides. The labeled mixtures are then combined, the peptides separated by multidimensional liquid chromatography and analyzed by tandem mass spectrometry. Isotope coded affinity tag (ICAT) reagents are the widely used isotope tags. In this method, the cysteine residues of proteins get covalently attached to the ICAT reagent, thereby reducing the complexity of the mixtures omitting the non-cysteine residues.

Quantitative proteomics using stable isotopic tagging is an increasingly useful tool in modern development. Firstly, chemical reactions have been used to introduce tags into specific sites or proteins for the purpose of probing specific protein functionalities. The isolation of phosphorylated peptides has been achieved using isotopic labeling and selective chemistries to capture the fraction of protein among the complex mixture. Secondly, the ICAT technology was used to differentiate between partially purified or purified macromolecular complexes such as large RNA polymerase II pre-initiation complex and the proteins complexed with yeast transcription factor. Thirdly, ICAT labeling was recently combined with chromatin isolation to identify and quantify chromatin-associated proteins. Finally ICAT reagents are useful for proteomic profiling of cellular organelles and specific cellular fractions.

Another quantitative approach is the Accurate Mass and Time (AMT) tag approach developed by Richard D. Smith and coworkers at Pacific Northwest National Laboratory. In this approach, increased throughput and sensitivity is achieved by avoiding the need for tandem mass spectrometry, and making use of precisely determined separation time information and highly accurate mass determinations for peptide and protein identifications.

Protein Chips

Balancing the use of mass spectrometers in proteomics and in medicine is the use of protein micro arrays. The aim behind protein micro arrays is to print thousands of protein detecting features for the interrogation of biological samples. Antibody arrays are an example in which a host of different antibodies are arrayed to detect their respective antigens from a sample of human blood. Another approach is the arraying of multiple protein types for the study of properties like protein-DNA, protein-protein and protein-ligand interactions. Ideally, the functional proteomic arrays would contain the entire complement of the proteins of a given organism. The first version of such arrays consisted of 5000 purified proteins from yeast deposited onto glass microscopic slides. Despite the success of first chip, it was a greater challenge for protein arrays to be implemented. Proteins are inherently much more difficult to work with than DNA. They have a broad dynamic range, are less stable than

DNA and their structure is difficult to preserve on glass slides, though they are essential for most assays. The global ICAT technology has striking advantages over protein chip technologies.

Reverse-phased Protein Microarrays

This is a promising and newer microarray application for the diagnosis, study and treatment of complex diseases such as cancer. The technology merges laser capture microdissection (LCM) with micro array technology, to produce reverse phase protein microarrays. In this type of microarrays, the whole collection of protein themselves are immobilized with the intent of capturing various stages of disease within an individual patient. When used with LCM, reverse phase arrays can monitor the fluctuating state of proteome among different cell population within a small area of human tissue. This is useful for profiling the status of cellular signaling molecules, among a cross section of tissue that includes both normal and cancerous cells. This approach is useful in monitoring the status of key factors in normal prostate epithelium and invasive prostate cancer tissues. LCM then dissects these tissue and protein lysates were arrayed onto nitrocellulose slides, which were probed with specific antibodies. This method can track all kinds of molecular events and can compare diseased and healthy tissues within the same patient enabling the development of treatment strategies and diagnosis. The ability to acquire proteomics snapshots of neighboring cell populations, using reverse phase microarrays in conjunction with LCM has a number of applications beyond the study of tumors. The approach can provide insights into normal physiology and pathology of all the tissues and is invaluable for characterizing developmental processes and anomalies.

Practical Applications of Proteomics

One major development to come from the study of human genes and proteins has been the identification of potential new drugs for the treatment of disease. This relies on genome and proteome information to identify proteins associated with a disease, which computer software can then use as targets for new drugs. For example, if a certain protein is implicated in a disease, its 3D structure provides the information to design drugs to interfere with the action of the protein. A molecule that fits the active site of an enzyme, but cannot be released by the enzyme, inactivates the enzyme. This is the basis of new drug-discovery tools, which aim to find new drugs to inactivate proteins involved in disease. As genetic differences among individuals are found, researchers expect to use these techniques to develop personalized drugs that are more effective for the individual.

Proteomics is also used to reveal complex plant-insect interactions that help identify candidate genes involved in the defensive response of plants to herbivory.

Interaction Proteomics and Protein Networks

Interaction proteomics is the analysis of protein interactions from scales of binary interactions to proteome- or network-wide. Most proteins function via protein-protein interactions, and one goal of interaction proteomics is to identify binary protein interactions, protein complexes, and interactomes.

Several methods are available to probe protein–protein interactions. While the most traditional method is yeast two-hybrid analysis, a powerful emerging method is affinity purification followed by protein mass spectrometry using tagged protein baits. Other methods include surface plasmon resonance (SPR), protein microarrays, dual polarisation interferometry, microscale thermophoresis and experimental methods such as phage display and *in silico* computational methods.

Knowledge of protein-protein interactions is especially useful in regard to biological networks and systems biology, for example in cell signaling cascades and gene regulatory networks (GRNs, where knowledge of protein-DNA interactions is also informative). Proteome-wide analysis of protein interactions, and integration of these interaction patterns into larger biological networks, is crucial towards understanding systems-level biology.

Expression Proteomics

Expression proteomics includes the analysis of protein expression at larger scale. It helps identify main proteins in a particular sample, and those proteins differentially expressed in related samples—such as diseased vs. healthy tissue. If a protein is found only in a diseased sample then it can be a useful drug target or diagnostic marker. Proteins with same or similar expression profiles may also be functionally related. There are technologies such as 2D-PAGE and mass spectrometry that are used in expression proteomics.

Biomarkers

The National Institutes of Health has defined a biomarker as "a characteristic that is objectively measured and evaluated as an indicator of normal biological processes, pathogenic processes, or pharmacologic responses to a therapeutic intervention."

Understanding the proteome, the structure and function of each protein and the complexities of protein–protein interactions is critical for developing the most effective diagnostic techniques and disease treatments in the future. For example, proteomics is highly useful in identification of candidate biomarkers (proteins in body fluids that are of value for diagnosis), identification of the bacterial antigens that are targeted by the immune response, and identification of possible immunohistochemistry markers of infectious or neoplastic diseases.

An interesting use of proteomics is using specific protein biomarkers to diagnose disease. A number of techniques allow to test for proteins produced during a particular disease, which helps to diagnose the disease quickly. Techniques include western blot, immunohistochemical staining, enzyme linked immunosorbent assay (ELISA) or mass spectrometry. Secretomics, a subfield of proteomics that studies secreted proteins and secretion pathways using proteomic approaches, has recently emerged as an important tool for the discovery of biomarkers of disease.

Proteogenomics

In proteogenomics, proteomic technologies such as mass spectrometry are used for improving gene annotations. Parallel analysis of the genome and the proteome facilitates discovery of post-translational modifications and proteolytic events, especially when comparing multiple species (comparative proteogenomics).

Structural Proteomics

Structural proteomics includes the analysis of protein structures at large-scale. It compares protein structures and helps identify functions of newly discovered genes. The structural analysis also helps to understand that where drugs bind to proteins and also show where proteins interact with each other. This understanding is achieved using different technologies such as X-ray crystallography and NMR spectroscopy.

Bioinformatics for Proteomics (Proteome Informatics)

Much proteomics data is collected with the help of high throughput technologies such as mass spectrometry and microarray. It would often take weeks or months to analyze the data and perform comparisons by hand. For this reason, biologists and chemists are collaborating with computer scientists and mathematicians to create programs and pipeline to computationally analyze the protein data. Using bioinformatics techniques, researchers are capable of faster analysis and data storage. A good place to find lists of current programs and databases is on the ExPASy bioinformatics resource portal. The applications of bioinformatics-based proteomics includes medicine, disease diagnosis, biomarker identification, and many more.

Protein Identification

Mass spectrometry and microarray produce peptide fragmentation information but do not give identification of specific proteins present in the original sample. Due to the lack of specific protein identification, past researchers were forced to decipher the peptide fragments themselves. However, there are currently programs available for protein identification. These programs take the peptide sequences output from mass spectrometry and microarray and return information about matching or similar proteins. This is done through algorithms implemented by the program which perform alignments with proteins from known databases such as UniProt and PROSITE to predict what proteins are in the sample with a degree of certainty.

Protein Structure

The biomolecular structure forms the 3D configuration of the protein. Understanding the protein's structure aids in identification of the protein's interactions and function. It used to be that the 3D structure of proteins could only be determined using X-ray crystallography and NMR spectroscopy. As of 2017, Cryo-electron microscopy is a leading technique, solving difficulties with crystallization (in X-ray crystallography) and conformational ambiguity (in NMR); resolution was 2.2Å as of 2015. Now, through bioinformatics, there are computer programs that can in some cases predict and model the structure of proteins. These programs use the chemical properties of amino acids and structural properties of known proteins to predict the 3D model of sample proteins. This also allows scientists to model protein interactions on a larger scale. In addition, biomedical engineers are developing methods to factor in the flexibility of protein structures to make comparisons and predictions.

Post-translational Modifications

Most programs available for protein analysis are not written for proteins that have undergone post-translational modifications. Some programs will accept post-translational modifications to aid in protein identification but then ignore the modification during further protein analysis. It is important to account for these modifications since they can affect the protein's structure. In turn, computational analysis of post-translational modifications has gained the attention of the scientific community. The current post-translational modification programs are only predictive. Chemists, biologists and computer scientists are working together to create and introduce new pipelines that allow for analysis of post-translational modifications that have been experimentally identified for their effect on the protein's structure and function.

Computational Methods in Studying Protein Biomarkers

One example of the use of bioinformatics and the use of computational methods is the study of protein biomarkers. Computational predictive models have shown that extensive and diverse feto-maternal protein trafficking occurs during pregnancy and can be readily detected non-invasively in maternal whole blood. This computational approach circumvented a major limitation, the abundance of maternal proteins interfering with the detection of fetal proteins, to fetal proteomic analysis of maternal blood. Computational models can use fetal gene transcripts previously identified in maternal whole blood to create a comprehensive proteomic network of the term neonate. Such work shows that the fetal proteins detected in pregnant woman's blood originate from a diverse group of tissues and organs from the developing fetus. The proteomic networks contain many biomarkers that are proxies for development and illustrate the potential clinical application of this technology as a way to monitor normal and abnormal fetal development.

An information theoretic framework has also been introduced for biomarker discovery, integrating biofluid and tissue information. This new approach takes advantage of functional synergy between certain biofluids and tissues with the potential for clinically significant findings not possible if tissues and biofluids were considered individually. By conceptualizing tissue-biofluid as information channels, significant biofluid proxies can be identified and then used for guided development of clinical diagnostics. Candidate biomarkers are then predicted based on information transfer criteria across the tissue-biofluid channels. Significant biofluid-tissue relationships can be used to prioritize clinical validation of biomarkers.

Emerging Trends in Proteomics

A number of emerging concepts have the potential to improve current features of proteomics. Obtaining absolute quantification of proteins and monitoring post-translational modifications are the two tasks that impact the understanding of protein function in healthy and diseased cells. For many cellular events, the protein concentrations do not change; rather, their function is modulated by post-translational modifications (PTM). Methods of monitoring PTM are an underdeveloped area in proteomics. Selecting a particular subset of protein for analysis substantially reduces protein complexity, making it advantageous for diagnostic purposes where blood is the starting material. Another important aspect of proteomics, yet not addressed, is that proteomics methods should focus on studying proteins in the context of the environment. The increasing use of chemical cross linkers, introduced into living cells to fix protein-protein, protein-DNA and other interactions, may ameliorate this problem partially. The challenge is to identify suitable methods of preserving relevant interactions. Another goal for studying protein is to develop more sophisticated methods to image proteins and other molecules in living cells and real time.

Proteomics for Systems Biology

Advances in quantitative proteomics would clearly enable more in-depth analysis of cellular systems. Biological systems are subject to a variety of perturbations (cell cycle, cellular differentiation, carcinogenesis, environment (biophysical), etc.). Transcriptional and translational responses to these perturbations results in functional changes to the proteome implicated in response to the stimulus. Therefore, describing and quantifying proteome-wide changes in protein abundance is crucial towards understanding biological phenomenon more holistically, on the level of the entire system. In this way, proteomics can be seen as complementary to genomics, transcriptomics, epigenomics, metabolomics, and other -omics approaches in integrative analyses attempting to define biological phenotypes more comprehensively. As an example, *The Cancer Proteome Atlas* provides quantitative protein expression data for ~200 proteins in over 4,000 tumor samples with matched transcriptomic and genomic data from The Cancer Genome Atlas. Similar datasets in other cell types, tissue types, and species, particularly using deep shotgun mass spectrometry, will be an immensely important resource for research in fields like cancer biology, developmental and stem cell biology, medicine, and evolutionary biology.

Human Plasma Proteome

Characterizing the human plasma proteome has become a major goal in the proteomics arena, but it is also the most challenging proteomes of all human tissues. It contains immunoglobulin, cytokines, protein hormones, and secreted proteins indicative of infection on top of resident, hemostatic proteins. It also contains tissue leakage proteins due to the blood circulation through different tissues in the body. The blood thus contains information on the physiological state of all tissues and, combined with its accessibility, makes the blood proteome invaluable for medical purposes. The blood plasma proteome is thought to be characterizing the proteome of blood plasma is a daunting challenge.

The depth of the plasma proteome encompassing a dynamic range of more than 10^{10} between the highest abundant protein (albumin) and the lowest (some cytokines) and is thought to be one of the main challenges for proteomics. Temporal and spatial dynamics further complicate the study of human plasma proteome. The turnover of some proteins is quite faster than others and the protein content of an artery may substantially vary from that of a vein. All these differences make even the simplest proteomic task of cataloging the proteome seem out of reach. To tackle this problem, priorities need to be established. Capturing the most meaningful subset of proteins among the entire proteome to generate a diagnostic tool is one such priority. Secondly, since cancer is associated with enhanced glycosylation of proteins, methods that focus on this part of proteins will also be useful. Again: multiparameter analysis best reveals a pathological state. As these technologies improve, the disease profiles should be continually related to respective gene expression changes. Due to the above-mentioned problems plasma proteomics remained challenging. However, technological advancements and continuous developments seem to result in a revival of plasma proteomics as it was shown recently by a technology called plasma proteome profiling. Due to such technologies researchers were able to investigate inflammation processes in mice, the heretability of plasma proteomes as well as to show the effect of such a common life style change like weight loss on the plasma proteome.

Protein Family

A protein family is a group of evolutionarily-related proteins. In many cases a protein family has a corresponding gene family, in which each gene encodes a corresponding protein with a 1:1 relationship. The term *protein family* should not be confused with *family* as it is used in taxonomy.

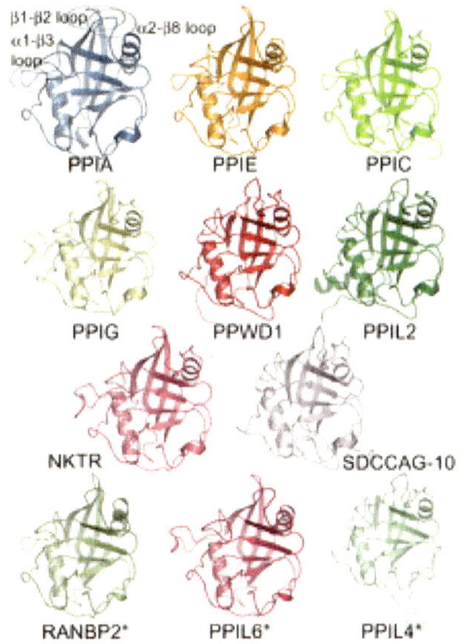

The human cyclophilin family, as represented by the structures of the isomerase domains of some of its members.

Proteins in a family descend from a common ancestor and typically have similar three-dimensional structures, functions, and significant sequence similarity. The most important of these is sequence similarity (usually amino acid sequence) since it is the strictest indicator of homology and therefore the clearest indicator of common ancestry. There is a fairly well developed framework for evaluating the significance of similarity between a group of sequences using sequence alignment methods. Proteins that do not share a common ancestor are very unlikely to show statistically significant sequence similarity, making sequence alignment a powerful tool for identifying the members of protein families.

Families are sometimes grouped together into larger clades called superfamilies based on structural and mechanistic similarity, even if there is no identifiable sequence homology.

Currently, over 60,000 protein families have been defined, although ambiguity in the definition of *protein family* leads different researchers to wildly varying numbers.

Terminology and usage

As with many biological terms, the use of *protein family* is somewhat context dependent; it may indicate large groups of proteins with the lowest possible level of detectable sequence similarity, or very narrow groups of proteins with almost identical sequence, function, and three-dimensional structure, or any kind of group in-between. To distinguish between these situations, the term protein superfamily is often used for distantly related proteins whose relatedness is not detectable by sequence similarity, but only from shared structural features. Other terms such as protein *class*, *group*, *clan* and *sub-family* have been coined over the years, but all suffer similar ambiguities of usage. A common usage is that superfamilies (structural homology) contain families (sequence homology) which contain sub-families. Hence a superfamily, such as the PA clan of proteases, has far lower sequence conservation than one of the families it contains, the C04 family. It is unlikely that an exact definition will be agreed and to it is up to the reader to discern exactly how these terms are being used in a particular context.

Above, sequence conservation of 250 members of the PA clan proteases (superfamily). Below, sequence conservation of 70 members of the C04 protease family. Arrows indicate catalytic triad residues. Aligned on the basis of structure by DALI

Protein Domains and Motifs

The concept of *protein family* was conceived at a time when very few protein structures or sequences were known; at that time, primarily small, single-domain proteins such as myoglobin, hemoglobin, and cytochrome c were structurally understood. Since that time, it was found that many proteins comprise multiple independent structural and functional units or *domains*. Due to evolutionary shuffling, different domains in a protein have evolved independently. This has led, in recent years, to a focus on families of protein domains. A number of online resources are devoted to identifying and cataloging such domains.

Regions of each protein have differing functional constraints (features critical to the structure and function of the protein). For example, the active site of an enzyme requires certain amino acid residues to be precisely oriented in three dimensions. On the other hand, a protein–protein binding interface may consist of a large surface with constraints on the hydrophobicity or polarity of the amino acid residues. Functionally constrained regions of proteins evolve more slowly than unconstrained regions such as surface loops, giving rise to discernible blocks of conserved sequence when the sequences of a protein family are compared. These blocks are most commonly referred to as *motifs*, although many other terms are used (blocks, signatures, fingerprints, etc.). Again, a large number of online resources are devoted to identifying and cataloging protein motifs.

Evolution of Protein Families

According to current consensus, protein families arise in two ways. Firstly, the separation of a parent species into two genetically isolated descendent species allows a gene/protein to independently accumulate variations (mutations) in these two lineages. This results in a family of orthologous proteins, usually with conserved sequence motifs. Secondly, a gene duplication may create a second copy of a gene (termed a paralog). Because the original gene is still able to perform its function, the duplicated gene is free to diverge and may acquire new functions (by random mutation). Certain gene/protein families, especially in eukaryotes, undergo extreme expansions and contractions in the course of evolution, sometimes in concert with whole genome duplications. This expansion and contraction of protein families is one of the salient features of genome evolution, but its importance and ramifications are currently unclear.

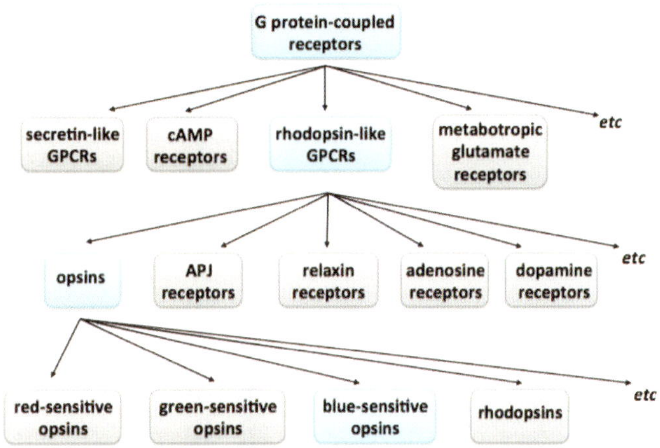

Use and Importance of Protein Families

As the total number of sequenced proteins increases and interest expands in proteome analysis, there is an ongoing effort to organize proteins into families and to describe their component domains and motifs. Reliable identification of protein families is critical to phylogenetic analysis, functional annotation, and the exploration of diversity of protein function in a given phylogenetic branch. The Enzyme Function Initiative (EFI) is using protein families and superfamilies as the basis for development of a sequence/structure-based strategy for large scale functional assignment of enzymes of unknown function.

The algorithmic means for establishing protein families on a large scale are based on a notion of similarity. Most of the time the only similarity we have access to is sequence similarity.

Protein Family Resources

There are many biological databases that record examples of protein families and allow users to identify if newly identified proteins belong to a known family. Here are a few examples:

- Pfam - Protein families database of alignments and HMMs
- PROSITE - Database of protein domains, families and functional sites
- PIRSF - SuperFamily Classification System
- PASS2 - Protein Alignment as Structural Superfamilies v2 - PASS2@NCBS
- SUPERFAMILY - Library of HMMs representing superfamilies and database of (superfamily and family) annotations for all completely sequenced organisms
- SCOP and CATH - classifications of protein structures into superfamilies, families and domains

Similarly many database-searching algorithms exist, for example:

- BLAST - DNA sequence similarity search

- BLASTp - Protein sequence similarity search
- OrthoFinder: a fast, scalable and accurate method for clustering proteins into families (orthogroups)

Protein Superfamily

A protein superfamily is the largest grouping (clade) of proteins for which common ancestry can be inferred. Usually this common ancestry is inferred from structural alignment and mechanistic similarity, even if no sequence similarity is evident. Sequence homology can then be deduced even if not apparent (due to low sequence similarity). Superfamilies typically contain several protein families which show sequence similarity within each family. The term *protein clan* is commonly used for protease superfamilies based on the MEROPS protease classification system.

Identification

Superfamilies of proteins are identified using a number of methods. Closely related members can be identified by different methods to those needed to group the most evolutionarily divergent members.

Above, secondary structural conservation of 80 members of the PA protease clan (superfamily). H indicates α-helix, E indicates β-sheet, L indicates loop. Below, sequence conservation for the same alignment. Arrows indicate catalytic triad residues. Aligned on the basis of structure by DALI

Sequence Similarity

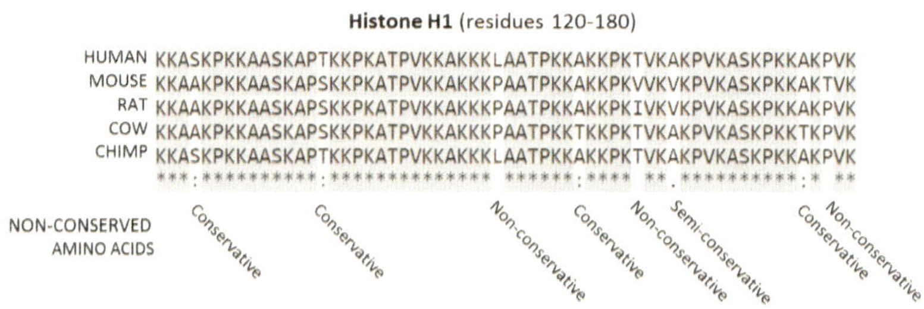

A sequence alignment of mammalian histone proteins. The similarity of the sequences implies that they evolved by gene duplication. Residues that are conserved across all sequences are highlighted in grey. Below the protein sequences is a key denoting conserved sequence (*), conservative mutations (:), semi-conservative mutations (.), and non-conservative mutations ( ).

Historically, the similarity of different amino acid sequences has been the most common method of inferring homology. Sequence similarity is considered a good predictor of relatedness, since similar sequences are more likely the result of gene duplication and divergent evolution, rather than the result of convergent evolution. Amino acid sequence is typically more conserved than DNA sequence (due to the degenerate genetic code), so is a more sensitive detection method. Since some of the amino acids have similar properties (e.g., charge, hydrophobicity, size), conservative mutations that interchange them are often neutral to function. The most conserved sequence regions of a protein often correspond to functionally important regions like catalytic sites and binding sites, since these regions are less tolerant to sequence changes.

Using sequence similarity to infer homology has several limitations. There is no minimum level of sequence similarity guaranteed to produce identical structures. Over long periods of evolution, related proteins may show no detectable sequence similarity to one another. Sequences with many insertions and deletions can also sometimes be difficult to align and so identify the homologous sequence regions. In the PA clan of proteases, for example, not a single residue is conserved through the superfamily, not even those in the catalytic triad. Conversely, the individual families that make up a superfamily are defined on the basis of their sequence alignment, for example the C04 protease family within the PA clan.

Nevertheless, sequence similarity is the most commonly used form of evidence to infer relatedness, since the number of known sequences vastly outnumbers the number of known tertiary structures. In the absence of structural information, sequence similarity constrains the limits of which proteins can be assigned to a superfamily.

Structural Similarity

Structure is much more evolutionarily conserved than sequence, such that proteins with highly similar structures can have entirely different sequences. Over very long evolutionary timescales, very few residues show detectable amino acid sequence conservation, however secondary structural elements and tertiary structural motifs are highly conserved. Conformational changes of the protein structure may also be conserved, as is seen in the serpin superfamily. Consequently, protein tertiary structure can be used to detect homology between proteins even when no evidence of relatedness remains in their sequences. Structural alignment programs, such as DALI, use the 3D structure of a protein of interest to find proteins with similar folds. However, on rare occasions, related proteins may evolve to be structurally dissimilar and relatedness can only be inferred by other methods.

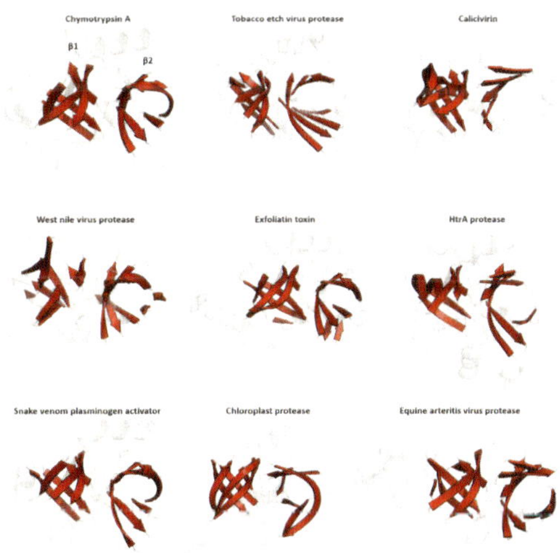

Structural homology in the PA superfamily (PA clan).

The double β-barrel that characterises the superfamily is highlighted in red. Shown are representative structures from several families within the PA superfamily. Note that some proteins show partially modified structural. Chymotrypsin (1gg6), tobacco etch virus protease (1lvm), calicivirin (1wqs), west nile virus protease (1fp7), exfoliatin toxin (1exf), HtrA protease (1l1j), snake venom plasminogen activator (1bqy), chloroplast protease (4fln) and equine arteritis virus protease (1mbm).

Mechanistic Similarity

The catalytic mechanism of enzymes within a superfamily is typically conserved, although substrate specificity may be significantly different. Catalytic residues also tend to occur in the same order in the protein sequence. The families within the PA clan of proteases, although there has been divergent evolution of the catalytic triad residues used to perform catalysis, all members use a similar mechanism to perform covalent, nucleophilic catalysis on proteins, peptides or amino acids. However, mechanism alone is not sufficient to infer relatedness, since some catalytic mechanisms have been convergently evolved multiple times independently, and so form separate superfamilies.

Evolutionary Significance

Protein superfamilies represent the current limits of our ability to identify common ancestry. They are the largest evolutionary grouping based on direct evidence that is currently possible. They are therefore amongst the most ancient evolutionary events currently studied. Some superfamilies have members present in all kingdoms of life, indicating that the last common ancestor of that superfamily was in the last universal common ancestor of all life (LUCA).

Superfamily members may be in different species, with the ancestral protein being the form of the protein that existed in the ancestral species (orthology). Conversely, the proteins may be in the same species, but evolved from a single protein whose gene was duplicated in the genome (paralogy).

Diversification

A majority of proteins contain multiple domains. Between 66-80% of eukaryotic proteins have multiple domains while about 40-60% of prokaryotic proteins have multiple domains. Over time, many of the superfamilies of domains have mixed together. In fact, it is very rare to find "consistently isolated superfamilies". When domains do combine, the N- to C- terminal domain order (the "domain architecture") is typically well conserved. Additionally, the number of domain combinations seen in nature is small compared to the number of possibilities, suggesting that selection acts on all combinations.

Examples

α/β hydrolase superfamily - Members share an α/β sheet, containing 8 strands connected by helices, with catalytic triad residues in the same order, activities include proteases, lipases, peroxidases, esterases, epoxide hydrolases and dehalogenases.

Alkaline phosphatase superfamily - Members share an αβα sandwich structure as well as performing common promiscuous reactions by a common mechanism.

Globin superfamily - Members share an 8-alpha helix globular globin fold.

Immunoglobulin superfamily - Members share a sandwich-like structure of two sheets of antiparallel β strands (Ig-fold), and are involved in recognition, binding, and adhesion.

PA clan - Members share a chymotrypsin-like double β-barrel fold and similar proteolysis mechanisms but sequence identity of <10%. The clan contains both cysteine and serine proteases (different nucleophiles).

Ras superfamily - Members share a common catalytic G domain of a 6-strand β sheet surrounded by 5 α-helices.

Serpin superfamily - Members share a high-energy, stressed fold which can undergo a large conformational change, which is typically used to inhibit serine and cysteine proteases by disrupting their structure.

TIM barrel superfamily - Members share a large $\alpha_8\beta_8$ barrel structure. It is one of the most common protein folds and the monophylicity of this superfamily is still contested.

Protein Metabolism

Protein metabolism denotes the various biochemical processes responsible for the synthesis of proteins and amino acids, and the breakdown of proteins (and other large molecules) by catabolism.

Dietary proteins are first broken down to individual amino acids by various enzymes and hydrochloric acid present in the gastro-intestinal tract. These amino acids are further broken down to α-keto acids which can be recycled in the body for generation of energy, and production of glucose or fat or other amino acids. This break-down of amino acids to α-keto acids occurs in the liver by a process known as transamination, which follows a bimolecular ping pong mechanism.

Protein Synthesis

Protein biosynthesis relies on four processes:

- amino acid synthesis
- RNA synthesis
- transcription
- translation

Protein anabolism is the process by which protein are formed from amino acids (a.k.a. anabolic amino acid synthesis).

Protein Breakdown

Protein catabolism is the process by which proteins are broken down to their amino acids. This is also called proteolysis.

This can be followed by further amino acid degradation.

Protein Biosynthesis

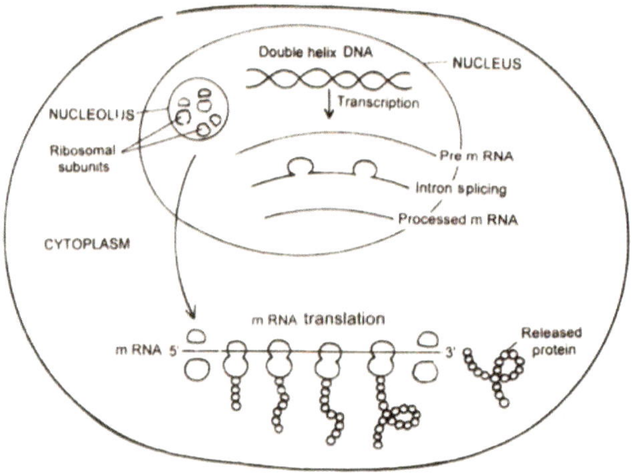

RNA is transcribed in the nucleus; once completely processed, it is transported to the cytoplasm and translated by the ribosome (shown in very pale grey behind the tRNA).

Protein synthesis is the process whereby biological cells generate new proteins; it is balanced by the loss of cellular proteins via degradation or export. Translation, the assembly of amino acids by ribosomes, is an essential part of the biosynthetic pathway, along with generation of messenger RNA (mRNA), aminoacylation of transfer RNA (tRNA), co-translational transport, and post-translational modification. Protein biosynthesis is strictly regulated at multiple steps. They are principally during transcription (phenomena of RNA synthesis from DNA template) and translation (phenomena of amino acid assembly from RNA).

The cistron DNA is transcribed into the first of a series of RNA intermediates. The last version is used as a template in synthesis of a polypeptide chain. Protein will often be synthesized directly from genes by translating mRNA. However, when a protein must be available on short notice or in large quantities, a protein precursor is produced. A proprotein is an inactive protein containing one or more inhibitory peptides that can be activated when the inhibitory sequence is removed by proteolysis during posttranslational modification. A preprotein is a form that contains a signal sequence (an N-terminal signal peptide) that specifies its insertion into or through membranes, i.e., targets them for secretion. The signal peptide is cleaved off in the endoplasmic reticulum. Preproproteins have both sequences (inhibitory and signal) still present.

In protein synthesis, a succession of tRNA molecules charged with appropriate amino acids are brought together with an mRNA molecule and matched up by base-pairing through the anti-codons of the tRNA with successive codons of the mRNA. The amino acids are then linked together to extend the growing protein chain, and the tRNAs, no longer carrying amino acids, are released. This whole complex of processes is carried out by the ribosome, formed of two main chains of RNA, called ribosomal RNA (rRNA), and more than 50 different proteins. The ribosome latches onto the end of an mRNA molecule and moves along it, capturing loaded tRNA molecules and joining together their amino acids to form a new protein chain.

Protein biosynthesis, although very similar, is different for prokaryotes and eukaryotes.

Transcription

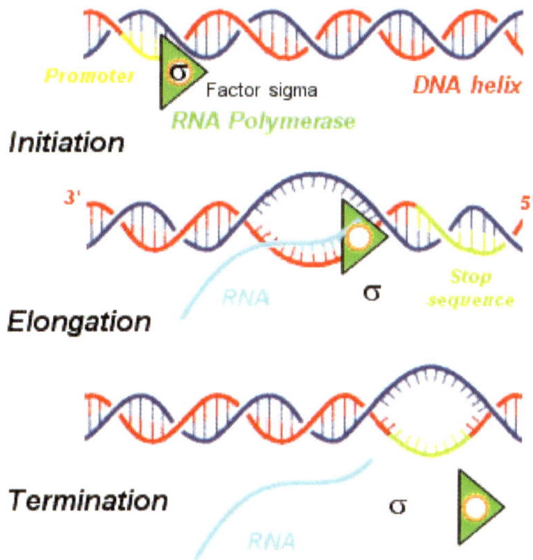

Diagram showing the process of transcription

Phenomena of RNA synthesis from DNA template In transcription an mRNA chain is generated, with one strand of the DNA double helix in the genome as a template. This strand is called the template strand. Transcription can be divided into 3 stages: initiation, elongation, and termination, each regulated by a large number of proteins such as transcription factors and coactivators that ensure that the correct gene is transcribed.

Transcription occurs in the cell nucleus, where the DNA is held. The DNA structure of the cell is made up of two helixes made up of sugar and phosphate held together by hydrogen bonds between the bases of opposite strands. The sugar and the phosphate in each strand are joined together by stronger phosphodiester covalent bonds. The DNA is "unzipped" (disruption of hydrogen bonds between different single strands) by the enzyme helicase, leaving the single nucleotide chain open to be copied. RNA polymerase reads the DNA strand from the 3-prime (3') end to the 5-prime (5') end, while it synthesizes a single strand of messenger RNA in the 5'-to-3' direction. The general RNA structure is very similar to the DNA structure, but in RNA the nucleotide uracil takes the place that thymine occupies in DNA. The single strand of mRNA leaves the nucleus through nuclear pores, and migrates into the cytoplasm.

The first product of transcription differs in prokaryotic cells from that of eukaryotic cells, as in prokaryotic cells the product is mRNA, which needs no post-transcriptional modification, whereas, in eukaryotic cells, the first product is called primary transcript, that needs post-transcriptional modification (capping with 7-methyl-guanosine, tailing with a poly A tail) to give hnRNA (heterogeneous nuclear RNA). hnRNA then undergoes splicing of introns (noncoding parts of the gene) via spliceosomes to produce the final mRNA.

Translation

Phenomena of amino acid assembly from RNA. The synthesis of proteins from RNA is known as translation. In eukaryotes, translation occurs in the cytoplasm, where the ribosomes are located. Ribosomes are made of a small and large subunit that surround the mRNA. In translation, messenger RNA (mRNA) is decoded to produce a specific polypeptide according to the rules specified by the trinucleotide genetic code. This uses an mRNA sequence as a template to guide the synthesis of a chain of amino acids that form a protein. Translation proceeds in four phases: activation, initiation, elongation, and termination (all describing the growth of the amino acid chain, or poly-peptide that is the product of translation).

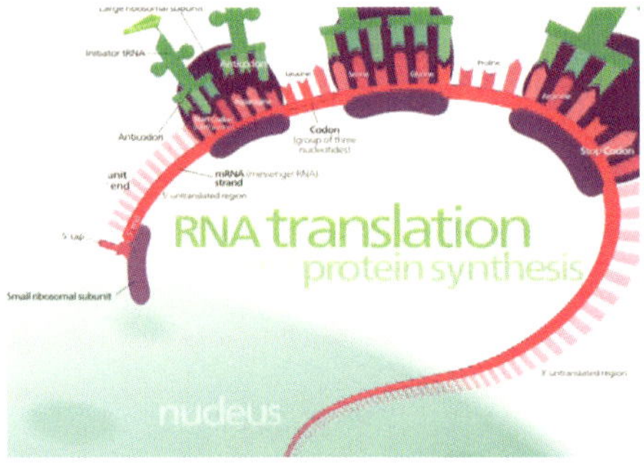

Diagram showing the process of translation

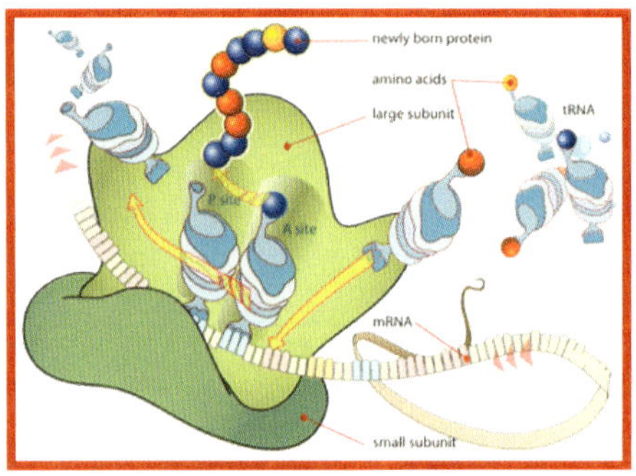

Diagram showing the translation of mRNA and the synthesis of proteins by a ribosome

In activation, the correct amino acid (AA) is joined to the correct transfer RNA (tRNA). While this is not, in the technical sense, a step in translation, it is required for translation to proceed. The AA is joined by its carboxyl group to the 3' OH of the tRNA by an ester bond. When the tRNA has an amino acid linked to it, it is termed "charged". Initiation involves the small subunit of the ribosome binding to 5' end of mRNA with the help of initiation factors (IF), other proteins that assist the process. Elongation occurs when the next aminoacyl-tRNA (charged tRNA) in line binds to the ribosome along with GTP and an elongation factor. Termination of the polypeptide happens when the A site of the ribosome faces a stop codon (UAA, UAG, or UGA). When this happens, no tRNA can recognize it, but releasing factor can recognize nonsense codons and causes the release of the polypeptide chain. The capacity of disabling or inhibiting translation in protein biosynthesis is used by some antibiotics such as anisomycin, cycloheximide, chloramphenicol, tetracycline, streptomycin, erythromycin, puromycin, etc.

Events During or Following Protein Translation

Events that occur during or following biosynthesis include proteolysis, post-translational modification and protein folding. Proteolysis may remove N-terminal, C-terminal or internal amino-acid residues or peptides from the polypeptide. The termini and side-chains of the polypeptide may be subjected to post-translational modification. These modifications may be required for correct cellular localisation or the natural function of the protein. During and after synthesis, polypeptide chains often fold to assume, so called, native secondary and tertiary structures. This is known as *protein folding* and is typically required for the natural function of the protein.

Proteolysis

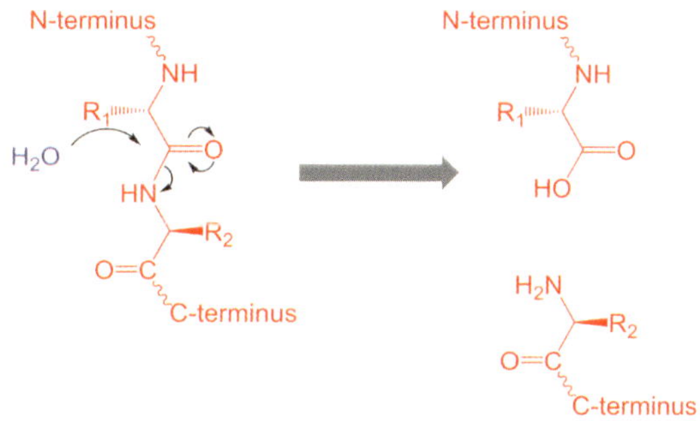

The hydrolysis of a protein (red) by the nucleophilic attack of water (blue).
The uncatalysed half-life is several hundred years.

Proteolysis is the breakdown of proteins into smaller polypeptides or amino acids. Uncatalysed, the hydrolysis of peptide bonds is extremely slow, taking hundreds of years. Proteolysis is typically catalysed by cellular enzymes called proteases, but may also occur by intra-molecular digestion. Low pH or high temperatures can also cause proteolysis non-enzymatically.

Proteolysis in organisms serves many purposes; for example, digestive enzymes break down proteins in food to provide amino acids for the organism, while proteolytic processing of a polypeptide chain after its synthesis may be necessary for the production of an active protein. It is also important in the regulation of some physiological and cellular processes, as well as preventing the accumulation of unwanted or abnormal proteins in cells. Consequently, dis-regulation of proteolysis can cause disease and is used by some venoms.

Proteolysis is important as an analytical tool for studying proteins in the laboratory, as well as industrially, for example in food processing and stain removal.

Biological Functions

Post-translational Proteolytic Processing

Limited proteolysis of a polypeptide during or after translation in protein synthesis often occurs for many proteins. This may involve removal of the N-terminal methionine, signal peptide, and/or the conversion of an inactive or non-functional protein to an active one. The precursor to the final functional form of protein is termed proprotein, and these proproteins may be first synthesized as preproprotein. For example, albumin is first synthesized as preproalbumin and contains an uncleaved signal peptide. This forms the proalbumin after the signal peptide is cleaved, and a further processing to remove the N-terminal 6-residue propeptide yields the mature form of the protein.

Removal of N-terminal Methionine

The initiating methionine (and, in prokaryotes, fMet) may be removed during translation of the nascent protein. For *E. coli*, fMet is efficiently removed if the second residue is small and uncharged, but not if the second residue is bulky and charged. In both prokaryotes and eukaryotes, the exposed N-terminal residue may determine the half-life of the protein according to the N-end rule.

Removal of the Signal Sequence

Proteins that are to be targeted to a particular organelle or for secretion have an N-terminal signal peptide that directs the protein to its final destination. This signal peptide is removed by proteolysis after their transport through a membrane.

Cleavage of Polyproteins

Some proteins and most eukaryotic polypeptide hormones are synthesized as a large precursor polypeptide known as a polyprotein that requires proteolytic cleavage into individual smaller polypeptide chains. The polyprotein pro-opiomelanocortin (POMC) contains many polypeptide hormones. The cleavage pattern of POMC, however, may vary between different tissues, yielding different sets of polypeptide hormones from the same polyprotein.

Many viruses also produce their proteins initially as a single polypeptide chain that were translated from a polycistronic mRNA. This polypeptide is subsequently cleaved into individual polypeptide chains.

Cleavage of Precursor Proteins

Many proteins and hormones are synthesized in the form of their precursors - zymogens, proenzymes, and prehormones. These proteins are cleaved to form their final active structures. Insulin, for example, is synthesized as preproinsulin, which yields proinsulin after the signal peptide has been cleaved. The proinsulin is then cleaved at two positions to yield two polypeptide chains linked by two disulfide bonds. Removal of two C-terminal residues from the B-chain then yields the mature insulin. Protein folding occurs in the single-chain Proinsulin form which facilitates formation of the ultimately inter-peptide disulfide bonds, and the ultimately intra-peptide disulfide bond, found in the native structure of insulin.

Proteases in particular are synthesized in the inactive form so that they may be safely stored in cells, and ready for release in sufficient quantity when required. This is to ensure that the protease is activated only in the correct location or context, as inappropriate activation of these proteases can be very destructive for an organism. Proteolysis of the zymogen yields an active protein; for example, when trypsinogen is cleaved to form trypsin, a slight rearrangement of the protein structure that completes the active site of the protease occurs, thereby activating the protein.

Proteolysis can, therefore, be a method of regulating biological processes by turning inactive proteins into active ones. A good example is the blood clotting cascade whereby an initial event triggers a cascade of sequential proteolytic activation of many specific proteases, resulting in blood coagulation. The complement system of the immune response also involves a complex sequential proteolytic activation and interaction that result in an attack on invading pathogens.

Protein Degradation

Protein degradation may take place intracellularly or extracellularly. In digestion of food, digestive enzymes may be released into the environment for extracellular digestion whereby proteolytic cleavage breaks down proteins into smaller peptides and amino acids so that they may be absorbed and used by an organism. In animals the food may be processed extracellularly in specialized digestive organs or guts, but in many bacteria the food may be internalized into the cell via phagocytosis. Microbial degradation of protein in the environment can be regulated by nutrient availability. For example, limitation for major elements in proteins (carbon, nitrogen, and sulfur) has been shown to induce proteolytic activity in the fungus *Neurospora crassa* as well as in whole communities of soil organisms.

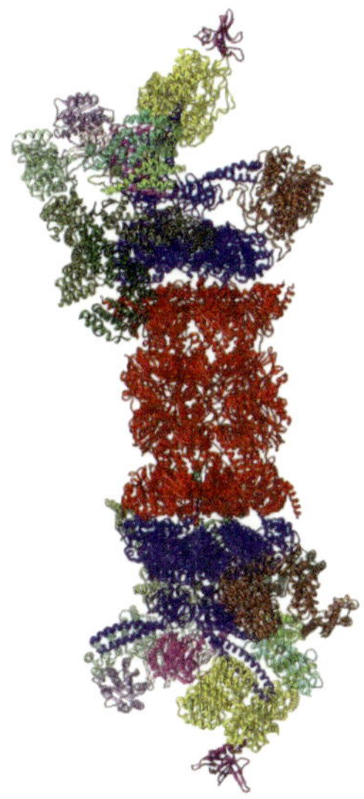

Structure of a proteasome. Its active sites are inside the tube (blue) where proteins are degraded.

Proteins in cells are also constantly being broken down into amino acids. This intracellular degradation of protein serves a number of functions: It removes damaged and abnormal protein and prevent their accumulation, and it also serves to regulate cellular processes by removing enzymes and regulatory proteins that are no longer needed. The amino acids may then be reused for protein synthesis.

Lysosome and Proteasome

The intracellular degradation of protein may be achieved in two ways - proteolysis in lysosome, or a ubiquitin-dependent process that targets unwanted proteins to proteasome. The autophagy-lysosomal pathway is normally a non-selective process, but it may become selective upon starvation whereby proteins with peptide sequence KFERQ or similar are selectively broken down. The lysosome contains a large number of proteases such as cathepsins.

The ubiquitin-mediated process is selective. Proteins marked for degradation are covalently linked to ubiquitin. Many molecules of ubiquitin may be linked in tandem to a protein destined for degradation. The polyubiquinated protein is targeted to an ATP-dependent protease complex, the proteasome. The ubiquitin is released and reused, while the targeted protein is degraded.

Rate of Intracellular Protein Degradation

Different proteins are degraded at different rates. Abnormal proteins are quickly degraded, whereas the rate of degradation of normal proteins may vary widely depending on their functions.

Enzymes at important metabolic control points may be degraded much faster than those enzymes whose activity is largely constant under all physiological conditions. One of the most rapidly degraded proteins is ornithine decarboxylase, which has a half-life of 11 minutes. In contrast, other proteins like actin and myosin have a half-life of a month or more, while, in essence, haemoglobin lasts for the entire life-time of an erythrocyte.

The N-end rule may partially determine the half-life of a protein, and proteins with segments rich in proline, glutamic acid, serine, and threonine (the so-called PEST proteins) have short half-life. Other factors suspected to affect degradation rate include the rate deamination of glutamine and asparagine and oxidation of cystein, histidine, and methionine, the absence of stabilizing ligands, the presence of attached carbohydrate or phosphate groups, the presence of free α-amino group, the negative charge of protein, and the flexibility and stability of the protein. Proteins with larger degrees of intrinsic disorder also tend to have short cellular half-life, with disordered segments having been proposed to facilitate efficient initiation of degradation by the proteasome.

The rate of proteolysis may also depend on the physiological state of the organism, such as its hormonal state as well as nutritional status. In time of starvation, the rate of protein degradation increases.

Digestion

In human digestion, proteins in food are broken down into smaller peptide chains by digestive enzymes such as pepsin, trypsin, chymotrypsin, and elastase, and into amino acids by various enzymes such as carboxypeptidase, aminopeptidase, and dipeptidase. It is necessary to break down proteins into small peptides (tripeptides and dipeptides) and amino acids so they can be absorbed by the intestines, and the absorbed tripeptides and dipeptides are also further broken into amino acids intracellularly before they enter the bloodstream. Different enzymes have different specificity for their substrate; trypsin, for example, cleaves the peptide bond after a positively charged residue (arginine and lysine); chymotrypsin cleaves the bond after an aromatic residue (phenylalanine, tyrosine, and tryptophan); elastase cleaves the bond after a small non-polar residue such as alanine or glycine.

In order to prevent inappropriate or premature activation of the digestive enzymes (they may, for example, trigger pancreatic self-digestion causing pancreatitis), these enzymes are secreted as inactive zymogen. The precursor of pepsin, pepsinogen, is secreted by the stomach, and is activated only in the acidic environment found in stomach. The pancreas secretes the precursors of a number of proteases such as trypsin and chymotrypsin. The zymogen of trypsin is trypsinogen, which is activated by a very specific protease, enterokinase, secreted by the mucosa of the duodenum. The trypsin, once activated, can also cleave other trypsinogens as well as the precursors of other proteases such as chymotrypsin and carboxypeptidase to activate them.

In bacteria, a similar strategy of employing an inactive zymogen or prezymogen is used. Subtilisin, which is produced by *Bacillus subtilis*, is produced as preprosubtilisin, and is released only if the signal peptide is cleaved and autocatalytic proteolytic activation has occurred.

Cellular Regulation

Proteolysis is also involved in the regulation of many cellular processes by activating or deactivating enzymes, transcription factors, and receptors, for example in the biosynthesis of cholesterol, or the mediation of thrombin signalling through protease-activated receptors.

Some enzymes at important metabolic control points such as ornithine decarboxylase is regulated entirely by its rate of synthesis and its rate of degradation. Other rapidly degraded proteins include the protein products of proto-oncogenes, which play central roles in the regulation of cell growth.

Cell Cycle Regulation

Cyclins are a group of proteins that activate kinases involved in cell division. The degradation of cyclins is the key step that governs the exit from mitosis and progress into the next cell cycle. Cyclins accumulate in the course the cell cycle, then abruptly disappear just before the anaphase of mitosis. The cyclins are removed via a ubiquitin-mediated proteolytic pathway.

Apoptosis

Caspases are an important group of proteases involved in apoptosis or programmed cell death. The precursors of caspase, procaspase, may be activated by proteolysis through its association with a protein complex that forms apoptosome, or by granzyme B, or via the death receptor pathways.

Proteolysis and Diseases

Abnormal proteolytic activity is associated with many diseases. In pancreatitis, leakage of proteases and their premature activation in the pancreas results in the self-digestion of the pancreas. People with diabetes mellitus may have increased lysosomal activity and the degradation of some proteins can increase significantly. Chronic inflammatory diseases such as rheumatoid arthritis may involve the release of lysosomal enzymes into extracellular space that break down surrounding tissues. Abnormal proteolysis and generation of peptides that aggregate in cells and their ineffective removal may result in many age-related neurological diseases such as Alzheimer's.

Proteases may be regulated by antiproteases or protease inhibitors, and imbalance between proteases and antiproteases can result in diseases, for example, in the destruction of lung tissues in emphysema brought on by smoking tobacco. Smoking is thought to increase the neutrophils and macrophages in the lung which release excessive amount of proteolytic enzymes such as elastase, such that they can no longer be inhibited by serpins such as α_1-antitrypsin, thereby resulting in the breaking down of connective tissues in the lung. Other proteases and their inhibitors may also be involved in this disease, for example matrix metalloproteinases (MMPs) and tissue inhibitors of metalloproteinases (TIMPs).

Other diseases linked to aberrant proteolysis include muscular dystrophy, degenerative skin disorders, respiratory and gastrointestinal diseases, and malignancy.

Non-enzymatic Proteolysis

Protein backbones are very stable in water at neutral pH and room temperature, although the rate of hydrolysis of different peptide bonds can vary. The half life of a peptide bond under normal conditions can range from 7 years to 350 years, even higher for peptides protected by modified terminus or within the protein interior. The rate of proteolysis however can be significantly increased by extremes of pH and heat.

Strong mineral acids can readily hydrolyse the peptide bonds in a protein (acid hydrolysis). The standard way to hydrolyze a protein or peptide into its constituent amino acids for analysis is to heat it to 105 °C for around 24 hours in 6M hydrochloric acid. However, some proteins are resistant to acid hydrolysis. One well-known example is ribonuclease A, which can be purified by treating crude extracts with hot sulphuric acid so that other proteins become degraded while ribonuclease A is left intact.

Certain chemicals cause proteolysis only after specific residues, and these can be used to selectively break down a protein into smaller polypeptides for laboratory analysis. For example, cyanogen bromide cleaves the peptide bond after a methionine. Similar methods may be used to specifically cleave tryptophanyl, aspartyl, cysteinyl, and asparaginyl peptide bonds. Acids such as trifluoroacetic acid and formic acid may also be used.

Like other biomolecules, proteins can also be broken down by high heat alone. At 250 °C, the peptide bond may be easily hydrolyzed, with its half-life dropping to about a minute. Protein may also be broken down without hydrolysis through pyrolysis; small heterocyclic compounds may start to form upon degradation, above 500 °C, polycyclic aromatic hydrocarbon may also form, which is of interest in the study of generation of carcinogens in tobacco smoke and cooking at high heat.

References

- Dobson CM (2000). "The nature and significance of protein folding". In Pain RH. Mechanisms of Protein Folding. Oxford, Oxfordshire: Oxford University Press. pp. 1–28. ISBN 0-19-963789-X
- Gutteridge A, Thornton JM (2005). "Understanding nature's catalytic toolkit". Trends in Biochemical Sciences. 30 (11): 622–29. doi:10.1016/j.tibs.2005.09.006. PMID 16214343
- EBI External Services (2010-01-20). "The Catalytic Site Atlas at The European Bioinformatics Institute". Ebi.ac.uk. Retrieved 2011-01-16
- Zagrovic B, Snow CD, Shirts MR, Pande VS (2002). "Simulation of folding of a small alpha-helical protein in atomistic detail using worldwide-distributed computing". Journal of Molecular Biology. 323 (5): 927–37. doi:10.1016/S0022-2836(02)00997-X. PMID 12417204
- Hey J, Posch A, Cohen A, Liu N, Harbers A (2008). "Fractionation of complex protein mixtures by liquid-phase isoelectric focusing". Methods in Molecular Biology. Methods in Molecular Biology™. 424: 225–39. doi:10.1007/978-1-60327-064-9_19. ISBN 978-1-58829-722-8. PMID 18369866
- Milo, Ron (2013-12-01). "What is the total number of protein molecules per cell volume? A call to rethink some published values". BioEssays. 35 (12): 1050–1055. doi:10.1002/bies.201300066. ISSN 1521-1878. PMC 3910158 . PMID 24114984
- Ward, J. J.; Sodhi, J. S.; McGuffin, L. J.; Buxton, B. F.; Jones, D. T. (2004). "Prediction and functional analysis of native disorder in proteins from the three kingdoms of life". Journal of Molecular Biology. 337 (3): 635–45. doi:10.1016/j.jmb.2004.02.002. PMID 15019783
- Walker JH, Wilson K (2000). Principles and Techniques of Practical Biochemistry. Cambridge, UK: Cambridge University Press. pp. 287–89. ISBN 0-521-65873-X
- "APAF - The Australian Proteome Analysis Facility - APAF - The Australian Proteome Analysis Facility". www.proteome.org.au. Retrieved 2017-02-06
- Brosnan J (June 2003). "Interorgan amino acid transport and its regulation". Journal of Nutrition. 133 (6 Suppl 1): 2068S–72S. PMID 12771367

- Sleator RD. (2012). "Prediction of protein functions". Methods in Molecular Biology. Methods in Molecular Biology. 815: 15–24. doi:10.1007/978-1-61779-424-7_2. ISBN 978-1-61779-423-0. PMID 22130980
- Schwarzer D, Cole P (2005). "Protein semisynthesis and expressed protein ligation: chasing a protein's tail". Current Opinion in Chemical Biology. 9 (6): 561–69. doi:10.1016/j.cbpa.2005.09.018. PMID 16226484
- Peter Tompa; Alan Fersht (18 November 2009). Structure and Function of Intrinsically Disordered Proteins. CRC Press. ISBN 978-1-4200-7893-0
- Terpe K (2003). "Overview of tag protein fusions: from molecular and biochemical fundamentals to commercial systems". Applied Microbiology and Biotechnology. 60 (5): 523–33. doi:10.1007/s00253-002-1158-6. PMID 12536251
- Reynolds JA, Tanford C (2003). Nature's Robots: A History of Proteins (Oxford Paperbacks). New York, New York: Oxford University Press. p. 15. ISBN 0-19-860694-X
- Görg A, Weiss W, Dunn MJ (2004). "Current two-dimensional electrophoresis technology for proteomics". Proteomics. 4 (12): 3665–85. doi:10.1002/pmic.200401031. PMID 15543535
- Watson, Tim D. G. (1998-12-01). "Diet and Skin Disease in Dogs and Cats". The Journal of Nutrition. 128 (12): 2783S–2789S. ISSN 0022-3166. PMID 9868266
- Ritchie DW (2008). "Recent progress and future directions in protein–protein docking". Current Protein and Peptide Science. 9 (1): 1–15. doi:10.2174/138920308783565741. PMID 18336319

Chapter 5

Nucleic Acids: Design, Structure and Metabolism

Nucleic acids are the most significant biopolymers present in all living organisms. They store information in the nucleus. The three primary types of nucleic acids are deoxyribonucleic acid, artificial nucleic acid and ribonucleic acid. The following chapter elucidates the diverse aspects related to the structure and metabolism of nucleic acids.

Nucleic acids are biopolymers, or small biomolecules, essential to all known forms of life. They are composed of nucleotides, which are monomers made of three components: a 5-carbon sugar, a phosphate group and a nitrogenous base. If the sugar is a simple ribose, the polymer is RNA (ribonucleic acid); if the sugar is derived from ribose as deoxyribose, the polymer is DNA (deoxyribonucleic acid).

Nucleic acids are the most important of all biomolecules. They are found in abundance in all living things, where they function to create and encode and then store information in the nucleus of every living cell of every life-form organism on Earth. In turn, they function to transmit and express that information inside and outside the cell nucleus—to the interior operations of the cell and ultimately to the next generation of each living organism. The encoded information is contained and conveyed via the nucleic acid sequence, which provides the 'ladder-step' ordering of nucleotides within the molecules of RNA and DNA.

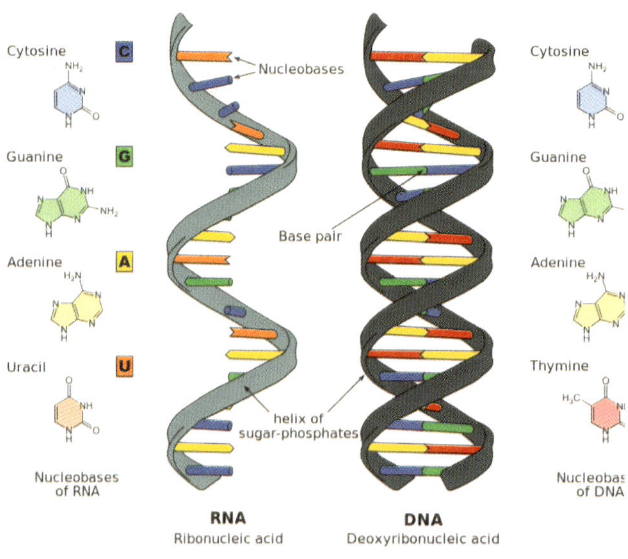

A comparison of the two principal nucleic acids: RNA (*left*) and DNA (*right*), showing the helices and nucleobases each employs.

The Swiss scientist Friedrich Miescher discovered nucleic acids (DNA). Later, he raised the idea that they could be involved in heredity.

Strings of nucleotides are bonded to form helical backbones—typically, one for RNA, two for DNA—and assembled into chains of base-pairs selected from the five primary, or canonical, nucleobases, which are: adenine, cytosine, guanine, thymine, and uracil; note, thymine occurs only in DNA and uracil only in RNA. Using amino acids and the process known as protein synthesis, the specific sequencing in DNA of these nucleobase-pairs enables storing and transmitting coded instructions as genes. In RNA, base-pair sequencing provides for manufacturing new proteins that determine the frames and parts and most chemical processes of all life forms.

Nucleic Acid Design

Nucleic acid design is the process of generating a set of nucleic acid base sequences that will associate into a desired conformation. Nucleic acid design is central to the fields of DNA nanotechnology and DNA computing. It is necessary because there are many possible sequences of nucleic acid strands that will fold into a given secondary structure, but many of these sequences will have undesired additional interactions which must be avoided. In addition, there are many tertiary structure considerations which affect the choice of a secondary structure for a given design.

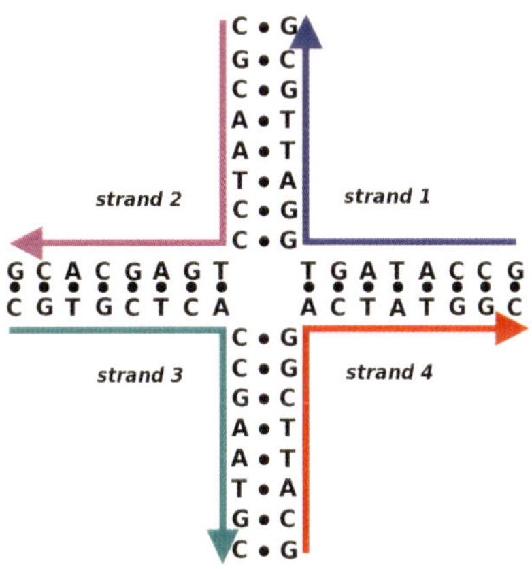

Nucleic acid design can be used to create nucleic acid complexes with complicated secondary structures such as this four-arm junction. These four strands associate into this structure because it maximizes the number of correct base pairs, with A's matched to T's and C's matched to G's. Image from Mao, 2004.

Nucleic acid design has similar goals to protein design: in both, the sequence of monomers is rationally designed to favor the desired folded or associated structure and to disfavor alternate structures. However, nucleic acid design has the advantage of being a much computationally simpler problem, since the simplicity of Watson-Crick base pairing rules leads to simple heuristic methods which yield experimentally robust designs. Computational models for protein folding require tertiary structure information whereas nucleic acid design can operate largely on the level of secondary structure. However, nucleic acid structures are less versatile than proteins in their functionality.

Nucleic acid design can be considered the inverse of nucleic acid structure prediction. In structure prediction, the structure is determined from a known sequence, while in nucleic acid design, a sequence is generated which will form a desired structure.

Fundamental Concepts

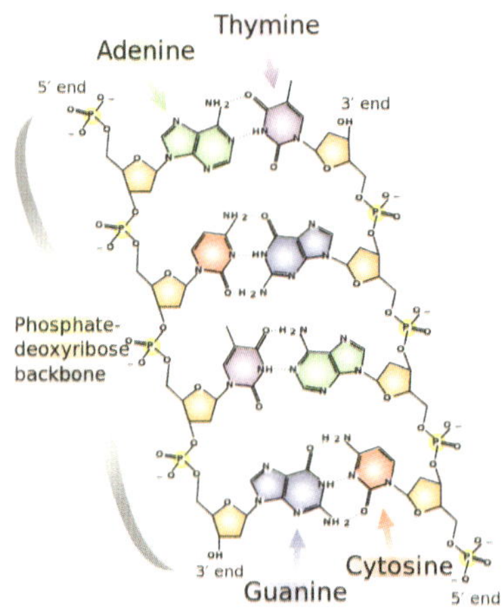

Chemical structure of DNA. Nucleic acid double helices will only form between two strands of complementary sequences, where the bases are matched into only A-T or G-C pairs.

The structure of nucleic acids consists of a sequence of nucleotides. There are four types of nucleotides distinguished by which of the four nucleobases they contain: in DNA these are adenine (A), cytosine (C), guanine (G), and thymine (T). Nucleic acids have the property that two molecules will bind to each other to form a double helix only if the two sequences are complementary, that is, they can form matching sequences of base pairs. Thus, in nucleic acids the sequence determines the pattern of binding and thus the overall structure.

Nucleic acid design is the process by which, given a desired target structure or functionality, sequences are generated for nucleic acid strands which will self-assemble into that target structure. Nucleic acid design encompasses all levels of nucleic acid structure:

- Primary structure—the raw sequence of nucleobases of each of the component nucleic acid strands;
- Secondary structure—the set of interactions between bases, i.e., which parts of which strands are bound to each other; and
- Tertiary structure—the locations of the atoms in three-dimensional space, taking into consideration geometrical and steric constraints.

One of the greatest concerns in nucleic acid design is ensuring that the target structure has the lowest free energy (i.e. is the most thermodynamically favorable) whereas misformed structures have higher values of free energy and are thus unfavored. These goals can be achieved through the use of a number of approaches, including heuristic, thermodynamic, and geometrical ones. Almost all nucleic acid design tasks are aided by computers, and a number of software packages are available for many of these tasks.

Two considerations in nucleic acid design are that desired hybridizations should have melting temperatures in a narrow range, and any spurious interactions should have very low melting temperatures (i.e. they should be very weak). There is also a contrast between affinity-optimizing "positive design", seeks to minimize the energy of the desired structure in an absolute sense, and specificity-optimizing "negative design", which considers the energy of the target structure relative to those of undesired structures. Algorithms which implement both kinds of design tend to perform better than those that consider only one type.

Approaches

Heuristic Methods

Heuristic methods use simple criteria which can be quickly evaluated to judge the suitability of different sequences for a given secondary structure. They have the advantage of being much less computationally expensive than the energy minimization algorithms needed for thermodynamic or geometrical modeling, and being easier to implement, but at the cost of being less rigorous than these models.

Sequence symmetry minimization is the oldest approach to nucleic acid design and was first used to design immobile versions of branched DNA structures. Sequence symmetry minimization divides the nucleic acid sequence into overlapping subsequences of a fixed length, called the criterion length. Each of the 4^N possible subsequences of length N is allowed to appear only once in the sequence. This ensures that no undesired hybridizations can occur which have a length greater than or equal to the criterion length.

A related heuristic approach is to consider the "mismatch distance", meaning the number of positions in a certain frame where the bases are not complementary. A greater mismatch distance lessens the chance that a strong spurious interaction can happen. This is related to the concept of Hamming distance in information theory. Another related but more involved approach is to use methods from coding theory to construct nucleic acid sequences with desired properties.

Thermodynamic Models

Information about the secondary structure of a nucleic acid complex along with its sequence can be used to predict the thermodynamic properties of the complex.

When thermodynamic models are used in nucleic acid design, there are usually two considerations: desired hybridizations should have melting temperatures in a narrow range, and any spurious interactions should have very low melting temperatures (i.e. they should be very weak). The Gibbs free energy of a perfectly matched nucleic acid duplex can be predicted using a nearest neighbor model. This model considers only the interactions between a nucleotide and its nearest neighbors on the nucleic acid strand, by summing the free energy of each of the overlapping two-nucleotide subwords of the duplex. This is then corrected for self-complementary monomers and for GC-content. Once the free energy is known, the melting temperature of the duplex can be determined. GC-content alone can also be used to estimate the free energy and melting temperature of a nucleic acid duplex. This is less accurate but also much less computationally costly.

Software for thermodynamic modeling of nucleic acids includes Nupack, mfold/UNAFold, and Vienna.

A related approach, inverse secondary structure prediction, uses stochastic local search which improves a nucleic acid sequence by running a structure prediction algorithm and the modifying the sequence to eliminate unwanted features.

Geometrical Models

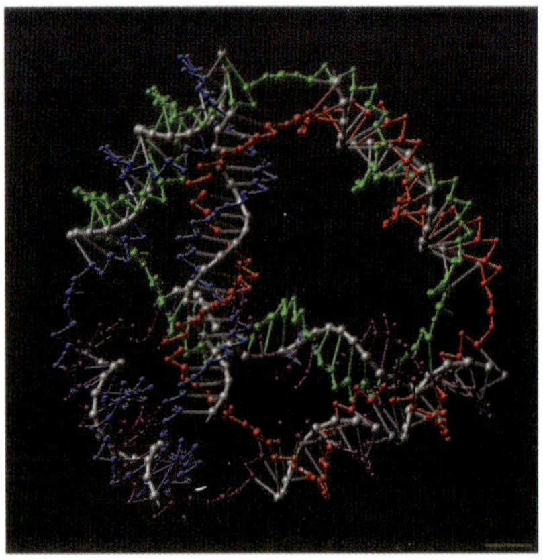

A geometrical model of a DNA tetrahedron described in Goodman. Models of this type are useful for ensuring that tertiary structure constraints do not cause excessive strain to the molecule.

Geometrical models of nucleic acids are used to predict tertiary structure. This is important because designed nucleic acid complexes usually contain multiple junction points, which introduces geometric constraints to the system. These constraints stem from the basic structure of nucleic acids, mainly that the double helix formed by nucleic acid duplexes has a fixed helicity of about 10.4 base pairs per turn, and is relatively stiff. Because of these constraints, the nucleic acid complexes are sensitive to the relative orientation of the major and minor grooves at junction points. Geometrical modeling can detect strain stemming from misalignments in the structure, which can then be corrected by the designer.

Geometric models of nucleic acids for DNA nanotechnology generally use reduced representations of the nucleic acid, because simulating every atom would be very computationally expensive for such large systems. Models with three pseudo-atoms per base pair, representing the two backbone sugars and the helix axis, have been reported to have a sufficient level of detail to predict experimental results. However, models with five pseudo-atoms per base pair, explicitly including the backbone phosphates, are also used.

Software for geometrical modeling of nucleic acids includes GIDEON, Tiamat, Nanoengineer-1, and UNIQUIMER 3D. Geometrical concerns are especially of interest in the design of DNA origami, because the sequence is predetermined by the choice of scaffold strand. Software specifically for DNA origami design has been made, including caDNAno and SARSE.

Applications

Nucleic acid design is used in DNA nanotechnology to design strands which will self-assemble into a desired target structure. These include examples such as DNA machines, periodic two- and three-dimensional lattices, polyhedra, and DNA origami. It can also be used to create sets of nucleic acid strands which are "orthogonal", or non-interacting with each other, so as to minimize or eliminate spurious interactions. This is useful in DNA computing, as well as for molecular barcoding applications in chemical biology and biotechnology.

Nucleic Acid Structure

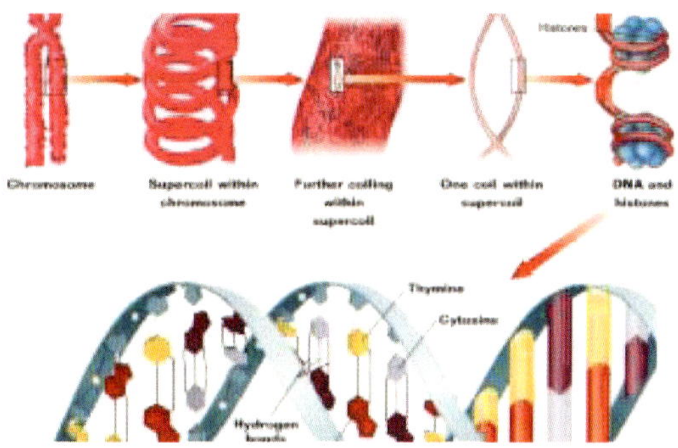

Interactive image of nucleic acid structure (primary, secondary, tertiary, and quaternary) using DNA helices and examples from the VS ribozyme and telomerase and nucleosome. (PDB: ADNA, 1BNA, 4OCB, 4R4V, 1YMO, 1EQZ)

Nucleic acid structure refers to the structure of nucleic acids such as DNA and RNA. Chemically speaking, DNA and RNA are very similar. Nucleic acid structure is often divided into four different levels: primary, secondary, tertiary and quaternary.

Primary Structure

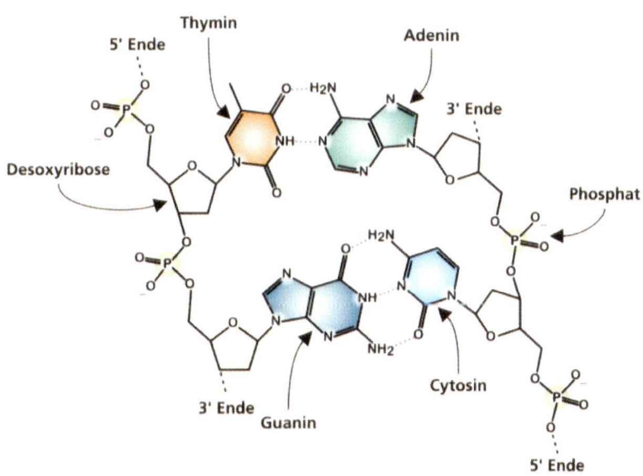

Chemical structure of DNA

Primary structure consists of a linear sequence of nucleotides that are linked together by phosphodiester bonds. It is this linear sequence of nucleotides that make up the Primary structure of DNA or RNA. Nucleotides consist of 3 components:

1. Nitrogenous base
    - Adenine
    - Guanine
    - Cytosine
    - Thymine (present in DNA only)
    - Uracil (present in RNA only)
2. 5-carbon sugar which is called deoxyribose (found in DNA) and ribose (found in RNA).
3. One or more phosphate groups.

The nitrogen bases adenine and guanine are purine in structure and form a glycosidic bond between their 9 nitrogen and the 1' -OH group of the deoxyribose. Cytosine, thymine and uracil are pyrimidines, hence the glycosidic bonds forms between their 1 nitrogen and the 1' -OH of the deoxyribose. For both the purine and pyrimidine bases, the phosphate group forms a bond with the deoxyribose sugar through an ester bond between one of its negatively charged oxygen groups and the 5' -OH of the sugar. The polarity in DNA and RNA is derived from the oxygen and nitrogen atoms in the backbone. Nucleic acids are formed when nucleotides come together through phosphodiester linkages between the 5' and 3' carbon atoms. A Nucleic acid sequence is the order of nucleotides within a DNA (GACT) or RNA (GACU) molecule that is determined by a series of letters. Sequences are presented from the 5' to 3' end and determine the covalent structure of the entire molecule. Sequences can be complementary to another sequence in that the base on each position is complementary as well as in the reverse order. An example of a complementary sequence to AGCT is TCGA. DNA is double-stranded containing both a sense strand and an antisense strand. Therefore, the complementary sequence will be to the sense strand.

Complexes with Alkali Metal Ions

There are three potential metal binding groups on nucleic acids: phosphate, sugar and base moieties. Solid-state structure of complexes with alkali metal ions have been reviewed.

Secondary Structure

Secondary structure is the set of interactions between bases, i.e., which parts of strands are bound to each other. In DNA double helix, the two strands of DNA are held together by hydrogen bonds. The nucleotides on one strand base pairs with the nucleotide on the other strand. The secondary structure is responsible for the shape that the nucleic acid assumes. The bases in the DNA are classified as purines and pyrimidines. The purines are adenine and guanine. Purines consist of a double ring structure, a six membered and a five membered ring containing nitrogen. The pyrimidines are cytosine and thymine. It has a single ringed structure, a six membered ring containing nitrogen. A purine base always pairs with a pyrimidine base (guanine (G) pairs with cytosine (C) and adenine (A) pairs with thymine (T) or uracil (U)). DNA's secondary structure is predominantly determined by base-pairing of the two polynucleotide strands wrapped around each other to form a double helix. Although the two strands are aligned by hydrogen bonds in base pairs, the stronger forces holding the two strands together are stacking interactions between the bases. These stacking interactions are stabilized by Van der Waals forces and hydrophobic interactions, and show a large amount of local structural variability. There are also two grooves in the double helix, which are called major groove and minor groove based on their relative size.

The secondary structure of RNA consists of a single polynucleotide. Base pairing in RNA occurs when RNA folds between complementarity regions. Both single- and double-stranded regions are often found in RNA molecules. The antiparallel strands form a helical shape. The four basic elements in the secondary structure of RNA are helices, loops, bulges, and junctions. Stem-loop or hairpin loop is the most common element of RNA secondary structure. Stem-loop is formed when the RNA chains fold back on themselves to form a double helical tract called the stem, the unpaired nucleotides forms single stranded region called the loop. Secondary structure of RNA can be predicted by experimental data on the secondary structure elements, helices, loops and bulges. Bulges and internal loops are formed by separation of the double helical tract on either one strand (bulge) or on both strands (internal loops) by unpaired nucleotides. A tetraloop is a four-base pairs hairpin RNA structure. There are three common families of tetraloop in ribosomal RNA: UNCG, GNRA, and CUUG (N is one of the four nucleotides and R is a purine).UNCG is the most stable tetraloop. Pseudoknot is a RNA secondary structure first identified in turnip yellow mosaic virus. Pseudoknots are formed when nucleotides from the hairpin loop pairs with a single stranded region outside of the hairpin to form a helical segment. H-type fold pseudoknots are best characterized. In H-type fold, nucleotides in the hairpin loop pairs with the bases outside the hairpin stem forming second stem and loop. This causes formation of pseudoknots with two stems and two loops. Pseudoknots are functional elements in RNA structure having diverse function and found in most classes of RNA. DotKnot-PW method is used for comparative pseudoknots prediction. The main points in the DotKnot-PW method is scoring the similarities found in stems, secondary elements and H-type pseudoknots.

Tertiary Structure

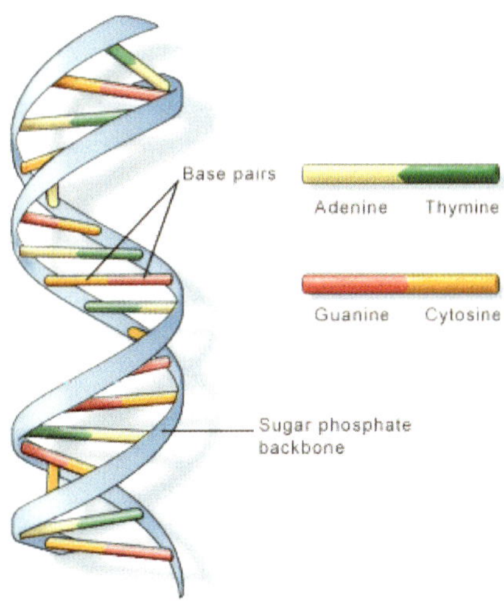

DNA structure and bases

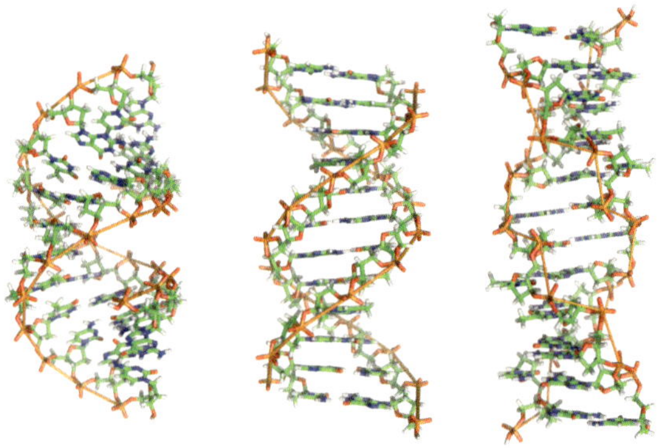

A-B-Z-DNA Side View

Tertiary structure refers to the locations of the atoms in three-dimensional space, taking into consideration geometrical and steric constraints. It is a higher order than the secondary structure, in which large-scale folding in a linear polymer occurs and the entire chain is folded into a specific 3-dimensional shape. There are 4 areas in which the structural forms of DNA can differ.

1. Handedness – right or left
2. Length of the helix turn
3. Number of base pairs per turn
4. Difference in size between the major and minor grooves

The tertiary arrangement of DNA's double helix in space includes B-DNA, A-DNA and Z-DNA.

B-DNA is the most common form of DNA in vivo and is a more narrow, elongated helix than A-DNA. Its wide major groove makes it more accessible to proteins. On the other hand, it has a narrow minor groove. B-DNA's favored conformations occur at high water concentrations; the hydration of the minor groove appears to favor B-DNA. B-DNA base pairs are nearly perpendicular to the helix axis. The sugar pucker which determines the shape of the a-helix, whether the helix will exist in the A-form or in the B-form, occurs at the C2'-endo.

A-DNA, is a form of the DNA duplex observed under dehydrating conditions. It is shorter and wider than B-DNA. RNA adopts this double helical form, and RNA-DNA duplexes are mostly A-form, but B-form RNA-DNA duplexes have been observed. In localized single strand dinucleotide contexts, RNA can also adopt the B-form without pairing to DNA. A-DNA has a deep, narrow major groove which does not make it easily accessible to proteins. On the other hand, its wide, shallow minor groove makes it accessible to proteins but with lower information content than the major groove. Its favored conformation is at low water concentrations. A-DNAs base pairs are tilted relative to the helix axis, and are displaced from the axis. The sugar pucker occurs at the C3'-endo and in RNA 2'-OH inhibits C2'-endo conformation. Long considered little more than a laboratory artifice, A-DNA is now known to have several biological functions.

Z-DNA is a relatively rare left-handed double-helix. Given the proper sequence and superhelical tension, it can be formed in vivo but its function is unclear. It has a more narrow, more elongated helix than A or B. Z-DNA's major groove is not really a groove, and it has a narrow minor groove.

The most favored conformation occurs when there are high salt concentrations. There are some base substitutions but they require an alternating purine-pyrimidine sequence. The N2-amino of G H-bonds to 5' PO, which explains the slow exchange of protons and the need for the G purine. Z-DNA base pairs are nearly perpendicular to the helix axis. Z-DNA does not contain single base-pairs but rather a GpC repeat with P-P distances varying for GpC and CpG. On the GpC stack there is good base overlap, whereas on the CpG stack there is less overlap. Z-DNA's zigzag backbone is due to the C sugar conformation compensating for G glycosidic bond conformation. The conformation of G is syn, C2'-endo and for C it is anti, C3'-endo.

A linear DNA molecule having free ends can rotate, to adjust to changes of various dynamic processes in the cell, by changing how many times the two chains of its double helix twist around each other. Some DNA molecules are circular and are topologically constrained. More recently circular RNA was described as well to be a natural pervasive class of nucleic acids, expressed in many organisms.

A covalently closed, circular DNA (also known as cccDNA) is topologically constrained as the number of times the chains coiled around one other cannot change. This cccDNA can be supercoiled, which is the tertiary structure of DNA. Supercoiling is characterized by the linking number, twist and writhe. The linking number (Lk) for circular DNA is defined as the number of times one strand would have to pass through the other strand to completely separate the two strands. The linking number for circular DNA can only be changed by breaking of a covalent bond in one of the two strands. Always an integer, the linking number of a cccDNA is the sum of two components: twists (Tw) and writhes (Wr).

$$Lk = Tw + Wr$$

Twists are the number of times the two strands of DNA are twisted around each other. Writhes are number of times the DNA helix crosses over itself. DNA in cells is negatively supercoiled and has the tendency to unwind. Hence the separation of strands is easier in negatively supercoiled DNA than in relaxed DNA. The two components of supercoiled DNA are solenoid and plectonemic. The plectonemic supercoil is found in prokaryotes, while the solenoidal supercoiling is mostly seen in eukaryotes.

Quaternary Structure

The quaternary structure of nucleic acids is similar to that of protein quaternary structure. Although some of the concepts are not exactly the same, the quaternary structure refers to a higher-level of organization of nucleic acids. Moreover, it refers to interactions of the nucleic acids with other molecules. The most commonly seen form of higher-level organization of nucleic acids is seen in the form of chromatin which leads to its interactions with the small proteins histones. Also, the quaternary structure refers to the interactions between separate RNA units in the ribosome or spliceosome.

Nucleic Acid Sequence

A nucleic acid sequence is a succession of letters that indicate the order of nucleotides within a DNA (using GACT) or RNA (GACU) molecule. By convention, sequences are usually presented from the 5' end to the 3' end. For DNA, the sense strand is used. Because nucleic acids are normally linear (unbranched) polymers, specifying the sequence is equivalent to defining the covalent structure of the entire molecule. For this reason, the nucleic acid sequence is also termed the primary structure.

Interactive image of nucleic acid structure (primary, secondary, tertiary, and quaternary) using DNA helices and examples from the VS ribozyme and telomerase and nucleosome. (PDB: ADNA, 1BNA, 4OCB, 4R4V, 1YMO, 1EQZ)

The sequence has capacity to represent information. Biological deoxyribonucleic acid represents the information which directs the functions of a living thing.

Nucleic acids also have a secondary structure and tertiary structure. Primary structure is sometimes mistakenly referred to as *primary sequence*. Conversely, there is no parallel concept of secondary or tertiary sequence.

Nucleotides

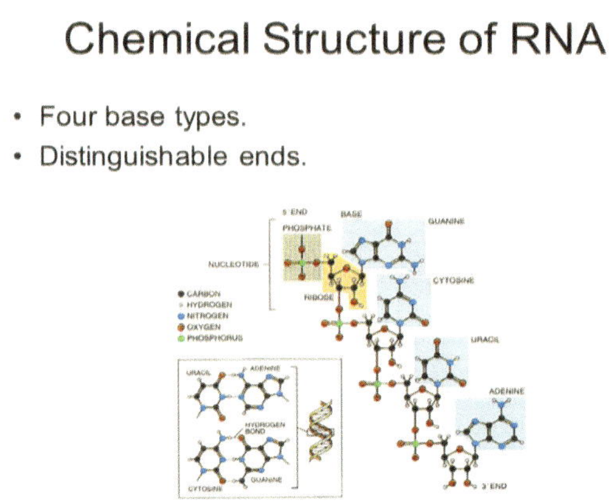

Chemical structure of RNA

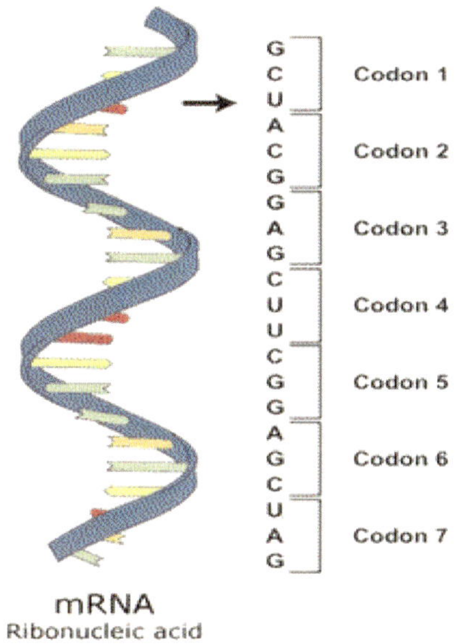

Image adapted from: National Human Genome Research Institute. Talking Glossary of Genetic Terms. Available at: www.genome.gov/ Pages/Hyperion/DIR/VIP/Glossary/Illustration/codon.shtml.

A series of codons in part of a mRNA molecule. Each codon consists of three nucleotides, usually representing a single amino acid.

Nucleic acids consist of a chain of linked units called nucleotides. Each nucleotide consists of three subunits: a phosphate group and a sugar (ribose in the case of RNA, deoxyribose in DNA) make up the backbone of the nucleic acid strand, and attached to the sugar is one of a set of nucleobases. The nucleobases are important in base pairing of strands to form higher-level secondary and tertiary structure such as the famed double helix.

The possible letters are *A*, *C*, *G*, and *T*, representing the four nucleotide bases of a DNA strand — adenine, cytosine, guanine, thymine — covalently linked to a phosphodiester backbone. In the typical case, the sequences are printed abutting one another without gaps, as in the sequence AAAGTCTGAC, read left to right in the 5' to 3' direction. With regards to transcription, a sequence is on the coding strand if it has the same order as the transcribed RNA.

One sequence can be complementary to another sequence, meaning that they have the base on each position in the complementary (i.e. A to T, C to G) and in the reverse order. For example, the complementary sequence to TTAC is GTAA. If one strand of the double-stranded DNA is considered the sense strand, then the other strand, considered the antisense strand, will have the complementary sequence to the sense strand.

Notation

Comparing and determining % difference between two nucleotide sequences.

- AATCCGCTAG
- AAACCCTTAG
- Given the two 10-nucleotide sequences, line them up and compare the differences between them. Calculate the percent similarity by taking the number of different DNA bases divided

by the total number of nucleotides. In the above case, there are three differences in the 10 nucleotide sequence. Therefore, divide 7/10 to get the 70% similarity and subtract that from 100% to get a 30% difference.

While A, T, C, and G represent a particular nucleotide at a position, there are also letters that represent ambiguity which are used when more than one kind of nucleotide could occur at that position. The rules of the International Union of Pure and Applied Chemistry (IUPAC) are as follows:

- A = adenine
- C = cytosine
- G = guanine
- T = thymine
- R = G A (purine)
- Y = T C (pyrimidine)
- K = G T (keto)
- M = A C (amino)
- S = G C (strong bonds)
- W = A T (weak bonds)
- B = G T C (all but A)
- D = G A T (all but C)
- H = A C T (all but G)
- V = G C A (all but T)
- N = A G C T (any)

These symbols are also valid for RNA, except with U (uracil) replacing T (thymine).

Apart from adenine (A), cytosine (C), guanine (G), thymine (T) and uracil (U), DNA and RNA also contain bases that have been modified after the nucleic acid chain has been formed. In DNA, the most common modified base is 5-methylcytidine (m5C). In RNA, there are many modified bases, including pseudouridine (Ψ), dihydrouridine (D), inosine (I), ribothymidine (rT) and 7-methylguanosine (m7G). Hypoxanthine and xanthine are two of the many bases created through mutagen presence, both of them through deamination (replacement of the amine-group with a carbonyl-group). Hypoxanthine is produced from adenine, xanthine from guanine. Similarly, deamination of cytosine results in uracil.

Biological Significance

In biological systems, nucleic acids contain information which is used by a living cell to construct specific proteins. The sequence of nucleobases on a nucleic acid strand is translated by cell machinery into a sequence of amino acids making up a protein strand. Each group of three bases, called a codon, corresponds to a single amino acid, and there is a specific genetic code by which each possible combination of three bases corresponds to a specific amino acid.

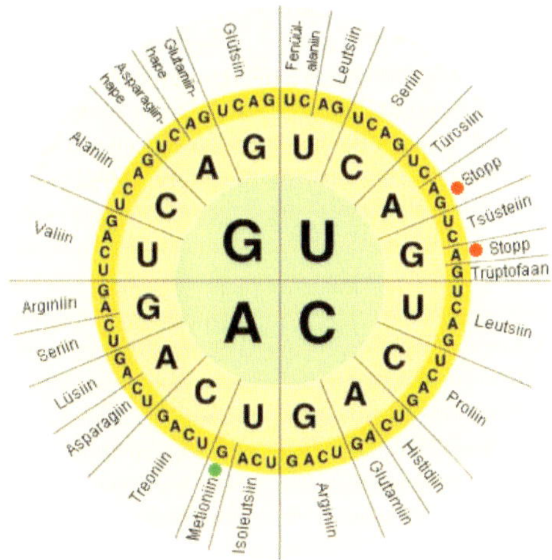

A depiction of the genetic code, by which the information contained in nucleic acids are translated into amino acid sequences in proteins.

The central dogma of molecular biology outlines the mechanism by which proteins are constructed using information contained in nucleic acids. DNA is transcribed into mRNA molecules, which travels to the ribosome where the mRNA is used as a template for the construction of the protein strand. Since nucleic acids can bind to molecules with complementary sequences, there is a distinction between "sense" sequences which code for proteins, and the complementary "antisense" sequence which is by itself nonfunctional, but can bind to the sense strand.

Sequence Determination

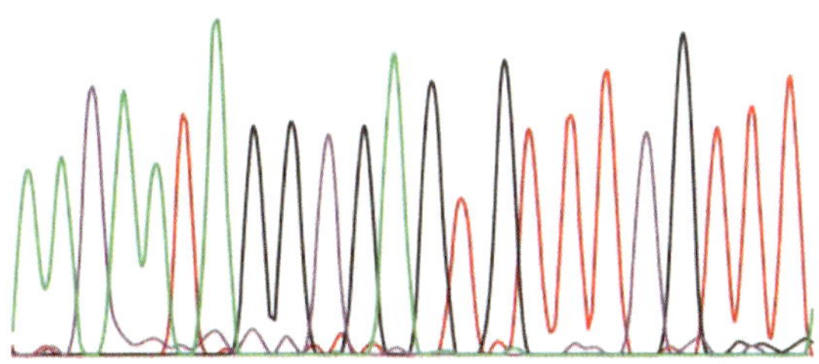

Electropherogram printout from automated sequencer for determining part of a DNA sequence

DNA sequencing is the process of determining the nucleotide sequence of a given DNA fragment. The sequence of the DNA of a living thing encodes the necessary information for that living thing to survive and reproduce. Therefore, determining the sequence is useful in fundamental research into why and how organisms live, as well as in applied subjects. Because of the importance of DNA to living things, knowledge of a DNA sequence may be useful in practically any biological research. For example, in medicine it can be used to identify, diagnose and potentially develop treatments for genetic diseases. Similarly, research into pathogens may lead to treatments for contagious diseases. Biotechnology is a burgeoning discipline, with the potential for many useful products and services.

RNA is not sequenced directly. Instead, it is copied to a DNA by reverse transcriptase, and this DNA is then sequenced.

Current sequencing methods rely on the discriminatory ability of DNA polymerases, and therefore can only distinguish four bases. An inosine (created from adenosine during RNA editing) is read as a G, and 5-methyl-cytosine (created from cytosine by DNA methylation) is read as a C. With current technology, it is difficult to sequence small amounts of DNA, as the signal is too weak to measure. This is overcome by polymerase chain reaction (PCR) amplification.

Digital Representation

Once a nucleic acid sequence has been obtained from an organism, it is stored *in silico* in digital format. Digital genetic sequences may be stored in sequence databases, be analyzed, be digitally altered and be used as templates for creating new actual DNA using artificial gene synthesis.

Sequence Analysis

Digital genetic sequences may be analyzed using the tools of bioinformatics to attempt to determine its function.

Genetic Testing

The DNA in an organism's genome can be analyzed to diagnose vulnerabilities to inherited diseases, and can also be used to determine a child's paternity (genetic father) or a person's ancestry. Normally, every person carries two variations of every gene, one inherited from their mother, the other inherited from their father. The human genome is believed to contain around 20,000 - 25,000 genes. In addition to studying chromosomes to the level of individual genes, genetic testing in a broader sense includes biochemical tests for the possible presence of genetic diseases, or mutant forms of genes associated with increased risk of developing genetic disorders.

Genetic testing identifies changes in chromosomes, genes, or proteins. Usually, testing is used to find changes that are associated with inherited disorders. The results of a genetic test can confirm or rule out a suspected genetic condition or help determine a person's chance of developing or passing on a genetic disorder. Several hundred genetic tests are currently in use, and more are being developed.

Sequence Alignment

In bioinformatics, a sequence alignment is a way of arranging the sequences of DNA, RNA, or protein to identify regions of similarity that may be due to functional, structural, or evolutionary relationships between the sequences. If two sequences in an alignment share a common ancestor, mismatches can be interpreted as point mutations and gaps as insertion or deletion mutations (indels) introduced in one or both lineages in the time since they diverged from one another. In sequence alignments of proteins, the degree of similarity between amino acids occupying a particular position in the sequence can be interpreted as a rough measure of how conserved a particular region or sequence motif is among lineages. The absence of substitutions, or the presence of only very conservative substitutions (that is, the substitution of amino acids whose side chains have similar biochemical properties) in a particular region of the sequence, suggest that this region has structural or functional importance. Although DNA and RNA nucleotide bases are more similar to each other than are amino acids, the conservation of base pairs can indicate a similar functional or structural role.

Computational phylogenetics makes extensive use of sequence alignments in the construction and interpretation of phylogenetic trees, which are used to classify the evolutionary relationships between homologous genes represented in the genomes of divergent species. The degree to which sequences in a query set differ is qualitatively related to the sequences' evolutionary distance from one another. Roughly speaking, high sequence identity suggests that the sequences in question have a comparatively young most recent common ancestor, while low identity suggests that the divergence is more ancient. This approximation, which reflects the "molecular clock" hypothesis that a roughly constant rate of evolutionary change can be used to extrapolate the elapsed time since two genes first diverged (that is, the coalescence time), assumes that the effects of mutation and selection are constant across sequence lineages. Therefore, it does not account for possible difference among organisms or species in the rates of DNA repair or the possible functional conservation of specific regions in a sequence. (In the case of nucleotide sequences, the molecular clock hypothesis in its most basic form also discounts the difference in acceptance rates between silent mutations that do not alter the meaning of a given codon and other mutations that result in a different amino acid being incorporated into the protein.) More statistically accurate methods allow the evolutionary rate on each branch of the phylogenetic tree to vary, thus producing better estimates of coalescence times for genes.

Sequence Motifs

Frequently the primary structure encodes motifs that are of functional importance. Some examples of sequence motifs are: the C/D and H/ACA boxes of snoRNAs, Sm binding site found in spliceosomal RNAs such as U1, U2, U4, U5, U6, U12 and U3, the Shine-Dalgarno sequence, the Kozak consensus sequence and the RNA polymerase III terminator.

Long Range Correlations

Peng found the existence of long-range correlations in the non-coding base pair sequences of DNA. In contrast, such correlations seem not to appear in coding DNA sequences.

Sequence Entropy

In Bioinformatics, a sequence entropy, also known as sequence complexity or information profile, is a numerical sequence providing a quantitative measure of the local complexity of a DNA sequence, independently of the direction of processing. The manipulations of the information profiles enable the analysis of the sequences using alignment-free techniques, such as for example in motif and rearrangements detection.

Nucleic Acid Secondary Structure

Nucleic acid secondary structure is the basepairing interactions within a single nucleic acid polymer or between two polymers. It can be represented as a list of bases which are paired in a nucleic acid molecule. The secondary structures of biological DNA's and RNA's tend to be different:

biological DNA mostly exists as fully base paired double helices, while biological RNA is single stranded and often forms complex and intricate base-pairing interactions due to its increased ability to form hydrogen bonds stemming from the extra hydroxyl group in the ribose sugar.

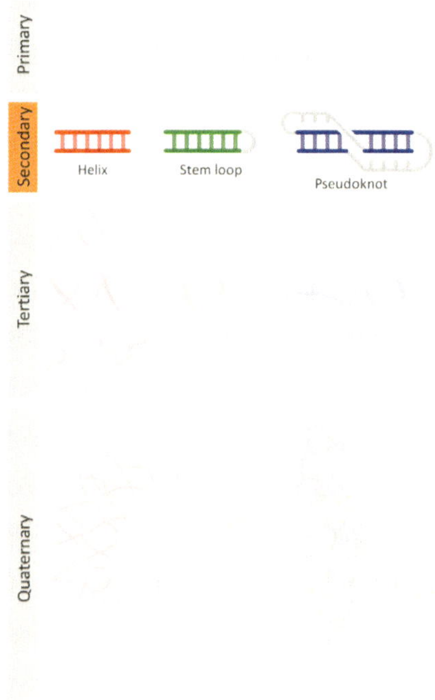

Interactive image of nucleic acid structure (primary, secondary, tertiary, and quaternary) using DNA helices and examples from the VS ribozyme and telomerase and nucleosome. (PDB: ADNA, 1BNA, 4OCB, 4R4V, 1YMO, 1EQZ)

In a non-biological context, secondary structure is a vital consideration in the nucleic acid design of nucleic acid structures for DNA nanotechnology and DNA computing, since the pattern of base-pairing ultimately determines the overall structure of the molecules.

Fundamental Concepts

Base Pairing

In molecular biology, two nucleotides on opposite complementary DNA or RNA strands that are connected via hydrogen bonds are called a base pair (often abbreviated bp). In the canonical Watson-Crick base pairing, adenine (A) forms a base pair with thymine (T) and guanine (G) forms one with cytosine (C) in DNA. In RNA, thymine is replaced by uracil (U). Alternate hydrogen bonding patterns, such as the wobble base pair and Hoogsteen base pair, also occur—particularly in RNA—giving rise to complex and functional tertiary structures. Importantly, pairing is the mechanism by which codons on messenger RNA molecules are recognized by anticodons on transfer RNA during protein translation. Some DNA- or RNA-binding enzymes can recognize specific base pairing patterns that identify particular regulatory regions of genes. Hydrogen bonding is the chemical mechanism that underlies the base-pairing rules described above. Appropriate geometrical correspondence of hydrogen bond donors and acceptors allows only the "right" pairs to form stably. DNA with high GC-content is more stable than DNA with low GC-content, but contrary to popular belief, the hydrogen bonds do not stabilize the DNA significantly and stabilization is mainly due to stacking interactions.

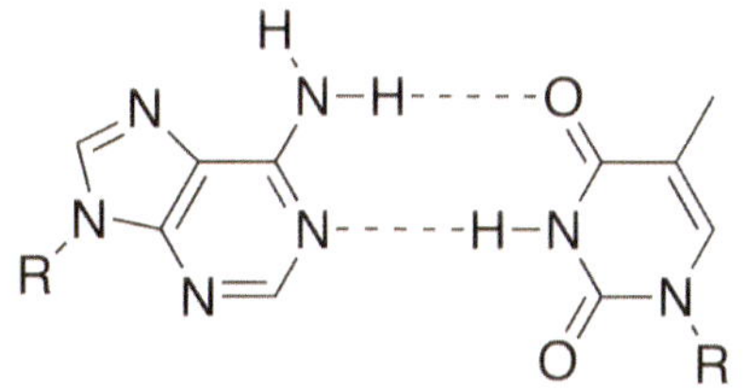

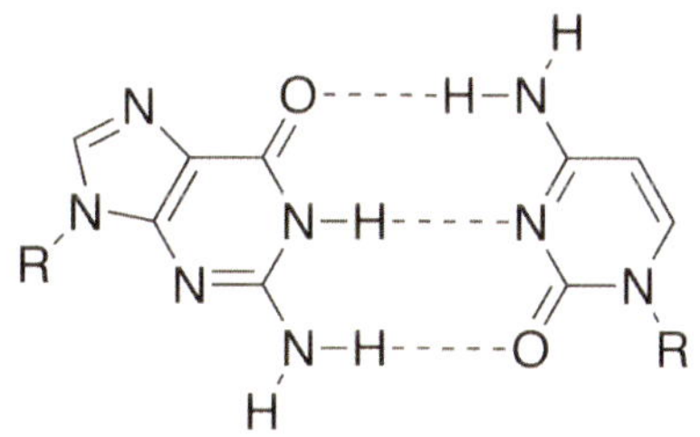

Top, an AT base pair demonstrating two intermolecular hydrogen bonds; *bottom*, a GC base pair demonstrating three intermolecular hydrogen bonds.

The larger nucleobases, adenine and guanine, are members of a class of doubly ringed chemical structures called purines; the smaller nucleobases, cytosine and thymine (and uracil), are members of a class of singly ringed chemical structures called pyrimidines. Purines are only complementary with pyrimidines: pyrimidine-pyrimidine pairings are energetically unfavorable because the molecules are too far apart for hydrogen bonding to be established; purine-purine pairings are energetically unfavorable because the molecules are too close, leading to overlap repulsion. The only other possible pairings are GT and AC; these pairings are mismatches because the pattern of hydrogen donors and acceptors do not correspond. The GU wobble base pair, with two hydrogen bonds, does occur fairly often in RNA.

Nucleic Acid Hybridization

Hybridization is the process of complementary base pairs binding to form a double helix. Melting is the process by which the interactions between the strands of the double helix are broken, separating the two nucleic acid strands. These bonds are weak, easily separated by gentle heating, enzymes, or physical force. Melting occurs preferentially at certain points in the nucleic acid. T and A rich sequences are more easily melted than C and G rich regions. Particular base steps are also susceptible to DNA melting, particularly T A and T G base steps. These mechanical features are reflected by the use of sequences such as TATAA at the start of many genes to assist RNA polymerase in melting the DNA for transcription.

Strand separation by gentle heating, as used in PCR, is simple providing the molecules have fewer than about 10,000 base pairs (10 kilobase pairs, or 10 kbp). The intertwining of the DNA strands makes long segments difficult to separate. The cell avoids this problem by allowing its DNA-melting enzymes (helicases) to work concurrently with topoisomerases, which can chemically cleave the phosphate backbone of one of the strands so that it can swivel around the other. Helicases unwind the strands to facilitate the advance of sequence-reading enzymes such as DNA polymerase.

Secondary Structure Motifs

Nucleic acid secondary structure is generally divided into helices (contiguous base pairs), and various kinds of loops (unpaired nucleotides surrounded by helices). Frequently these elements, or combinations of them, are further classified into additional categories including, for example, tetraloops, pseudoknots, and stem-loops.

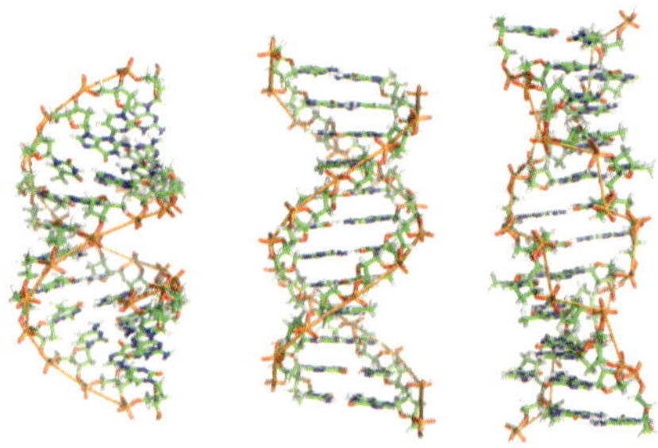

The main nucleic acid helix structures (A-, B- and Z-form)

Double Helix

The double helix is an important tertiary structure in nucleic acid molecules which is intimately connected with the molecule's secondary structure. A double helix is formed by regions of many consecutive base pairs.

The nucleic acid double helix is a spiral polymer, usually right-handed, containing two nucleotide strands which base pair together. A single turn of the helix constitutes about ten nucleotides, and contains a major groove and minor groove, the major groove being wider than the minor groove. Given the difference in widths of the major groove and minor groove, many proteins which bind to DNA do so through the wider major groove. Many double-helical forms are possible; for DNA the three biologically relevant forms are A-DNA, B-DNA, and Z-DNA, while RNA double helices have structures similar to the A form of DNA.

Stem-loop Structures

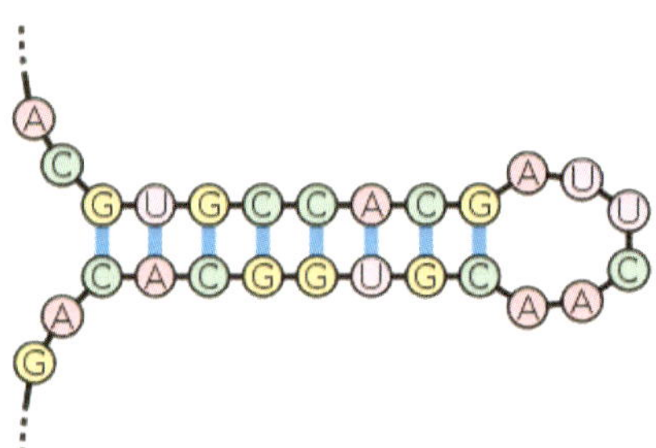

An RNA stem-loop secondary structure

The secondary structure of nucleic acid molecules can often be uniquely decomposed into stems and loops. The stem-loop structure (also often referred to as an "hairpin"), in which a base-paired helix ends in a short unpaired loop, is extremely common and is a building block for larger structural motifs such as cloverleaf structures, which are four-helix junctions such as those found in transfer RNA. Internal loops (a short series of unpaired bases in a longer paired helix) and bulges (regions in which one strand of a helix has "extra" inserted bases with no counterparts in the opposite strand) are also frequent.

There are many secondary structure elements of functional importance to biological RNA's; some famous examples are the Rho-independent terminator stem-loops and the tRNA cloverleaf. Active research is on-going to determine the secondary structure of RNA molecules, with approaches including both experimental and computational methods

Pseudoknots

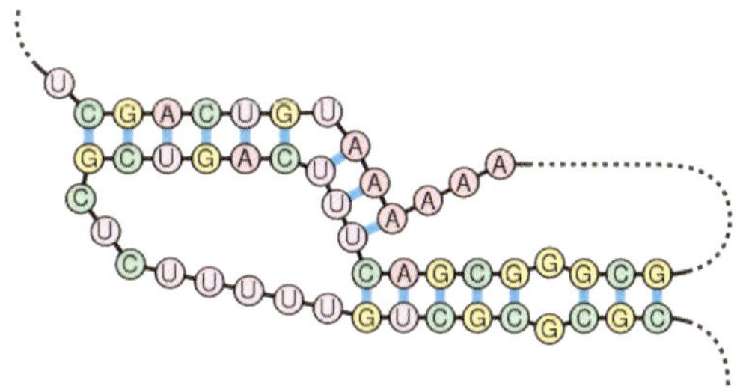

An RNA pseudoknot structure. For example, the RNA component of human telomerase.

A pseudoknot is a nucleic acid secondary structure containing at least two stem-loop structures in which half of one stem is intercalated between the two halves of another stem. Pseudoknots fold into knot-shaped three-dimensional conformations but are not true topological knots. The base pairing in pseudoknots is not well nested; that is, base pairs occur that "overlap" one another in sequence position. This makes the presence of general pseudoknots in nucleic acid sequences impossible to predict by the standard method of dynamic programming, which uses a recursive scoring system to identify paired stems and consequently cannot detect non-nested base pairs with common algorithms. However, limited subclasses of pseudoknots can be predicted using modified dynamic programs. Newer structure prediction techniques such as stochastic context-free grammars are also unable to consider pseudoknots.

Pseudoknots can form a variety of structures with catalytic activity and several important biological processes rely on RNA molecules that form pseudoknots. For example, the RNA component of the human telomerase contains a pseudoknot that is critical for its activity. The hepatitis delta virus ribozyme is a well known example of a catalytic RNA with a pseudoknot in its active site. Though DNA can also form pseudoknots, they are generally not present in standard physiological conditions.

Secondary Structure Prediction

Most methods for nucleic acid secondary structure prediction rely on a nearest neighbor thermodynamic model. A common method do determine the most probable structures given a sequence of nucleotides makes use of a dynamic programming algorithm that seeks to find structures with low free energy. Dynamic programming algorithms often forbid pseudoknots, or other cases in which base pairs are not fully nested, as considering these structures becomes computationally very expensive for even small nucleic acid molecules. Other methods, such as stochastic context-free grammars can also be used to predict nucleic acid secondary structure.

For many RNA molecules, the secondary structure is highly important to the correct function of the RNA — often more so than the actual sequence. This fact aids in the analysis of non-coding RNA sometimes termed "RNA genes". One application of bioinformatics uses predicted RNA secondary structures in searching a genome for noncoding but functional forms of RNA. For example, microRNAs have canonical long stem-loop structures interrupted by small internal loops.

RNA secondary structure applies in RNA splicing in certain species. In humans and other tetrapods, it has been shown that without the U2AF2 protein, the splicing process is inhibited. However, in zebrafish and other teleosts the RNA splicing process can still occur on certain genes in the absence of U2AF2. This may be because 10% of genes in zebrafish have alternating TG and AC base pairs at the 3' splice site (3'ss) and 5' splice site (5'ss) respectively on each intron, which alters the secondary structure of the RNA. This suggests that secondary structure of RNA can influence splicing, potentially without the use of proteins like U2AF2 that have been thought to be required for splicing to occur.

Nucleic Acid Tertiary Structure

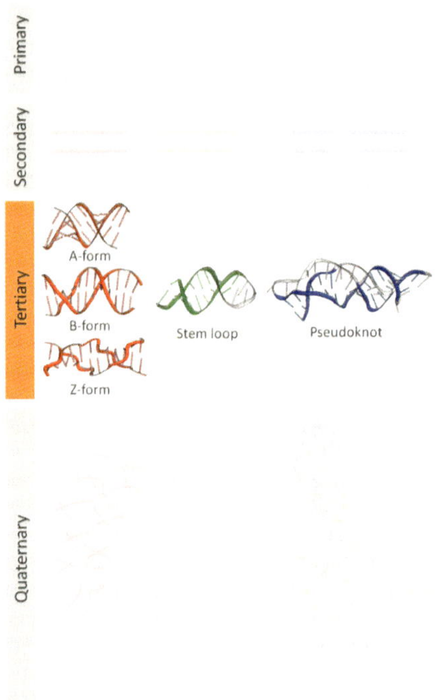

Interactive image of nucleic acid structure (primary, secondary, tertiary, and quaternary) using DNA helices and examples from the VS ribozyme and telomerase and nucleosome. (PDB: ADNA, 1BNA, 4OCB, 4R4V, 1YMO, 1EQZ)

Nucleic acid tertiary structure is the three-dimensional shape of a nucleic acid polymer. RNA and DNA molecules are capable of diverse functions ranging from molecular recognition to catalysis. Such functions require a precise three-dimensional tertiary structure. While such structures are diverse and seemingly complex, they are composed of recurring, easily recognizable tertiary structure motifs that serve as molecular building blocks. Some of the most common motifs for RNA and DNA tertiary structure are described below, but this information is based on a limited number of solved structures. Many more tertiary structural motifs will be revealed as new RNA and DNA molecules are structurally characterized.

Helical Structures

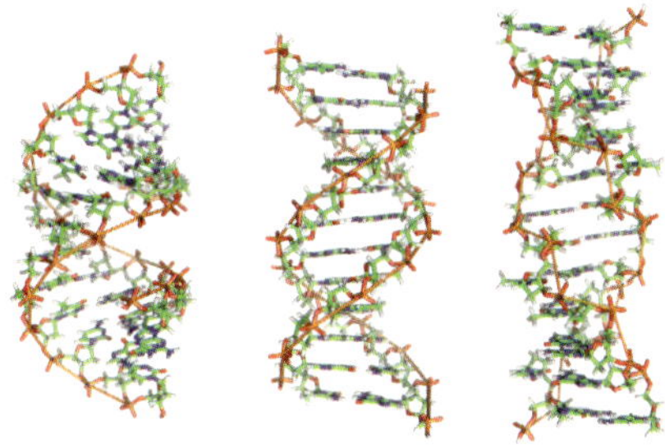

The structures of the A-, B-, and Z-DNA double helix structures.

Double Helix

The double helix is the dominant tertiary structure for biological DNA, and is also a possible structure for RNA. Three DNA conformations are believed to be found in nature, A-DNA, B-DNA, and Z-DNA. The "B" form described by James D. Watson and Francis Crick is believed to predominate in cells. James D. Watson and Francis Crick described this structure as a double helix with a radius of 10 Å and pitch of 34 Å, making one complete turn about its axis every 10 bp of sequence. The double helix makes one complete turn about its axis every 10.4–10.5 base pairs in solution. This frequency of twist (known as the helical *pitch*) depends largely on stacking forces that each base exerts on its neighbours in the chain. Double-helical RNA adopts a conformation similar to the A-form structure.

Other conformations are possible; in fact, only the letters F, Q, U, V, and Y are now available to describe any new DNA structure that may appear in the future. However, most of these forms have been created synthetically and have not been observed in naturally occurring biological systems.

RNA Triplexes

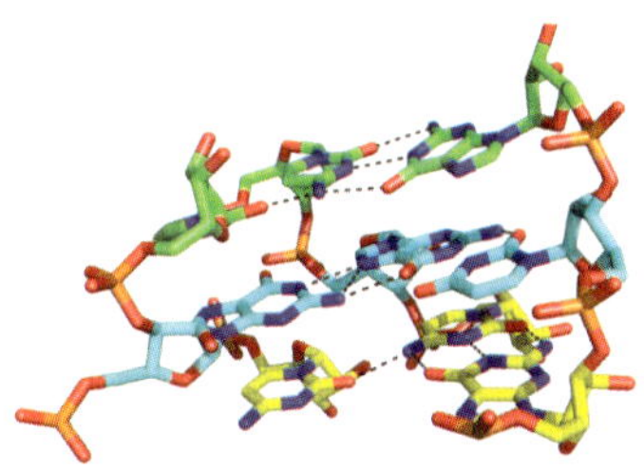

Major groove triples in the group II intron in *Oceanobacillus Iheyensis*. Each stacked layer is formed by one triplex with a different color scheme. Hydrogen bonds between triplexes are shown in black dashed lines. "N" atoms are colored in blue and "O" atoms in red. From top to bottom, the residues on the left side are G288, C289, and C377.

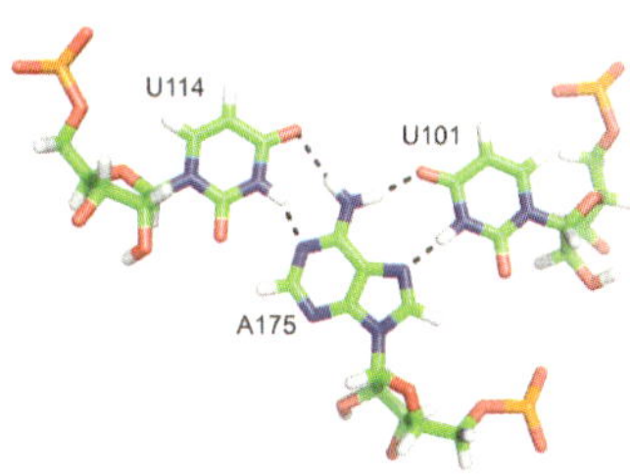

Close-up rendering of the U114:A175-U101 major groove (Hoogsteen base) triplex formed within the wild type pseudoknot of Human Telomerase RNA. Hydrogen bonds are shown in black dashed lines. "N" atoms are colored in blue and "o" atoms in red.

Major and Minor Groove Triplexes

The minor groove triple is a ubiquitous RNA structural motif. Because interactions with the minor groove are often mediated by the 2'-OH of the ribose sugar, this RNA motif looks very different from its DNA equivalent. The most common example of a minor loop triple is the A-minor motif, or the insertion of adenosine bases into the minor groove. However, this motif is not restricted to adenosines, as other nucleobases have also been observed to interact with the RNA minor groove.

The minor groove presents a near-perfect complement for an inserted base. This allows for optimal van der Waals contacts, extensive hydrogen bonding and hydrophobic surface burial, and creates a highly energetically favorable interaction. Because minor groove triples are capable of stably packing a free loop and helix, they are key elements in the structure of large ribonucleotides, including the group I intron, the group II intron, and the ribosome.

Although the major groove of standard A-form RNA is fairly narrow and therefore less available for triplex interaction than the minor groove, major groove triplex interactions can be observed in several RNA structures. These structures consist of several combinations of base pair and Hoogsteen interactions. For example, the GGC triplex (GGC amino(N-2)-N-7, imino-carbonyl, carbonyl-amino(N-4); Watson-Crick) observed in the 50S ribosome, composed of a Watson-Crick type G-C pair and an incoming G which forms a pseudo-Hoogsteen network of hydrogen bonding interactions between both bases involved in the canonical pairing. Other notable examples of major groove triplexes include (i) the catalytic core of the group II intron shown in the figure at left (ii) a catalytically essential triple helix observed in human telomerase RNA and (iii) the SAM-II riboswitch.

Triple-stranded DNA is also possible from Hoogsteen or reversed Hoogsteen hydrogen bonds in the major groove of B-form DNA.

Quadruplexes

Besides double helices and the above-mentioned triplexes, RNA and DNA can both also form quadruple helices. There are diverse structures of RNA base quadruplexes. Four consecutive guanine residues can form a quadruplex in RNA by Hoogsteen hydrogen bonds to form a "Hoogsteen ring". G-C and A-U pairs can also form base quadruplex with a combination of Watson-Crick pairing and noncanonical pairing in the minor groove.

The core of malachite green aptamer is also a kind of base quadruplex with a different hydrogen bonding pattern. The quadruplex can repeat several times consecutively, producing an immensely stable structure.

The unique structure of quadruplex regions in RNA may serve different functions in a biological system. Two important functions are the binding potential with ligands or proteins, and its ability to stabilize the whole tertiary structure of DNA or RNA. The strong structure can inhibit or modulate transcription and replication, such as in the telomeres of chromosomes and the UTR of mRNA. The base identity is important towards ligand binding. The G-quartet typically binds monovalent cations such as potassium, while other bases can bind numerous other ligands such as hypoxanthine in a U-U-C-U quadruplex.

Along with these functions, the G-quadruplex in the mRNA around the ribosome binding regions could serve as a regulator of gene expression in bacteria. There may be more interesting structures and functions yet to be discovered *in vivo*.

Quadruplexes

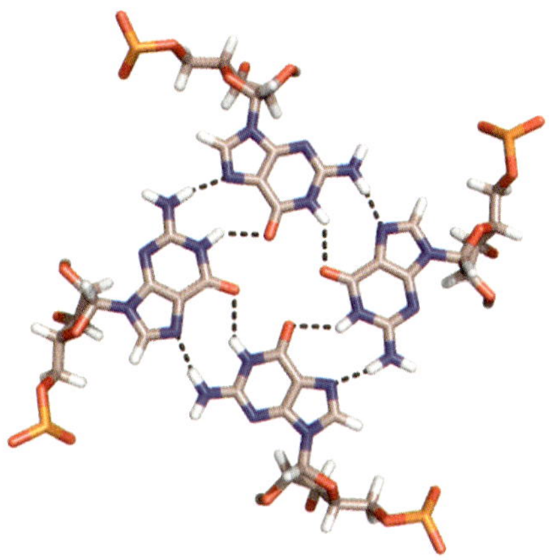

Above: Typical Ring Structure of a Hoogsteen paired G-quartet.

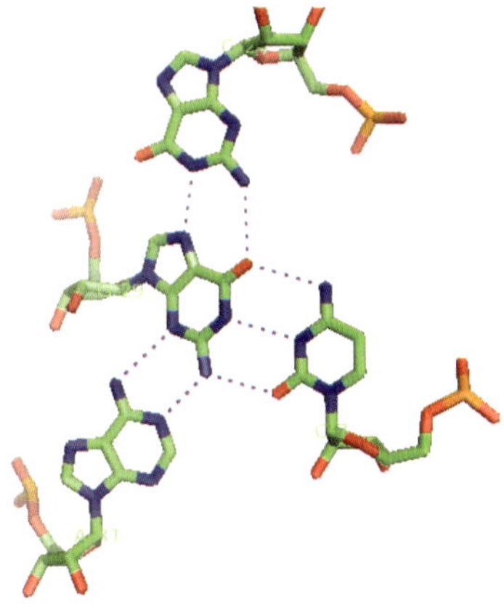

Above: Quadruplex seen in crystal structure of Malachite Green RNA aptamer. G29 involved in major groove, minor groove, and Watson-Crick hydrogen-bonding with three other bases.

Coaxial Stacking

Coaxial stacking, otherwise known as helical stacking, is a major determinant of higher order RNA tertiary structure. Coaxial stacking occurs when two RNA duplexes form a contiguous helix, which is stabilized by base stacking at the interface of the two helices. Coaxial stacking was noted in the crystal structure of tRNAPhe. More recently, coaxial stacking has been observed in higher order structures of many ribozymes, including many forms of the self-splicing group I and group II introns. Common coaxial stacking motifs include the kissing loop interaction and the pseudoknot. The stability of these interactions can be predicted by an adaptation of "Turner's rules".

In 1994, Walter and Turner determined the free energy contributions of nearest neighbor stacking interactions within a helix-helix interface by using a model system that created a helix-helix interface between a short oligomer and a four-nucleotide overhang at the end of a hairpin stem . Their experiments confirmed that the thermodynamic contribution of base-stacking between two helical secondary structures closely mimics the thermodynamics of standard duplex formation (nearest neighbor interactions predict the thermodynamic stability of the resulting helix). The relative stability of nearest neighbor interactions can be used to predict favorable coaxial stacking based on known secondary structure. Walter and Turner found that, on average, prediction of RNA structure improved from 67% to 74% accuracy when coaxial stacking contributions were included. Turner and colleagues later used an explicit quantification of stacking based on x-ray crystal structure databases to define such stacking and study kinetics in molecular dynamics simulations. Theories of coaxial stacking can be tested using the technique of helical fusion. This approach was used by Murphy and Cech to confirm a coaxial stacking interaction between the P4 and P6 helices within the catalytic center of the Tetrahymena group I intron.

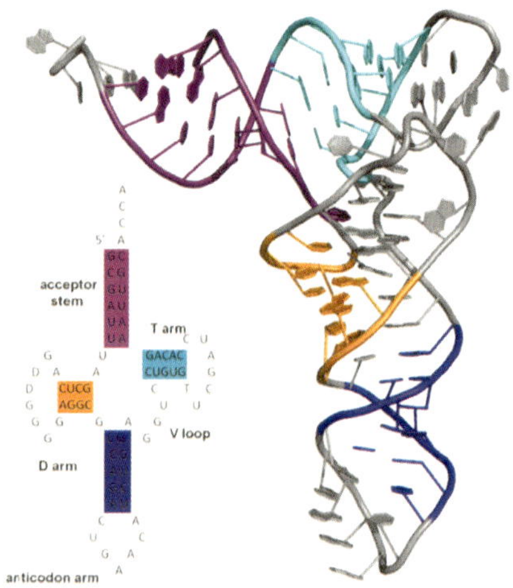

Secondary (inset) and tertiary structure of tRNA demonstrating coaxial stacking.

Most well-studied RNA tertiary structures contain examples of coaxial stacking. Some prominent examples are tRNA-Phe, group I introns, group II introns, and ribosomal RNAs. Crystal structures of tRNA revealed the presence of two extended helices that result from coaxial stacking of the amino-acid acceptor stem with the T-arm, and stacking of the D- and anticodon-arms. These interactions within tRNA orient the anticodon stem perpendicularly to the amino-acid stem, leading to the functional L-shaped tertiary structure. In group I introns, the P4 and P6 helices were shown to coaxially stack using a combination of biochemical and crystallographic methods. The P456 crystal structure provided a detailed view of how coaxial stacking stabilizes the packing of RNA helices into tertiary structures. In the self-splicing group II intron from Oceanobacillus iheyensis, the IA and IB stems coaxially stack and contribute to the relative orientation of the constituent helices of a five-way junction. This orientation facilitates proper folding of the active site of the functional ribozyme. The ribosome contains numerous examples of coaxial stacking, including stacked segments as long as 70 bp.

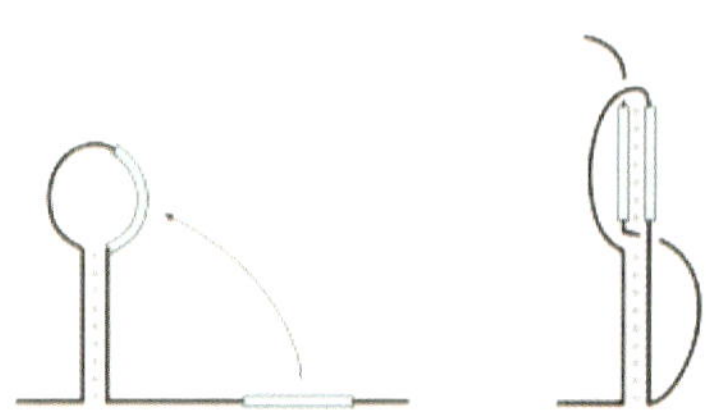

formation of a pseudoknot with coaxial stacking of the two helices

Two common motifs involving coaxial stacking are kissing loops and pseudoknots. In kissing loop interactions, the single-stranded loop regions of two hairpins interact through base pairing, forming a composite, coaxially stacked helix. Notably, this structure allows all of the nucleotides in each loop to participate in base-pairing and stacking interactions. This motif was visualized and studied using NMR analysis by Lee and Crothers. The pseudoknot motif occurs when a single stranded region of a hairpin loop basepairs with an upstream or downstream sequence within the same RNA strand. The two resulting duplex regions often stack upon one another, forming a stable coaxially stacked composite helix. One example of a pseudoknot motif is the highly stable Hepatitis Delta virus ribozyme, in which the backbone shows an overall double pseudoknot topology.

An effect similar to coaxial stacking has been observed in rationally designed DNA structures. DNA origami structures contain a large number of double helixes with exposed blunt ends. These structures were observed to stick together along the edges that contained these exposed blunt ends, due to the hydrophobic stacking interactions.

Other Motifs

Tetraloop-receptor Interactions

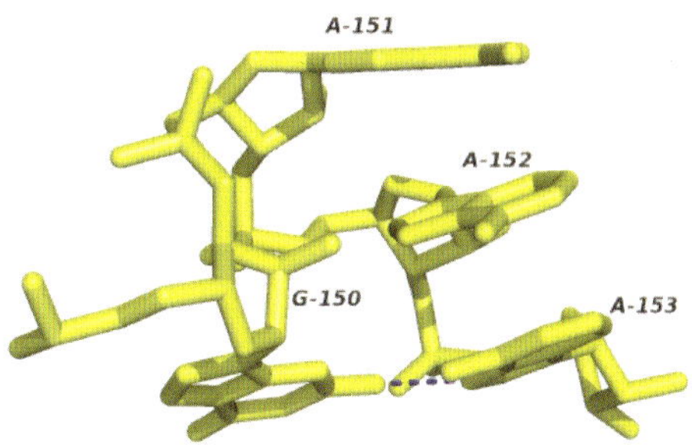

Stick representation of a GAAA tetraloop - an example from the GNRA tetraloop family.

Tetraloop-receptor interactions combine base-pairing and stacking interactions between the loop nucleotides of a tetraloop motif and a receptor motif located within an RNA duplex, creating a tertiary contact that stabilizes the global tertiary fold of an RNA molecule. Tetraloops are also possible structures in DNA duplexes.

Stem-loops can vary greatly in size and sequence, but tetraloops of four nucleotides are very common and they usually belong to one of three categories, based on sequence. These three families are the CUYG, UNCG and GNRA tetraloops. In each of these tetraloop families, the second and third nucleotides form a turn in the RNA strand and a base-pair between the first and fourth nucleotides stabilizes the stemloop structure. It has been determined, in general, that the stability of the tetraloop depends on the composition of bases within the loop and on the composition of this "closing base pair". The GNRA family of tetraloops is the most commonly observed within Tetraloop-receptor interactions.

"Tetraloop receptor motifs" are long-range tertiary interactions consisting of hydrogen bonding between the bases in the tetraloop to stemloop sequences in distal sections of the secondary RNA structure. In addition to hydrogen bonding, stacking interactions are an important component of these tertiary interactions. For example, in GNRA-tetraloop interactions, the second nucleotide of the tetraloop stacks directly on an A-platform motif within the receptor. The sequence of the tetraloop and its receptor often covary so that the same type of tertiary contact can be made with different isoforms of the tetraloop and its cognate receptor.

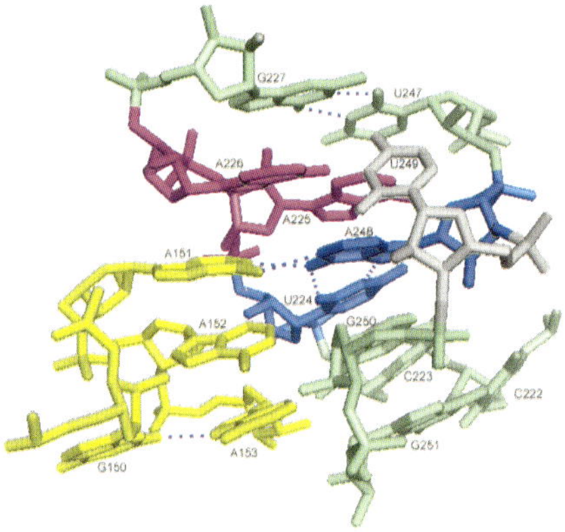

GAAA Tetraloop and Receptor: Stick representation of tetraloop (yellow) and its receptor, showing both Watson-Crick and Hoogsteen base-pairing.

For example, the self-splicing group I intron relies on tetraloop receptor motifs for its structure and function. Specifically, the three adenine residues of the canonical GAAA motif stack on top of the receptor helix and form multiple stabilizing hydrogen bonds with the receptor. The first adenine of the GAAA sequence forms a triple base-pair with the receptor AU bases. The second adenine is stabilized by hydrogen bonds with the same uridine, as well as via its 2'-OH with the receptor and via interactions with the guanine of the GAAA tetraloop. The third adenine forms a triple base pair.

A-minor Motif

A-minor Interactions

The A-minor motif is a ubiquitous RNA tertiary structural motif. It is formed by the insertion of an unpaired nucleoside into the minor groove of an RNA duplex. As such it is an example of a minor groove triple. Although guanosine, cytosine and uridine can also form minor groove triple interactions, minor groove interactions by adenine are very common. In the case of adenine, the N1-C2-N3 edge of the inserting base forms hydrogen bonds with one or both of the 2'-OH's of the duplex, as well as the bases of the duplex. The host duplex is often a G-C basepair.

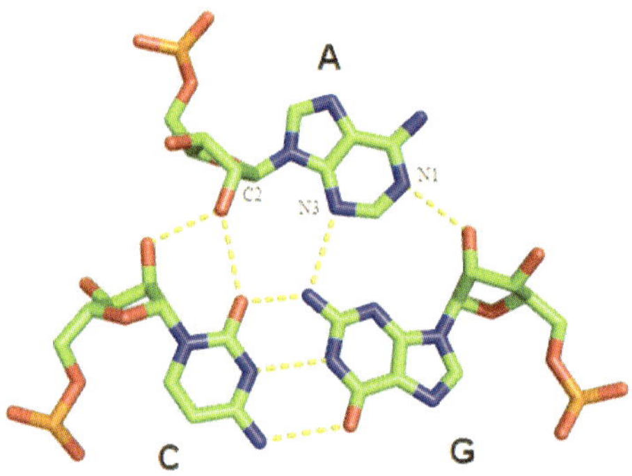

Type I A-minor interaction: Type I interactions are the most common, strongest A-minor interactions, as they involve numerous hydrogen bonds, and bury the incoming A base in the minor groove.

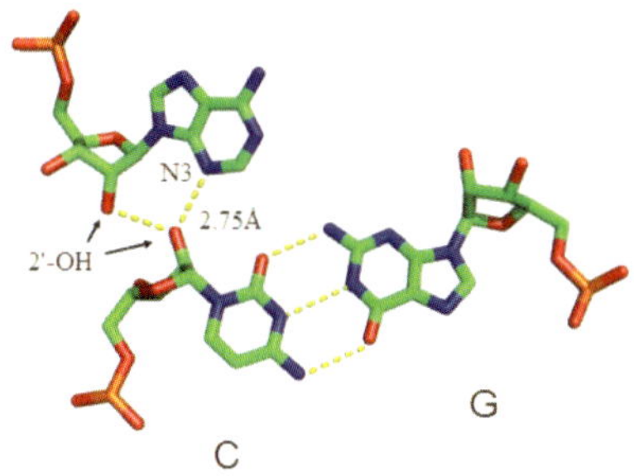

Type II A-minor interaction: Type II interactions involve the 2'-OH group and N3 of the adenosine. The adenosine interacts with the cytosine's 2'-OH group in the minor groove. The strength of this interaction is on the order of the Type I interaction.

A-minor motifs have been separated into four classes, types 0 to III, based upon the position of the inserting base relative to the two 2'-OH's of the Watson-Crick base pair. In type I and II A-minor motifs, N3 of adenine is inserted deeply within the minor groove of the duplex, and there is good shape complementarity with the base pair. Unlike types 0 and III, type I and II interactions are specific for adenine due to hydrogen bonding interactions. In the type III interaction, both the O2' and N3 of the inserting base are associated less closely with the minor groove of the duplex. Type 0 and III motifs are weaker and non-specific because they are mediated by interactions with a single 2'-OH.

The A-minor motif is among the most common RNA structural motifs in the ribosome, where it contributes to the binding of tRNA to the 23S subunit. They most often stabilize RNA duplex interactions in loops and helices, such as in the core of group II introns.

An interesting example of A-minor is its role in anticodon recognition. The ribosome must discriminate between correct and incorrect codon-anticodon pairs. It does so, in part, through the insertion of adenine bases into the minor groove. Incorrect codon-anticodon pairs will present distorted helical geometry, which will prevent the A-minor interaction from stabilizing the binding, and increase the dissociation rate of the incorrect tRNA.

An analysis of A-minor motifs in the 23S ribosomal RNA has revealed a hierarchical network of structural dependencies, suggested to be related to ribosomal evolution and to the order of events that led to the development of the modern bacterial large subunit.

Ribose Zipper

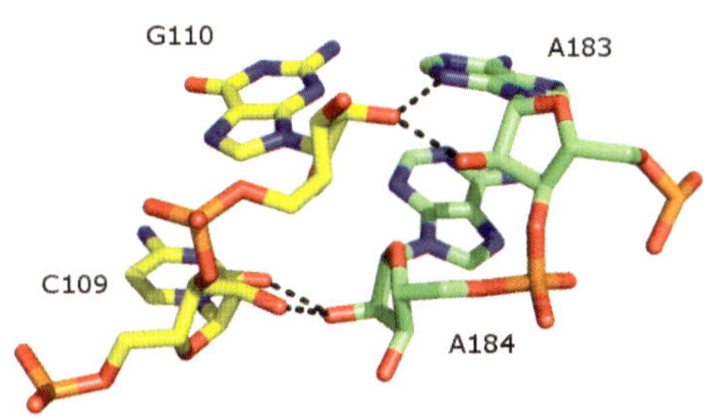

Ribose Zippers: View of a canonical ribose zipper between two RNA backbones.

The ribose zipper is an RNA tertiary structural element in which two RNA chains are held together by hydrogen bonding interactions involving the 2'OH of ribose sugars on different strands. The

2'OH can behave as both hydrogen bond donor and acceptor, which allows formation of bifurcated hydrogen bonds with another 2' OH.

Numerous forms of ribose zipper have been reported, but a common type involves four hydrogen bonds between 2'-OH groups of two adjacent sugars. Ribose zippers commonly occur in arrays that stabilize interactions between separate RNA strands. Ribose zippers are often observed as Stem-loop interactions with very low sequence specificity. However, in the small and large ribosomal subunits, there exists a propensity for ribose zippers of the CC/AA sequence- two cytosines on the first chain paired to two adenines on the second chain.

Role of Metal Ions

Functional RNAs are often folded, stable molecules with three-dimensional shapes rather than floppy, linear strands. Cations are essential for thermodynamic stabilization of RNA tertiary structures. Metal cations that bind RNA can be monovalent, divalent or trivalent. Potassium (K^+) is a common monovalent ion that binds RNA. A common divalent ion that binds RNA is magnesium (Mg^{2+}). Other ions including sodium (Na^+), calcium (Ca^{2+}) and manganese (Mn^{2+}) have been found to bind RNA *in vivo* and *in vitro*. Multivalent organic cations such as spermidine or spermine are also found in cells and these make important contributions to RNA folding. Trivalent ions such as cobalt hexamine or lanthanide ions such as terbium (Tb^{3+}) are useful experimental tools for studying metal binding to RNA.

A metal ion can interact with RNA in multiple ways. An ion can associate diffusely with the RNA backbone, shielding otherwise unfavorable electrostatic interactions. This charge screening is often fulfilled by monovalent ions. Site-bound ions stabilize specific elements of RNA tertiary structure. Site-bound interactions can be further subdivided into two categories depending on whether water mediates the metal binding. "Outer sphere" interactions are mediated by water molecules that surround the metal ion. For example, magnesium hexahydrate interacts with and stabilizes specific RNA tertiary structure motifs via interactions with guanosine in the major groove. Conversely, "inner sphere" interactions are directly mediated by the metal ion. RNA often folds in multiple stages and these steps can be stabilized by different types of cations. In the early stages, RNA forms secondary structures stabilized through the binding of monovalent cations, divalent cations and polyanionic amines in order to neutralize the polyanionic backbone. The later stages of this process involve the formation of RNA tertiary structure, which is stabilized almost largely through the binding of divalent ions such as magnesium with possible contributions from potassium binding.

Metal-binding sites are often localized in the deep and narrow major groove of the RNA duplex, coordinating to the Hoogsteen edges of purines. In particular, metal cations stabilize sites of backbone twisting where tight packing of phosphates results in a region of dense negative charge. There are several metal ion-binding motifs in RNA duplexes that have been identified in crystal structures. For instance, in the P4-P6 domain of the *Tetrahymena thermophila* group I intron, several ion-binding sites consist of tandem G-U wobble pairs and tandem G-A mismatches, in which divalent cations interact with the Hoogsteen edge of guanosine via O6 and N7. Another ion-binding motif in the *Tetrahymena* group I intron is the A-A platform motif, in which consecutive adenosines in the same strand of RNA form a non-canonical pseudobase pair. Unlike the tandem G-U motif, the A-A platform motif binds preferentially to monovalent cations. In many of these motifs, absence of the monovalent or divalent cations results in either greater flexibility or loss of tertiary structure.

Metal Ion Binding in the Group I Intron

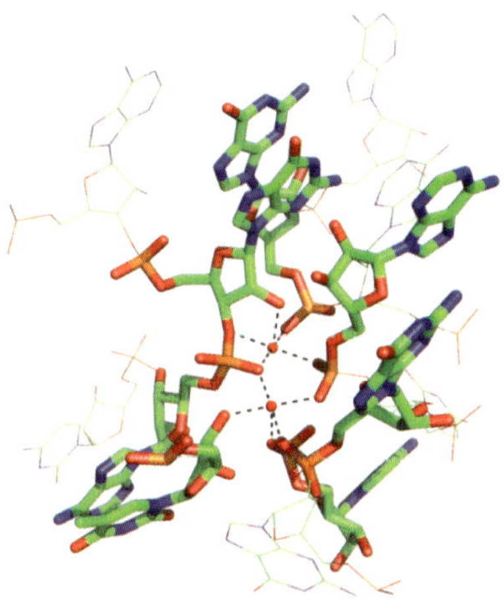

PDB rendering of Group I intron inner sphere magnesium coordination. The two red balls indicate magnesium ions and dashed lines coming from the ions indicate coordination with the respective groups on nucleotides. The color-coding scheme is as follows: green=carbon, orange=phosphate, pink=oxygen, blue=nitrogen.

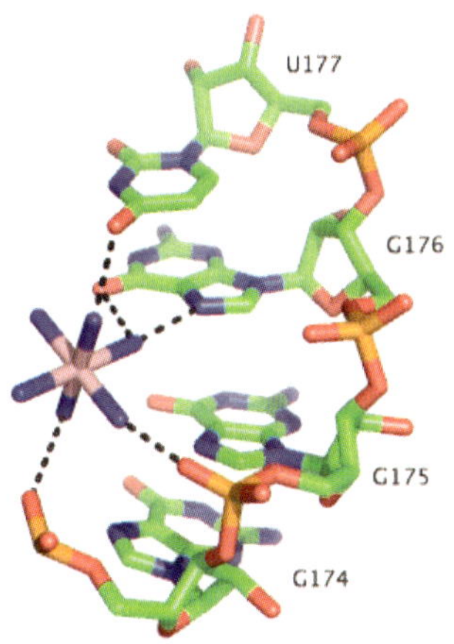

PDB rendering of Group 1 intron P5c binding pocket demonstrating outer sphere coordination. Here, the six amines of osmium hexamine(III) fulfill the role generally served by water molecules and mediate the ion's interaction with the major groove. Coordination via hydrogen bonds is indicated by dashed lines and osmium is rendered in pink, all other colors are as above.

Divalent metal ions, especially magnesium, have been found to be important for the structure of DNA junctions such as the Holliday junction intermediate in genetic recombination. The magnesium ion shields the negatively charged phosphate groups in the junction and allows them to be positioned closer together, allowing a stacked conformation rather than an unstacked conformation. Magnesium is vital in stabilizing these kinds of junctions in artificially designed structures used in DNA nanotechnology, such as the double crossover motif.

Nucleic Acid Quaternary Structure

Nucleic acid quaternary structure refers to the interactions between separate nucleic acid molecules, or between nucleic acid molecules and proteins. The concept is analogous to protein quaternary structure, but as the analogy is not perfect, the term is used to refer to a number of different concepts in nucleic acids and is less commonly encountered.

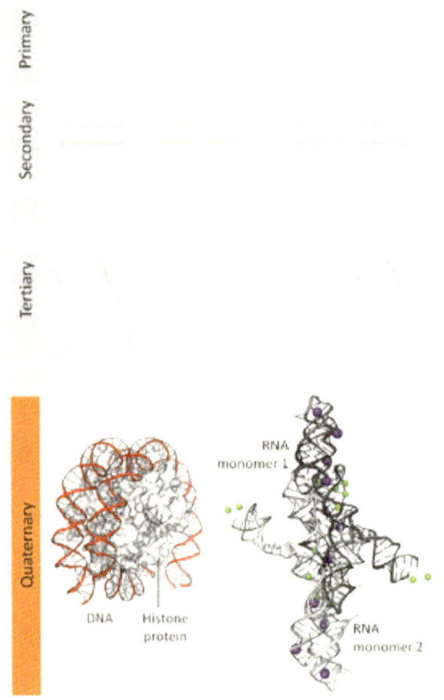

Interactive image of nucleic acid structure (primary, secondary, tertiary, and quaternary) using DNA helices and examples from the VS ribozyme and telomerase and nucleosome. (PDB: ADNA, 1BNA, 4OCB, 4R4V, 1YMO, 1EQZ)

DNA

DNA quaternary structure is used to refer to the binding of DNA to histones to form nucleosomes, and then their organisation into higher-order chromatin fibres. The quaternary structure of DNA strongly affects how accessible the DNA sequence is to the transcription machinery for expression of genes. DNA quaternary structure varies over time, as regions of DNA are condensed or exposed for transcription. The term has also been used to describe the hierarchical assembly of artificial nucleic acid building blocks used in DNA nanotechnology.

RNA

Symmetrical complexes of RNA molecules are extremely uncommon compared to protein oligomers. One example of an RNA homodimer is the VS ribozyme from Neurospora, with its two active sites consisting of nucleotides from both monomers.

The best known example of RNA forming quaternary structures with proteins is the ribosome, which consists of multiple rRNAs, supported by rProteins. Similar RNA-Protein complexes are also found in the spliceosome.

Nucleic Acid Structure Determination

Structure probing of nucleic acids is the process by which biochemical techniques are used to determine nucleic acid structure. This analysis can be used to define the patterns which can infer the molecular structure, experimental analysis of molecular structure and function, and further understanding on development of smaller molecules for further biological research. Structure probing analysis can be done through many different methods, which include chemical probing, hydroxyl radical probing, SHAPE, nucleotide analog interference mapping (NAIM), and in-line probing.

Physical Methods

Nuclear Magnetic Resonance Spectroscopy

Nucleic acid NMR is the use of NMR spectroscopy to obtain information about the structure and dynamics of nucleic acid molecules, such as DNA or RNA. As of 2003, nearly half of all known RNA structures had been determined by NMR spectroscopy.

Nucleic acid NMR uses similar techniques as protein NMR, but has several differences. Nucleic acids have a smaller percentage of hydrogen atoms, which are the atoms usually observed in NMR, and because nucleic acid double helices are stiff and roughly linear, they do not fold back on themselves to give "long-range" correlations. The types of NMR usually done with nucleic acids are ^{1}H or proton NMR, ^{13}C NMR, ^{15}N NMR, and ^{31}P NMR. Two-dimensional NMR methods are almost always used, such as correlation spectroscopy (COSY) and total coherence transfer spectroscopy (TOCSY) to detect through-bond nuclear couplings, and nuclear Overhauser effect spectroscopy (NOESY) to detect couplings between nuclei that are close to each other in space.

Parameters taken from the spectrum, mainly NOESY cross-peaks and coupling constants, can be used to determine local structural features such as glycosidic bond angles, dihedral angles (using the Karplus equation), and sugar pucker conformations. For large-scale structure, these local parameters must be supplemented with other structural assumptions or models, because errors add up as the double helix is traversed, and unlike with proteins, the double helix does not have a compact interior and does not fold back upon itself. NMR is also useful for investigating nonstandard geometries such as bent helices, non-Watson–Crick basepairing, and coaxial stacking. It has been especially useful in probing the structure of natural RNA oligonucleotides, which tend to adopt complex conformations such as stem-loops and pseudoknots. NMR is also useful for probing the binding of nucleic acid molecules to other molecules, such as proteins or drugs, by seeing which resonances are shifted upon binding of the other molecule.

RNA Sequencing

While methods for each type of probing differ in steps, the probing of secondary structure involves certain steps in order to determine the structure. The first step involved is submitting the structural RNA to the probe of interest and incubating over a certain amount of time to allow the reaction to occur. The RNA is then transcribed using reverse transcriptase PCR, where this results in different lengths of bands due to the modification of the RNA at specific sites, which causes the reverse transcriptase to fall off. These bands are then run on a gel with sequencing data determined from RNA sequencing.

Chemical Methods

RNA chemical probing can involve many different chemicals which serve to modify specific bases at certain sites to show certain locations available for specific modification type.

Hydroxyl Radical Probing

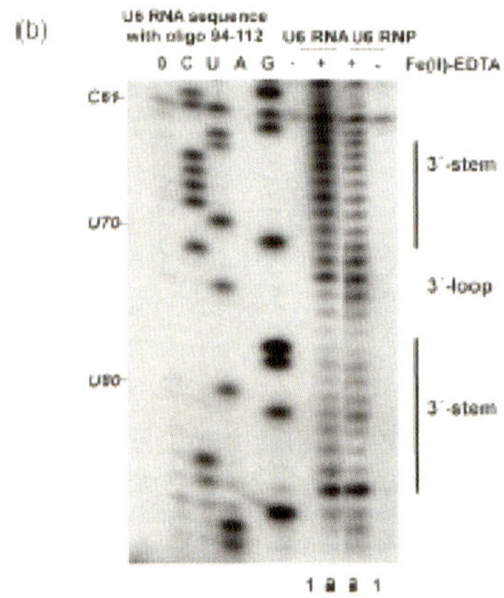

Hydroxyl radical probing gel showing bands at positions and dots indicating strength of protection.

Probing with hydroxyl radicals involves an additional step, as hydroxyl radicals are short lived in solution they need to be generated. This can be done using H_2O_2, ascorbic acid, and Fe(II)-EDTA complex which is attached to the backbone through EDTA. The Fe(II) along with ascorbic acid generates hydroxyl radicals which then can react with the nucleic acid molecules. Hydroxyl radical probing is often used in conjunction with chemical probing of nucleic acid molecules that are thought to associate with proteins, this is due to the modification done by hydroxyl radicals. Hydroxyl radicals attack the ribose/deoxyribose ring and this results in breaking of the phosphate backbone, which is independent of secondary structure, as all backbone is accessible, but is instead resultant of protein or tertiary structure protection. Probing with hydroxyl radicals shows the protection of structured nucleic acids by the proteins thought to be associated or folding on itself, where cleavage again results in a band formed through gel electrophoresis (after RT-PCR in the case of RNA) that is shorter than the full nucleic acid depending upon where it is cleaved. In this case since the last nucleotide is not modified the band length is indicative of the base that was cleaved. When examining the gel produced by running the gels on a band the areas of various strength of protection where areas of stronger protection for hydroxyl radicals can be said to have tighter association with a protein, or if no protein associates with the nucleic acid it can be caused by the tertiary fold.

DMS

Dimethyl sulfate, known as DMS, is a chemical that can be used to modify nucleic acids in order to determine secondary structure. DMS modifies certain bases through methylation. One set of methylation products that is used for RNA is the methylation of N1 adenosine and N3 of cytosine which prevents the natural hydrogen bonds to form between bases modified by DMS. This enables modification sites to be detected by RT-PCR, as modified sites cannot be basepaired, which results in the reverse transcriptase to fall off and produce different band sizes. DMS may modify adenines and cytosines that are single stranded, base paired at the end of a helix, or in a base pair next to a GU pair. Detection of these modifications is done by examination of a gel and looking at bands present. One thing of note is that since modification prevents the addition of nucleotides at the modification site each of the bands generated is shifted by one base down on the gel. Nucleotides may be protected from DMS modification by base pairing, tertiary contacts, or protein-RNA interactions. This can be used to begin to develop a model of secondary structure for the RNA molecule.

Structure probing by DMS allows for detection of secondary structure changes due to binding of RNA molecules along with detecting changes in tertiary structure.

DMS modification can also be used for DNA, for example in footprinting DNA-protein interactions.

CMCT

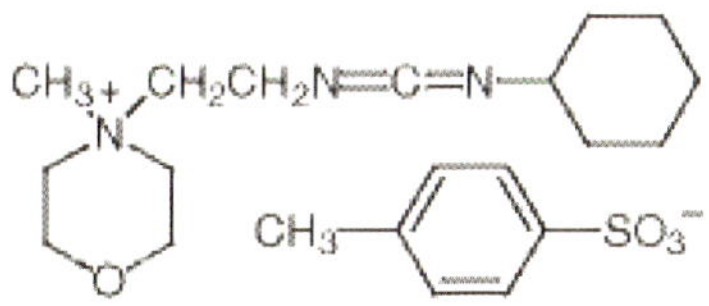

Structure of CMCT used in RNA structure probing

1-cyclohexyl-(2-morpholinoethyl)carbodiimide metho-p-toluene sulfonate known as CMCT is another chemical that is used in structure probing. CMCT like DMS serves to modify the exposed bases of specific nucleotides, which are uridine, and to a smaller extent guanine. CMCT reacts primarily with N3 of uridine and N1 of guanine modifying two sites responsible for hydrogen bonding on the bases. Modification using CMCT is analogous in detection to DMS as modification prevents basepairing at specific nucleotides which are then detected using rt-PCR and running of the PCR products on an agarose gel as shown in Figure. Structure probing of RNA by CMCT indicates the presence of uridine and guanine in single stranded regions by accessibility to modification or the presence of uridine and guanine in double stranded regions by protection from CMCT, which is the absence of a band. CMCT, like DMS can detect secondary and tertiary structure changes, but still has the same weaknesses as the method of modification is the same as DMS.

Kethoxal

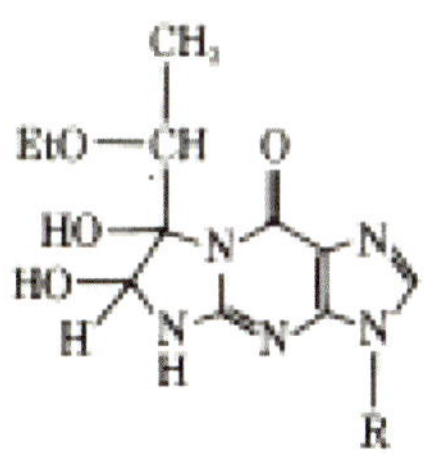

II

Binding of kethoxal to modified guanine preventing basepairing.

1,1-Dihydroxy-3-ethoxy-2-butanone, known as kethoxal, is used like DMS and CMCT, where treatment with kethoxal causes the modification of guanine, specifically altering the N1 and the exocyclic amino group simultaneously by covalent interaction. The modification by kethoxal prevents single stranded guanine nucleotides to become modified, so that when reverse transcriptase reaches the modified guanine it falls off resulting in a band. After rt-PCR the products will be of different lengths where a band present indicates modification of a guanine base, and the absence of a band indicates that the base was not available for modification. Using this gel in combination with DMS and CMCT a model for structural RNAs can be formed through comparison between the gels which indicate protected and accessible positions in all the bases, which can be used to form a preliminary model.

SHAPE

Selective 2′-hydroxyl acylation analyzed by primer extension, or SHAPE, takes advantage of reagents that preferentially modify the backbone of RNA in structurally flexible regions.

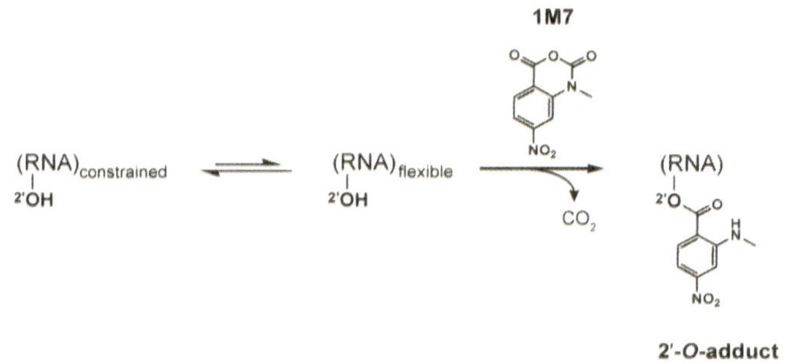

1-methyl-7-nitroisatoic anhydride (1M7) undergoes hydrolysis to form adducts on the backbone of unpaired RNA nucleotides.

Reagents such as N-methylisotoic anhydride (NMIA) and 1-methyl-7-nitroisatoic anhydride (1M7) react with the 2'-hydroxyl group to form adducts on the 2'-hydroxyl of the RNA backbone. Compared to the chemicals used in other RNA probing techniques, these reagents have the advantage of being largely unbiased to base identity, while remaining very sensitive to conformational dynamics. Nucleotides which are constrained (usually by base-pairing) show less adduct formation than nucleotides which are unpaired. Adduct formation is quantified for each nucleotide in a given RNA by extension of a complementary DNA primer with reverse transcriptase and comparison of the resulting fragments with those from an unmodified control. SHAPE therefore reports on RNA structure at the individual nucleotide level. This data can be used as input to generate highly accurate secondary structure models. SHAPE has been used to analyze diverse RNA structures, including that of an entire HIV-1 genome. The best approach is to use a combination of chemical probing reagents and experimental data. In SHAPE-Seq SHAPE is extended by bar-code based multiplexing combined with RNA-Seq and can be performed in a high-throughput fashion.

LASER

Light-Activated Structural Examination of RNA (LASER) probing utilizes UV light to active nicotinoyl azide (NAz), generating nitrenium cation, which reacts with guanosine and adenosine of RNA at C-8 position through a barrierless Friedel-Crafts reaction. This chemical probing method is light-controllable, and probes solvent accessbility of nucleobase, which has been shown to footprint RNA binding proteins inside cells.

In-line Probing

In-line probing does not involve treatment with any type of chemical or reagent to modify RNA structures. This type of probing assay uses the structure dependent cleavage of RNA; single stranded regions are more flexible and unstable and will degrade over time. The process of in-line probing is often used to determine changes in structure due to ligand binding. Binding of a ligand can result in different cleavage patterns. The process of in-line probing involves incubation of structural or functional RNAs over a long period of time. This period can be several days, but varies in each experiment. The incubated products are then run on a gel to visualize the bands. This experiment is often done using two different conditions: 1) with ligand and 2) in the absence of ligand. Cleavage results in shorter band lengths and is indicative of areas that are not basepaired, as basepaired regions tend to be less sensitive to spontaneous cleavage. In-line probing is a functional assay that can be used to determine structural changes in RNA in response to ligand binding. It can directly show the change in flexibility and binding of regions of RNA in response to a ligand, as well as compare that response to analogous ligands. This assay is commonly used in dynamic studies, specifically when examining riboswitches.

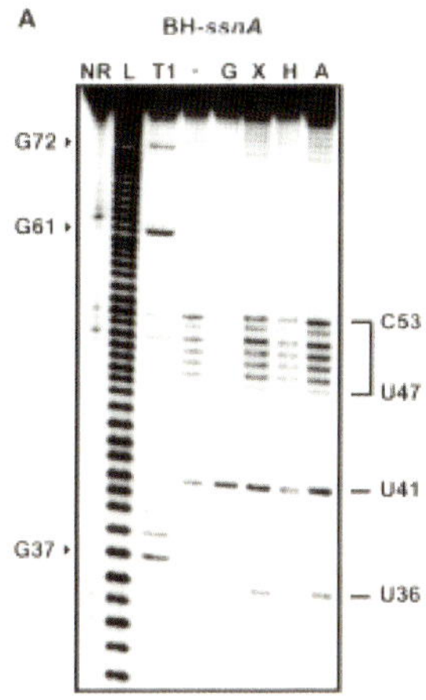

In-line probing assay of guanine riboswitches showing change in flexibility in response to various nucleotide ligands

Nucleotide Analog Interference Mapping

Nucleotide analog interference mapping (NAIM) is the process of using nucleotide analogs, molecules that are similar in some ways to nucleotides but lack function, to determine the importance of a functional group at each location of an RNA molecule. The process of NAIM is to insert a single nucleotide analog into a unique site. This can be done by transcribing a short RNA using T7 RNA polymerase, then synthesizing a short oligonucleotide containing the analog in a specific position, then ligating them together on the DNA template using a ligase. The nucleotide analogs are tagged with a phosphorothioate, the active members of the RNA population are then distinguished from the inactive members, the inactive members then have the phosphorothioate tag removed and the analog sites are identified using gel electrophoresis and autoradiography. This indicates a functionally important nucleotide, as cleavage of the phosphorothioate by iodine results in an RNA that is cleaved at the site of the nucleotide analog insert. By running these truncated RNA molecules on a gel, the nucleotide of interest can be identified against a sequencing experiment Site directed incorporation results indicate positions of importance where when running on a gel, functional RNAs that have the analog incorporated at that position will have a band present, but if the analog results in non-functionality, when the functional RNA molecules are run on a gel there will be no band corresponding to that position on the gel. This process can be used to evaluate an entire area, where analogs are placed in site specific locations, differing by a single nucleotide, then when functional RNAs are isolated and run on a gel, all areas where bands are produced indicate non-essential nucleotides, but areas where bands are absent from the functional RNA indicate that inserting a nucleotide analog in that position caused the RNA molecule to become non-functional.

Nucleic Acid Structure Prediction

Nucleic acid structure prediction is a computational method to determine *secondary* and *tertiary* nucleic acid structure from its sequence. Secondary structure can be predicted from one or several nucleic acid sequences. Tertiary structure can be predicted from the sequence, or by comparative modeling (when the structure of a homologous sequence is known).

The problem of predicting nucleic acid secondary structure is dependent mainly on base pairing and base stacking interactions; many molecules have several possible three-dimensional structures, so predicting these structures remains out of reach unless obvious sequence and functional similarity to a known class of nucleic acid molecules, such as transfer RNA (tRNA) or microRNA (miRNA), is observed. Many secondary structure prediction methods rely on variations of dynamic programming and therefore are unable to efficiently identify pseudoknots.

While the methods are similar, there are slight differences in the approaches to RNA and DNA structure prediction. *In vivo*, DNA structures are more likely to be duplexes with full complementarity between two strands, while RNA structures are more likely to fold into complex secondary and tertiary structures such as in the ribosome, spliceosome, or transfer RNA. This is partly because the extra oxygen in RNA increases the propensity for hydrogen bonding in the nucleic acid backbone. The energy parameters are also different for the two nucleic acids.

Single Sequence Structure Prediction

A common problem for researchers working with RNA is to determine the three-dimensional structure of the molecule given only a nucleic acid sequence. However, in the case of RNA much of the final structure is determined by the secondary structure or intra-molecular base pairing interactions of the molecule. This is shown by the high conservation of base pairings across diverse species.

The Most Stable Structure

Secondary structure of small RNA molecules is largely determined by strong, local interactions such as hydrogen bonds and base stacking. Summing the free energy for such interactions should provide an approximation for the stability of a given structure. To predict the folding free energy of a given secondary structure, an empirical nearest-neighbor model is used. In the nearest neighbor model the free energy change for each motif depends on the sequence of the motif and of its closest base-pairs. The model and parameters of minimal energy for Watson–Crick pairs, GU pairs and loop regions were derived from empirical calorimetric experiments, the most up-to-date parameters were published in 2004, although most software packages use the prior set assembled in 1999.

The simplest way to find the lowest free energy structure would be to generate all possible structures and calculate the free energy for it, but the number of possible structures for a sequence increases exponentially with the length of RNA: number of secondary structures = $(1,8)^N$, N- number of nucleotides. For longer molecules, the number of possible secondary structures is huge: a sequence of 100 nucleotides has more than 10^{25} possible secondary structures.

Dynamic Programming Algorithms

The first and the most popular method for finding the most stable structure is a dynamic programming algorithm. One of the first attempts to predict RNA secondary structure was made by Ruth Nussinov and co-workers who used the dynamic programming method for maximizing the number of base-pairs. However, there are several issues with this approach: most importantly, the solution is not unique. Nussinov et al. published an adaptation of their approach using a simple nearest-neighbour energy model in 1980. In 1981, Michael Zuker and Patrick Stiegler proposed using a slightly refined dynamic programming approach to modelling nearest neighbor energy interactions that directly incorporates stacking into the prediction.

Dynamic programming algorithms provide a means to implicitly check all variants of possible RNA secondary structures without explicitly generating the structures. First, the lowest conformational free energy is determined for each possible sequence fragment starting with the shortest fragments and then for longer fragments. For longer fragments, recursion on the optimal free energy changes determined for shorter sequences speeds the determination of the lowest folding free energy. Once the lowest free energy of the complete sequence is calculated, the exact structure of RNA molecule is determined.

Dynamic programming algorithms are commonly used to detect base pairing patterns that are "well-nested", that is, form hydrogen bonds only to bases that do not overlap one another in sequence position. Secondary structures that fall into this category include double helices, stem-loops, and variants of the "cloverleaf" pattern found in transfer RNA molecules. These methods rely on pre-calculated parameters which estimate the free energy associated with certain types of base-pairing interactions, including Watson-Crick and Hoogsteen base pairs. Depending on the complexity of the method, single base pairs may be considered, and short two- or three-base segments, to incorporate the effects of base stacking. This method cannot identify pseudoknots, which are not well nested, without substantial algorithmic modifications that are computationally very costly.

Suboptimal Structures

The accuracy of RNA secondary structure prediction from one sequence by free energy minimization is limited by several factors:

1. The free energy value's list in nearest neighbor model is incomplete.
2. Not all known RNA folds in such a way as to conform with the thermodynamic minimum.
3. Some RNA sequences have more than one biologically active conformation (i.e., riboswitches).

For this reason, the ability to predict structures which have similar low free energy can provide significant information. Such structures are termed suboptimal structures. MFOLD is one program that generates suboptimal structures.

Predicting Pseudoknots

One of the issues when predicting RNA secondary structure is that the standard free energy minimization and statistical sampling methods can not find pseudoknots. The major problem is that the usual dynamic programing algorithms, when predicting secondary structure, consider only the interactions between the closest nucleotides, while pseudoknotted structures are formed due to interactions between distant nucleotides. Rivas and Eddy published a dynamic programming algorithm for predicting pseudoknots. However, this dynamic programming algorithm is very slow. The standard dynamic programming algorithm for free energy minimization scales $O(N^3)$ in time (N is the number of nucleotides in the sequence), while the Rivas and Eddy algorithm scales $O(N^6)$ in time. This has prompted several researchers to implement versions of the algorithm that restrict classes of pseudoknots, resulting in performance gains. For example, pknotsRG tool includes only the class of simple recursive pseudoknots and scales O(N4) in time.

Other Approaches for RNA Secondary Structure Prediction

Another approach for RNA secondary structure determination is to sample structures from the Boltzmann ensemble, as exemplified by the program SFOLD. The program generates a statistical sample of all possible RNA secondary structures. The algorithm samples secondary structures according to the Boltzmann distribution. The sampling method offers an appealing solution to the problem of uncertainties in folding.

Comparative Secondary Structure Prediction

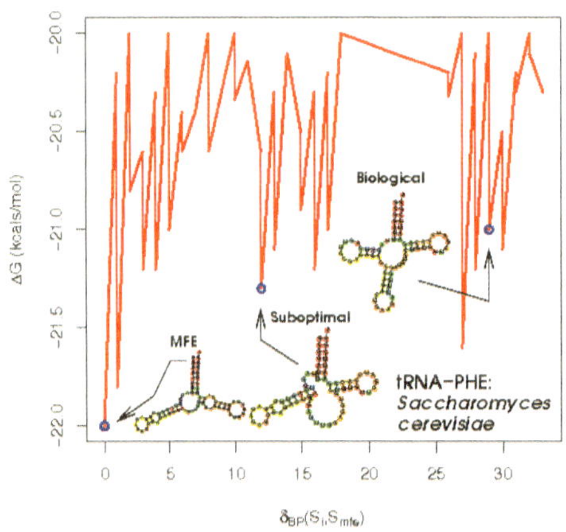

S. cerevisiae tRNA-PHE structure space: the energies and structures were calculated using RNAsubopt and the structure distances computed using RNAdistance.

Sequence covariation methods rely on the existence of a data set composed of multiple homologous RNA sequences with related but dissimilar sequences. These methods analyze the covariation of individual base sites in evolution; maintenance at two widely separated sites of a pair of base-pairing nucleotides indicates the presence of a structurally required hydrogen bond between those positions. The general problem of pseudoknot prediction has been shown to be NP-complete.

In general, the problem of alignment and consensus structure prediction are closely related. Three different approaches to the prediction of consensus structures can be distinguished:

1. Folding of alignment
2. Simultaneous sequence alignment and folding
3. Alignment of predicted structures

Align then Fold

A practical heuristic approach is to use multiple sequence alignment tools to produce an alignment of several RNA sequences, to find consensus sequence and then fold it. The quality of the alignment determines the accuracy of the consensus structure model. Consensus sequences are folded using various approaches similarly as in individual structure prediction problem. The thermodynamic folding approach is exemplified by RNAalifold program. The different approaches are exemplified by Pfold and ILM programs. Pfold program implements a SCFGs. ILM (iterated loop matching) unlike the other algorithms for folding of alignments, can return pseudoknoted structures. It uses combination of thermodynamics and mutual information content scores.

Align and Fold

Evolution frequently preserves functional RNA structure better than RNA sequence. Hence, a common biological problem is to infer a common structure for two or more highly diverged but homologous RNA sequences. In practice, sequence alignments become unsuitable and do not help to improve the accuracy of structure prediction, when sequence similarity of two sequences is less than 50%.

Structure-based alignment programs improves the performance of these alignments and most of them are variants of the Sankoff algorithm. Basically, Sankoff algorithm is a merger of sequence alignment and Nussinov (maximal-pairing) folding dynamic programming method. Sankoff algorithm itself is a theoretical exercise because it requires extreme computational resources ($O^{(n3m)}$ in time, and $O^{(n2m)}$ in space, where n is the sequence length and m is the number of sequences). Some notable attempts at implementing restricted versions of Sankoff's algorithm are Foldalign, Dynalign, PMmulti/PMcomp, Stemloc, and Murlet. In these implementations the maximal length of alignment or variants of possible consensus structures are restricted. For example, Foldalign focuses on local alignments and restricts the possible length of the sequences alignment.

Fold then Align

A less widely used approach is to fold the sequences using single sequence structure prediction methods and align the resulting structures using tree-based metrics. The fundamental weakness with this approach is that single sequence predictions are often inaccurate, thus all further analyses are affected.

Tertiary Structure Prediction

Once secondary structure of RNA is known, the next challenge is to predict tertiary structure. The biggest problem is to determine the structure of regions between double stranded helical regions. Also RNA molecules often contain posttranscriptionally modified nucleosides, which because of new possible non-canonical interactions, cause a lot of troubles for tertiary structure prediction.

Nucleic Acid Quantitation

In molecular biology, quantitation of nucleic acids is commonly performed to determine the average concentrations of DNA or RNA present in a mixture, as well as their purity. Reactions that use nucleic acids often require particular amounts and purity for optimum performance. To date, there are two main approaches used by scientists to quantitate, or establish the concentration, of nucleic acids (such as DNA or RNA) in a solution. These are spectrophotometric quantification and UV fluorescence tagging in presence of a DNA dye.

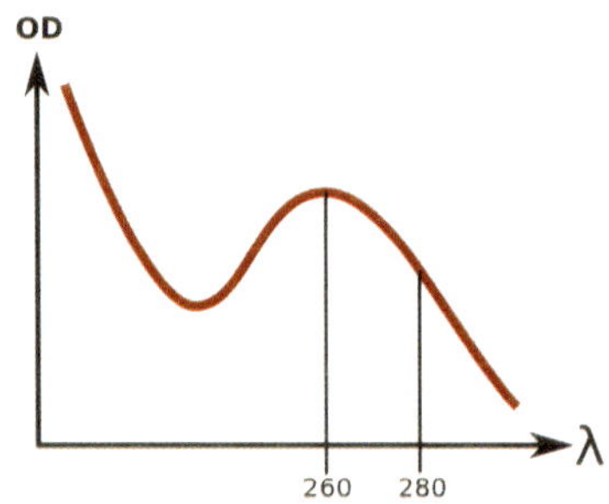

Optical density of ribosome sample. The important wavelengths of 260nm and 280nm are labeled.

Spectrophotometric Analysis

One of the more commonly used practices to quantitate DNA or RNA is the use of spectrophotometric analysis using a spectrophotometer. A spectrophotometer is able to determine the average concentrations of the nucleic acids DNA or RNA present in a mixture, as well as their purity.

Spectrophotometric analysis is based on the principles that nucleic acids absorb ultraviolet light in a specific pattern. In the case of DNA and RNA, a sample is exposed to ultraviolet light at a wavelength of 260 nanometres (nm) and a photo-detector measures the light that passes through the sample. Some of the ultraviolet light will pass through and some will be absorbed by the DNA / RNA. The more light absorbed by the sample, the higher the nucleic acid concentration in the sample. The resulting effect is that less light will strike the photodetector and this will produce a higher optical density (OD).

Using the Beer-Lambert Law it is possible to relate the amount of light absorbed to the concentration of the absorbing molecule. At a wavelength of 260 nm, the average extinction coefficient for double-stranded DNA is 0.020 $(\mu g/ml)^{-1}\ cm^{-1}$, for single-stranded DNA it is 0.027 $(\mu g/ml)^{-1}\ cm^{-1}$, for single-stranded RNA it is 0.025 $(\mu g/ml)^{-1}\ cm^{-1}$ and for short single-stranded oligonucleotides it is dependent on the length and base composition. Thus, an Absorbance (A) of 1 corresponds to a concentration of 50 μg/ml for double-stranded DNA. This method of calculation is valid for up to an A of at least 2. A more accurate extinction coefficient may be needed for oligonucleotides; these can be predicted using the nearest-neighbor model.

Calculations

The optical density is generated from equation:

Optical Density= Log (Intensity of Incident Light / Intensity of Transmitted Light)

In practical terms, a sample that contains no DNA or RNA should not absorb any of the ultraviolet light and therefore produce an OD of 0

Optical Density=Log (100/100)=0

When using spectrophotometric analysis to determine the concentration of DNA or RNA, the Beer-Lambert law is used to determine unknown concentrations without the need for standard curves. In essence, the Beer Lambert Law makes it possible to relate the amount of light absorbed to the concentration of the absorbing molecule. The following absorbance units to nucleic acid concentration conversion factors are used to convert OD to concentration of unknown nucleic acid samples:

A260 dsDNA = 50 μg/ml

A260 ssDNA = 33 μg/ml

A260 ssRNA = 40 μg/ml

Conversion Factors

When using a 10 mm path length, simply multiply the OD by the conversion factor to determine the concentration. Example, a 2.0 OD dsDNA sample corresponds to a sample with a 100 ug/ml concentration.

When using a path length that is shorter than 10mm, the resultant OD will be reduced by a factor of 10/path length. Using the example above with a 3 mm path length, the OD for the 100 ug/ml sample would be reduced to 0.6. To normalize the concentration to a 10mm equivalent, the following is done:

0.6 OD X (10/3) * 50 ug/ml=100 ug/ml

Most spectrophotometers allow selection of the nucleic acid type and path length such that resultant concentration is normalized to the 10 mm path length which is based on the principles of Beer's law.

A260 as Quantity Measurement

The "A260 unit" is used as a quantity measure for nucleic acids. One A260 unit is the amount of nucleic acid contained in 1 mL and producing an OD of 1. The same conversion factors apply, and therefore, in such contexts:

1 A260 unit dsDNA = 50 μg

1 A260 unit ssDNA = 33 μg

1 A260 unit ssRNA = 40 μg

Sample Purity (260:280 / 260:230 Ratios)

It is common for nucleic acid samples to be contaminated with other molecules (i.e. proteins, organic compounds, other). The secondary benefit of using spectrophotometric analysis for nucleic acid quantitation is the ability to determine sample purity using the 260 nm:280 nm calculation. The ratio of the absorbance at 260 and 280 nm ($A_{260/280}$) is used to assess the purity of nucleic acids. For pure DNA, $A_{260/280}$ is widely considered ~1.8 but has been argued to translate - due to numeric errors in the original Warburg paper - into a mix of 60% protein and 40% DNA. The ratio for pure RNA $A_{260/280}$ is ~2.0. These ratios are commonly used to assess the amount of protein contamination that is left from the nucleic acid isolation process since proteins absorb at 280 nm.

The ratio of absorbance at 260 nm vs 280 nm is commonly used to assess DNA contamination of protein solutions, since proteins (in particular, the aromatic amino acids) absorb light at 280 nm. The reverse, however, is not true — it takes a relatively large amount of protein contamination to significantly affect the 260:280 ratio in a nucleic acid solution.

260:280 ratio has high sensitivity for nucleic acid contamination in protein:

% protein	% nucleic acid	260:280 ratio
100	0	0.57
95	5	1.06
90	10	1.32
70	30	1.73

260:230 ratio lacks sensitivity for protein contamination in nucleic acids (table shown for RNA, 100% DNA is approximately 1.8):

% nucleic acid	% protein	260:230 ratio
100	0	2.00
95	5	1.99
90	10	1.98
70	30	1.94

This difference is due to the much higher mass attenuation coefficient nucleic acids have at 260 nm and 280 nm, compared to that of proteins. Because of this, even for relatively high concentrations of protein, the protein contributes relatively little to the 260 and 280 absorbance. While the protein contamination cannot be reliably assessed with a 260:280 ratio, this also means that it contributes little error to DNA quantity estimation.

Contamination Identification

Examination of sample spectra may be useful in identifying that a problem with sample purity exists.

Table of Potential Contamination Factors		
	Low Reading	High Reading
A260/A230 Ratio	• Carbohydrate carryover (often a problem with plants). • Residual phenol from nucleic acid extraction. • Residual guanidine (often used in column based kits). • Glycogen used for precipitation.	• Making a Blank measurement on a dirty pedestal. • Using an inappropriate solution for the Blank measurement. The blank solution should be the same pH and of a similar ionic strength as the sample solution. Example: Using water for the Blank measurement for samples dissolved in TE may result in low 260/230 ratios.

A260/A280 Ratio	• Residual phenol or other reagent associated with the extraction protocol. • A very low concentration(< 10 ng/ul).of nucleic acid.	• Residual RNA from nucleic acid extraction. * High 260/280 purity ratios are not normally indicative of any issues.

Other Common Contaminants

- Contamination by phenol, which is commonly used in nucleic acid purification, can significantly throw off quantification estimates. Phenol absorbs with a peak at 270 nm and a $A_{260/280}$ of 1.2. Nucleic acid preparations uncontaminated by phenol should have a $A_{260/280}$ of around 2. Contamination by phenol can significantly contribute to overestimation of DNA concentration.
- Absorption at 230 nm can be caused by contamination by phenolate ion, thiocyanates, and other organic compounds. For a pure RNA sample, the $A_{230:260:280}$ should be around 1:2:1, and for a pure DNA sample, the $A_{230:260:280}$ should be around 1:1.8:1.
- Absorption at 330 nm and higher indicates particulates contaminating the solution, causing scattering of light in the visible range. The value in a pure nucleic acid sample should be zero.
- Negative values could result if an incorrect solution was used as blank. Alternatively, these values could arise due to fluorescence of a dye in the solution.

Analysis with Fluorescent Dye Tagging

An alternative method to assess DNA and RNA concentration is to tag the sample with a Fluorescent tag, which is a fluorescent dye used to measure the intensity of the dyes that bind to nucleic acids and selectively fluoresce when bound (e.g. Ethidium bromide). This method is useful for cases where concentration is too low to accurately assess with spectrophotometry and in cases where contaminants absorbing at 260 nm make accurate quantitation by that method impossible. The benefit of fluorescence quantitation of DNA and RNA is the improved sensitivity over spectrophotometric analysis. Although, that increase in sensitivity comes at the cost of a higher price per sample and a lengthier sample preparation process.

There are two main ways to approach this. "Spotting" involves placing a sample directly onto an agarose gel or plastic wrap. The fluorescent dye is either present in the agarose gel, or is added in appropriate concentrations to the samples on the plastic film. A set of samples with known concentrations are spotted alongside the sample. The concentration of the unknown sample is then estimated by comparison with the fluorescence of these known concentrations. Alternatively, one may run the sample through an agarose or polyacrylamide gel, alongside some samples of known concentration. As with the spot test, concentration is estimated through comparison of fluorescent intensity with the known samples.

If the sample volumes are large enough to use microplates or cuvettes, the dye-loaded samples can also be quantified with a fluorescence photometer. Minimum sample volume starts at 0.3 μl.

To date there is no fluorescence method to determine protein contamination of a DNA sample that is similar to the 260 nm/280 nm spectrophotometric version.

Nucleic Acid Double Helix

In molecular biology, the term double helix refers to the structure formed by double-stranded molecules of nucleic acids such as DNA. The double helical structure of a nucleic acid complex arises as a consequence of its secondary structure, and is a fundamental component in determining its tertiary structure.

The DNA double helix polymer of nucleic acid, held together by nucleotides which base pair together. In B-DNA, the most common double helical structure found in nature, the double helix is right-handed with about 10–10.5 base pairs per turn. The double helix structure of DNA contains a *major groove* and *minor groove*. In B-DNA the major groove is wider than the minor groove. Given the difference in widths of the major groove and minor groove, many proteins which bind to B-DNA do so through the wider major groove.

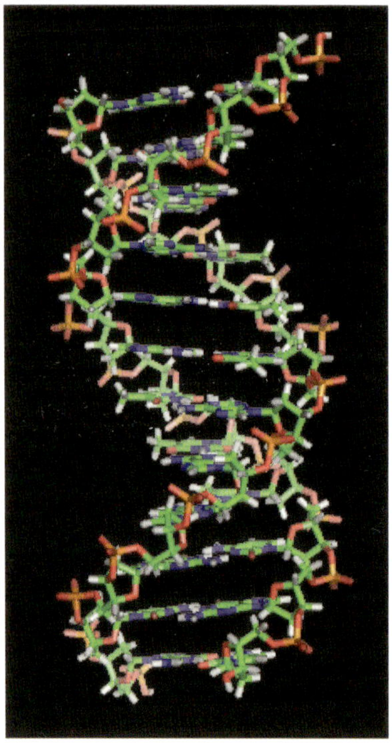

Two complementary regions of nucleic acid molecules will bind and form a double helical structure held together by base pairs.

Nucleic Acid Hybridization

Hybridization is the process of complementary base pairs binding to form a double helix. Melting is the process by which the interactions between the strands of the double helix are broken, separating the two nucleic acid strands. These bonds are weak, easily separated by gentle heating, enzymes, or physical force. Melting occurs preferentially at certain points in the nucleic acid. *T* and *A* rich regions are more easily melted than *C* and *G* rich regions. Some base steps (pairs) are also susceptible to DNA melting, such as *T A* and *T G*. These mechanical features are reflected by the use of sequences such as *TATA* at the start of many genes to assist RNA polymerase in melting the DNA for transcription.

Strand separation by gentle heating, as used in polymerase chain reaction (PCR), is simple, providing the molecules have fewer than about 10,000 base pairs (10 kilobase pairs, or 10 kbp). The intertwining of the DNA strands makes long segments difficult to separate. The cell avoids this problem by allowing its DNA-melting enzymes (helicases) to work concurrently with topoisomerases, which can chemically cleave the phosphate backbone of one of the strands so that it can swivel around the other. Helicases unwind the strands to facilitate the advance of sequence-reading enzymes such as DNA polymerase.

Base Pair Geometry

The geometry of a base, or base pair step can be characterized by 6 coordinates: shift, slide, rise, tilt, roll, and twist. These values precisely define the location and orientation in space of every base or base pair in a nucleic acid molecule relative to its predecessor along the axis of the helix. Together, they characterize the helical structure of the molecule. In regions of DNA or RNA where the *normal* structure is disrupted, the change in these values can be used to describe such disruption.

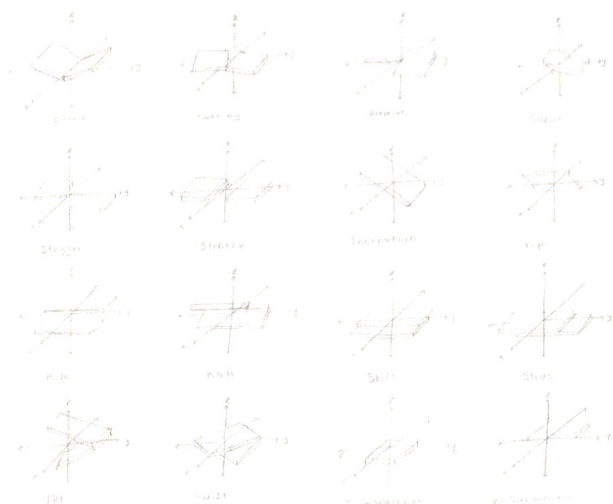

Base pair geometries

For each base pair, considered relative to its predecessor, there are the following base pair geometries to consider:

- Shear
- Stretch
- Stagger
- Buckle
- Propeller: rotation of one base with respect to the other in the same base pair.
- Opening
- Shift: displacement along an axis in the base-pair plane perpendicular to the first, directed from the minor to the major groove.
- Slide: displacement along an axis in the plane of the base pair directed from one strand to the other.

- Rise: displacement along the helix axis.
- Tilt: rotation around the shift axis.
- Roll: rotation around the slide axis.
- Twist: rotation around the rise axis.
- x-displacement
- y-displacement
- inclination
- tip
- pitch: the number of base pairs per complete turn of the helix.

Rise and twist determine the handedness and pitch of the helix. The other coordinates, by contrast, can be zero. Slide and shift are typically small in B-DNA, but are substantial in A- and Z-DNA. Roll and tilt make successive base pairs less parallel, and are typically small.

Note that "tilt" has often been used differently in the scientific literature, referring to the deviation of the first, inter-strand base-pair axis from perpendicularity to the helix axis. This corresponds to slide between a succession of base pairs, and in helix-based coordinates is properly termed "inclination".

Helix Geometries

At least three DNA conformations are believed to be found in nature, A-DNA, B-DNA, and Z-DNA. The *B* form described by James Watson and Francis Crick is believed to predominate in cells. It is 23.7 Å wide and extends 34 Å per 10 bp of sequence. The double helix makes one complete turn about its axis every 10.4-10.5 base pairs in solution. This frequency of twist (termed the helical *pitch*) depends largely on stacking forces that each base exerts on its neighbours in the chain. The absolute configuration of the bases determines the direction of the helical curve for a given conformation.

A-DNA and Z-DNA differ significantly in their geometry and dimensions to B-DNA, although still form helical structures. It was long thought that the A form only occurs in dehydrated samples of DNA in the laboratory, such as those used in crystallographic experiments, and in hybrid pairings of DNA and RNA strands, but DNA dehydration does occur in vivo, and A-DNA is now known to have biological functions. Segments of DNA that cells have been methylated for regulatory purposes may adopt the Z geometry, in which the strands turn about the helical axis the opposite way to A-DNA and B-DNA. There is also evidence of protein-DNA complexes forming Z-DNA structures.

Other conformations are possible; A-DNA, B-DNA, C-DNA, E-DNA, L-DNA (the enantiomeric form of D-DNA), P-DNA, S-DNA, Z-DNA, etc. have been described so far. In fact, only the letters F, Q, U, V, and Y are now available to describe any new DNA structure that may appear in the future. However, most of these forms have been created synthetically and have not been observed in naturally occurring biological systems. There are also triple-stranded DNA forms and quadruplex forms such as the G-quadruplex.

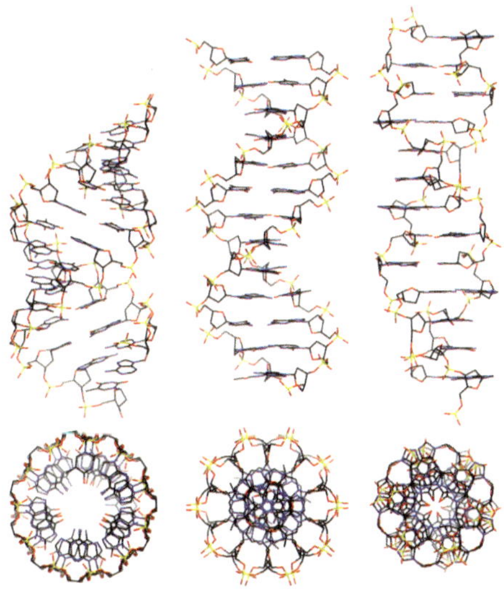

The structures of A-, B-, and Z-DNA.

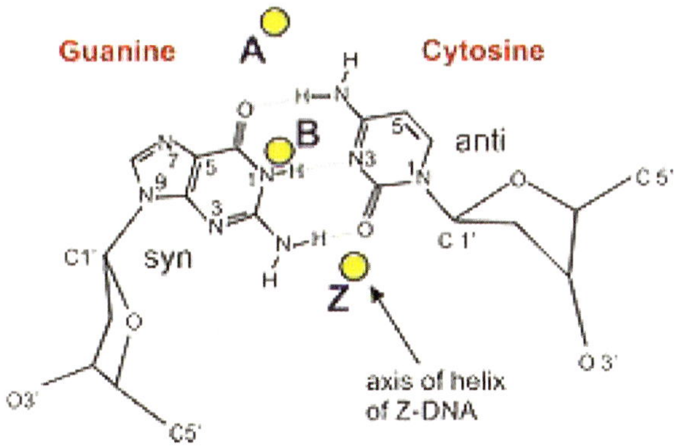

The helix axis of A-, B-, and Z-DNA.

Structural features of the three major forms of DNA			
Geometry attribute	A-DNA	B-DNA	Z-DNA
Helix sense	right-handed	right-handed	left-handed
Repeating unit	1 bp	1 bp	2 bp
Rotation/bp	32.7°	34.3°	60°/2
bp/turn	11	10.5	12
Inclination of bp to axis	+19°	−1.2°	−9°
Rise/bp along axis	2.3 Å (0.23 nm)	3.32 Å (0.332 nm)	3.8 Å (0.38 nm)
Pitch/turn of helix	28.2 Å (2.82 nm)	33.2 Å (3.32 nm)	45.6 Å (4.56 nm)
Mean propeller twist	+18°	+16°	0°
Glycosyl angle	anti	anti	C: anti, G: syn
Sugar pucker	C3'-endo	C2'-endo	C: C2'-endo, G: C2'-exo
Diameter	23 Å (2.3 nm)	20 Å (2.0 nm)	18 Å (1.8 nm)

Grooves

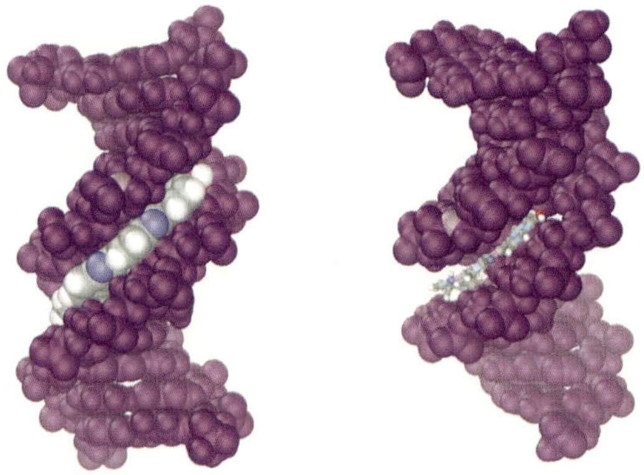

Major and minor grooves of DNA. Minor groove is a binding site for the dye Hoechst 33258.

Twin helical strands form the DNA backbone. Another double helix may be found by tracing the spaces, or grooves, between the strands. These voids are adjacent to the base pairs and may provide a binding site. As the strands are not directly opposite each other, the grooves are unequally sized. One groove, the major groove, is 22 Å wide and the other, the minor groove, is 12 Å wide. The narrowness of the minor groove means that the edges of the bases are more accessible in the major groove. As a result, proteins like transcription factors that can bind to specific sequences in double-stranded DNA usually make contacts to the sides of the bases exposed in the major groove. This situation varies in unusual conformations of DNA within the cell, but the major and minor grooves are always named to reflect the differences in size that would be seen if the DNA is twisted back into the ordinary B form.

Non-double Helical Forms

Alternative non-helical models were briefly considered in the late 1970s as a potential solution to problems in DNA replication in plasmids and chromatin. However, the models were set aside in favor of the double-helical model due to subsequent experimental advances such as X-ray crystallography of DNA duplexes and later the nucleosome core particle, and the discovery of topoisomerases. Also, the non-double-helical models are not currently accepted by the mainstream scientific community.

Single-stranded nucleic acids (ssDNA) do not adopt a helical formation, and are described by models such as the random coil or worm-like chain.

Bending

DNA is a relatively rigid polymer, typically modelled as a worm-like chain. It has three significant degrees of freedom; bending, twisting, and compression, each of which cause certain limits on what is possible with DNA within a cell. Twisting-torsional stiffness is important for the circularisation of DNA and the orientation of DNA bound proteins relative to each other and bending-axial stiffness is important for DNA wrapping and circularisation and protein interactions. Compression-extension is relatively unimportant in the absence of high tension.

Persistence Length, Axial Stiffness

Example sequences and their persistence lengths (B DNA)	
Sequence	Persistence length / base pairs
Random	154±10
$(CA)_{repeat}$	133±10
$(CAG)_{repeat}$	124±10
$(TATA)_{repeat}$	137±10

DNA in solution does not take a rigid structure but is continually changing conformation due to thermal vibration and collisions with water molecules, which makes classical measures of rigidity impossible. Hence, the bending stiffness of DNA is measured by the persistence length, defined as:

> "The length of DNA over which the time-averaged orientation of the polymer becomes uncorrelated by a factor of *e*".

This value may be directly measured using an atomic force microscope to directly image DNA molecules of various lengths. In an aqueous solution, the average persistence length is 46-50 nm or 140-150 base pairs (the diameter of DNA is 2 nm), although can vary significantly. This makes DNA a moderately stiff molecule.

The persistence length of a section of DNA is somewhat dependent on its sequence, and this can cause significant variation. The variation is largely due to base stacking energies and the residues which extend into the minor and major grooves.

Models for DNA Bending

Stacking stability of base steps (B DNA)	
Step	Stacking ΔG /kcal mol^{-1}
T A	-0.19
T G or C A	-0.55
C G	-0.91
A G or C T	-1.06
A A or T T	-1.11
A T	-1.34
G A or T C	-1.43
C C or G G	-1.44
A C or G T	-1.81
G C	-2.17

The entropic flexibility of DNA is remarkably consistent with standard polymer physics models, such as the *Kratky-Porod* worm-like chain model. Consistent with the worm-like chain model is the observation that bending DNA is also described by Hooke's law at very small (sub-piconewton) forces. However, for DNA segments less than the persistence length, the bending force is approximately constant and behaviour deviates from the worm-like chain predictions.

This effect results in unusual ease in circularising small DNA molecules and a higher probability of finding highly bent sections of DNA.

Bending Preference

DNA molecules often have a preferred direction to bend, i.e., anisotropic bending. This is, again, due to the properties of the bases which make up the DNA sequence - a random sequence will have no preferred bend direction, i.e., isotropic bending.

Preferred DNA bend direction is determined by the stability of stacking each base on top of the next. If unstable base stacking steps are always found on one side of the DNA helix then the DNA will preferentially bend away from that direction. As bend angle increases then steric hindrances and ability to roll the residues relative to each other also play a role, especially in the minor groove. *A* and *T* residues will be preferentially be found in the minor grooves on the inside of bends. This effect is particularly seen in DNA-protein binding where tight DNA bending is induced, such as in nucleosome particles.

DNA molecules with exceptional bending preference can become intrinsically bent. This was first observed in trypanosomatid kinetoplast DNA. Typical sequences which cause this contain stretches of 4-6 *T* and *A* residues separated by *G* and *C* rich sections which keep the A and T residues in phase with the minor groove on one side of the molecule. For example:

```
                                                                  AATGCCAA
G ATTCCCAAAAATGTCAAAA AATAGGCAAAA AAAATCCC
                                                                    AAAC
```

The intrinsically bent structure is induced by the 'propeller twist' of base pairs relative to each other allowing unusual bifurcated Hydrogen-bonds between base steps. At higher temperatures this structure is denatured, and so the intrinsic bend, is lost.

All DNA which bends anisotropically has, on average, a longer persistence length and greater axial stiffness. This increased rigidity is required to prevent random bending which would make the molecule act isotropically.

Circularization

DNA circularization depends on both the axial (bending) stiffness and torsional (rotational) stiffness of the molecule. For a DNA molecule to successfully circularize it must be long enough to easily bend into the full circle and must have the correct number of bases so the ends are in the correct rotation to allow bonding to occur. The optimum length for circularization of DNA is around 400 base pairs (136 nm), with an integral number of turns of the DNA helix, i.e., multiples of 10.4 base pairs. Having a non integral number of turns presents a significant energy barrier for circularization, for example a 10.4 x 30 = 312 base pair molecule will circularize hundreds of times faster than 10.4 x 30.5 ≈ 317 base pair molecule.

Stretching

Elastic Stretching Regime

Longer stretches of DNA are entropically elastic under tension. When DNA is in solution, it undergoes continuous structural variations due to the energy available in the thermal bath of the solvent. This is due to the thermal vibration of the molecule combined with continual collisions with water molecules. For entropic reasons, more compact relaxed states are thermally accessible than stretched out states, and so DNA molecules are almost universally found in a tangled relaxed layouts. For this reason, one molecule of DNA will stretch under a force, straightening it out. Using optical tweezers, the entropic stretching behavior of DNA has been studied and analyzed from a polymer physics perspective, and it has been found that DNA behaves largely like the *Kratky-Porod* worm-like chain model under physiologically accessible energy scales.

Phase Transitions Under Stretching

Under sufficient tension and positive torque, DNA is thought to undergo a phase transition with the bases splaying outwards and the phosphates moving to the middle. This proposed structure for overstretched DNA has been called *P-form DNA*, in honor of Linus Pauling who originally presented it as a possible structure of DNA.

Evidence from mechanical stretching of DNA in the absence of imposed torque points to a transition or transitions leading to further structures which are generally referred to as *S-form DNA*. These structures have not yet been definitively characterised due to the difficulty of carrying out atomic-resolution imaging in solution while under applied force although many computer simulation studies have been made.

Proposed S-DNA structures include those which preserve base-pair stacking and hydrogen bonding (GC-rich), while releasing extension by tilting, as well as structures in which partial melting of the base-stack takes place, while base-base association is nonetheless overall preserved (AT-rich).

Periodic fracture of the base-pair stack with a break occurring once per three bp (therefore one out of every three bp-bp steps) has been proposed as a regular structure which preserves planarity of the base-stacking and releases the appropriate amount of extension , with the term "Σ-DNA" introduced as a mnemonic, with the three right-facing points of the Sigma character serving as a reminder of the three grouped base pairs. The Σ form has been shown to have a sequence preference for GNC motifs which are believed under the GNC_hypothesis to be of evolutionary importance .

Supercoiling and Topology

The B form of the DNA helix twists 360° per 10.4-10.5 bp in the absence of torsional strain. But many molecular biological processes can induce torsional strain. A DNA segment with excess or insufficient helical twisting is referred to, respectively, as positively or negatively *supercoiled*. DNA *in vivo* is typically negatively supercoiled, which facilitates the unwinding (melting) of the double-helix required for RNA transcription.

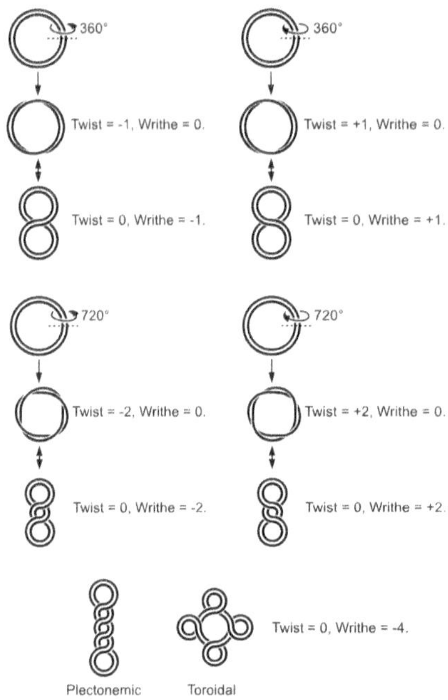

Supercoiled structure of circular DNA molecules with low writhe. The helical aspect of the DNA duplex is omitted for clarity.

Within the cell most DNA is topologically restricted. DNA is typically found in closed loops (such as plasmids in prokaryotes) which are topologically closed, or as very long molecules whose diffusion coefficients produce effectively topologically closed domains. Linear sections of DNA are also commonly bound to proteins or physical structures (such as membranes) to form closed topological loops.

Francis Crick was one of the first to propose the importance of linking numbers when considering DNA supercoils. In a paper published in 1976, Crick outlined the problem as follows:

In considering supercoils formed by closed double-stranded molecules of DNA certain mathematical concepts, such as the linking number and the twist, are needed. The meaning of these for a closed ribbon is explained and also that of the writhing number of a closed curve. Some simple examples are given, some of which may be relevant to the structure of chromatin.

Analysis of DNA topology uses three values:

L = linking number - the number of times one DNA strand wraps around the other. It is an integer for a closed loop and constant for a closed topological domain.

T = twist - total number of turns in the double stranded DNA helix. This will normally tend to approach the number of turns that a topologically open double stranded DNA helix makes free in solution: number of bases/10.5, assuming there are no intercalating agents (e.g., ethidium bromide) or other elements modifying the stiffness of the DNA.

W = writhe - number of turns of the double stranded DNA helix around the superhelical axis

$L = T + W$ and $\Delta L = \Delta T + \Delta W$

Any change of T in a closed topological domain must be balanced by a change in W, and vice versa. This results in higher order structure of DNA. A circular DNA molecule with a writhe of 0 will be circular. If the twist of this molecule is subsequently increased or decreased by supercoiling then the writhe will be appropriately altered, making the molecule undergo plectonemic or toroidal superhelical coiling.

When the ends of a piece of double stranded helical DNA are joined so that it forms a circle the strands are topologically knotted. This means the single strands cannot be separated any process that does not involve breaking a strand (such as heating). The task of un-knotting topologically linked strands of DNA falls to enzymes termed topoisomerases. These enzymes are dedicated to un-knotting circular DNA by cleaving one or both strands so that another double or single stranded segment can pass through. This un-knotting is required for the replication of circular DNA and various types of recombination in linear DNA which have similar topological constraints.

The Linking Number Paradox

For many years, the origin of residual supercoiling in eukaryotic genomes remained unclear. This topological puzzle was referred to by some as the "linking number paradox". However, when experimentally determined structures of the nucleosome displayed an over-twisted left-handed wrap of DNA around the histone octamer, this *paradox* was considered to be solved by the scientific community.

Nucleic Acid Metabolism

Nucleic acid metabolism is the process by which nucleic acids (DNA and RNA) are synthesized and degraded. Nucleic acids are polymers of nucleotides. Nucleotide synthesis is an anabolic mechanism generally involving the chemical reaction of phosphate, pentose sugar, and a nitrogenous base. Destruction of nucleic acid is a catabolic reaction. Additionally, parts of the nucleotides or nucleobases can be salvaged to recreate new nucleotides. Both synthesis and degradation reactions require enzymes to facilitate the event. Defects or deficiencies in these enzymes can lead to a variety of diseases.

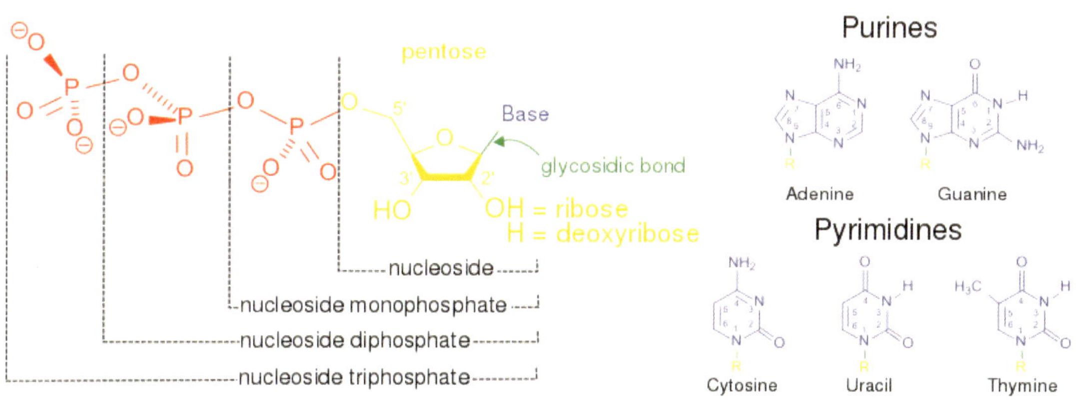

Composition of nucleotides, which make up nucleic acids.

Synthesis of Nucleic Acids

Nucleotides can be separated into purines and pyrimidines. They are both primarily produced in the liver. They both contain a sugar and a phosphate, but have nitrogenous bases that are different sizes. Because of this, the two different groups are synthesized in different ways. However, all nucleotide synthesis requires the use of phosphoribosyl pyrophosphate (PRPP) which donates the ribose and phosphate necessary to create a nucleotide.

Purine Synthesis

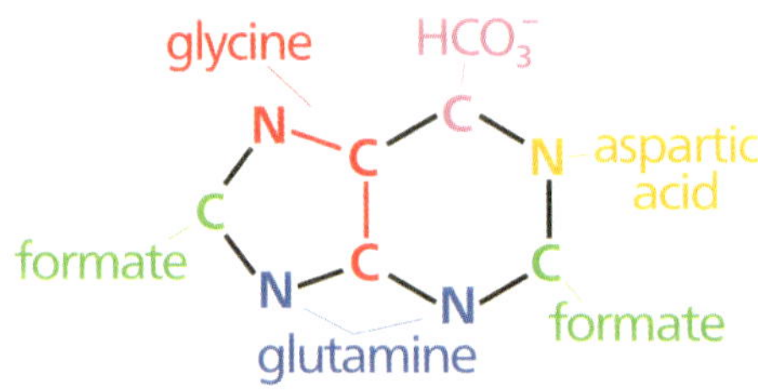

The origin of atoms that make up purine bases.

Adenine and guanine are the two nucleotides classified as purines. In purine synthesis, PRPP is turned into inosine monophosphate, or IMP. Production of IMP from PRPP requires glutamine, glycine, aspartate, and 6 ATP, among other things. IMP is then converted to AMP (adenosine monophosphate) using GTP and aspartate, which is converted into fumarate. While IMP can be directly converted to AMP, synthesis of GMP (guanosine monophosphate) requires an intermediate step, in which NAD+ is used to form the intermediate xanthosine monophosphate, or XMP. XMP is then converted into GMP by using the hydrolysis of 1 ATP and the conversion of glutamine to glutamate. AMP and GMP can then be converted into ATP and GTP, respectively, by kinases that add additional phosphates.

ATP stimulates production of GTP, while GTP stimulates production of ATP. This cross regulation keeps the relative amounts of ATP and GTP the same. Excess of either nucleotide could increase the likelihood of DNA mutations, where the wrong purine nucleotide is inserted.

Lesch-Nyhan syndrome is caused by a deficiency in hypoxanthine-guanine phosphoribosyltransferase or HGPRT, the enzyme that catalyzes the reversible reaction of producing guanine from GMP. This is a sex-linked congenital defect that causes overproduction of uric acid along with mental retardation, spasticity, and an urge to self-mutilate.

Pyrimidine Synthesis

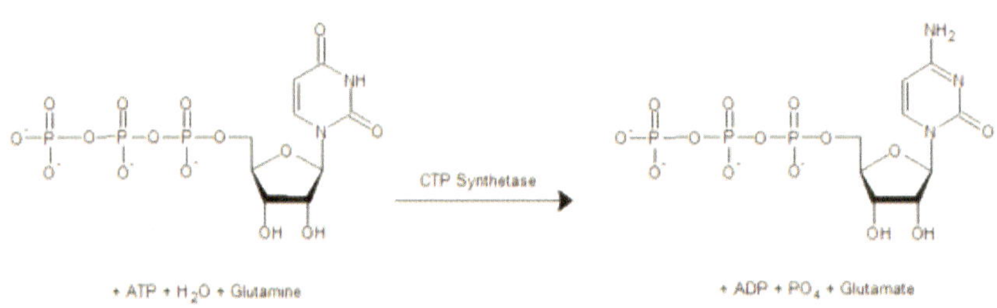

Uridine-triphosphate (UTP), at left, reacts with glutamine and other chemicals to form cytidine-triphosphate (CTP), on the right.

Pyrimidine nucleotides include cytidine, uridine, and thymidine. The synthesis of any pyrimidine nucleotide begins with the formation of uridine. This reaction requires aspartate, glutamine, bicarbonate, and 2 ATP molecules (to provide energy), as well as PRPP which provides the ribose-monophosphate. Unlike in purine synthesis, the sugar/phosphate group from PRPP is not added to the nitrogenous base until towards the end of the process. After uridine-monophosphate is synthesized, it can react with 2 ATP to form uridine-triphosphate or UTP. UTP can be converted to CTP (cytidine-triphosphate) in a reaction catalyzed by CTP synthetase. Thymidine synthesis first requires reduction of the uridine to deoxyuridine, before the base can be methylated to produce thymidine.

ATP, a purine nucleotide, is an activator of pyrimidine synthesis, while CTP, a pyrimidine nucleotide, is an inhibitor of pyrimidine synthesis. This regulation helps to keep the purine/pyrimidine amounts similar, which is beneficial because equal amounts of purines and pyrimidines are required for DNA synthesis.

Deficiencies of enzymes involved in pyrimidine synthesis can lead to the genetic disease Orotic aciduria which causes excessive excretion of orotic acid in the urine.

Converting Nucleotides to Deoxynucleotides

Nucleotides are initially made with ribose as the sugar component, which is a feature of RNA. DNA, however, requires *deoxy*ribose, which is missing the 2'-hydroxyl (-OH group) on the ribose. The reaction to remove this -OH is catalyzed by ribonucleotide reductase. This enzyme converts NDPs (nucleoside-diphosphate) to dNDPs (deoxynucleoside-diphosphate). The nucleotides must be in the diphosphate form for the reaction to occur.

In order to synthesize thymidine, a component of DNA which only exists in the deoxy form, uridine is converted to deoxyuridine (by ribonucleotide reductase), and then is methylated by thymidylate synthase to create thymidine.

Degradation of Nucleic Acids

The breakdown of DNA and RNA is occurring continuously in the cell. Purine and pyrimidine nucleosides can either be degraded to waste products and excreted or can be salvaged as nucleotide components.

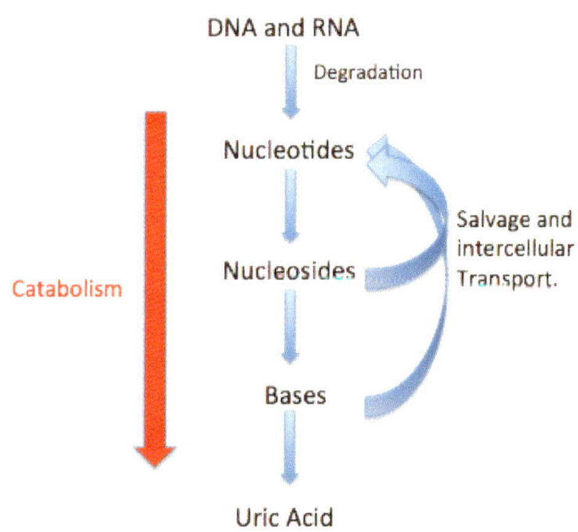

General outline of nucleic acid degradation for purines.

Pyrimidine Catabolism

Cytosine and uracil are converted into beta-alanine and later to malonyl-CoA which is needed for fatty acid synthesis, among other things. Thymine, on the other hand, is converted into β-aminoisobutyric acid which is then used to form methylmalonyl-CoA. The leftover carbon skeletons such as acetyl-CoA and Succinyl-CoA can then be oxidized by the citric acid cycle. Pyrimidine degradation ultimately ends in the formation of ammonium, water, and carbon dioxide. The ammonium can then enter the urea cycle which occurs in the cytosol and the mitochondria of cells.

Pyrimidine bases can also be salvaged. For example, the uracil base can be combined with ribose-1-phosphate to create uridine monophosphate or UMP. A similar reaction can also be done with thymine and deoxyribose-1-phosphate.

Deficiencies in enzymes involved in pyrimidine catabolism can lead to diseases such as Dihydropyrimidine dehydrogenase deficiency which has negative neurological effects.

Purine Catabolism

Purine degradation takes place mainly in the liver of humans and requires an assortment of enzymes to degrade purines to uric acid. First, the nucleotide will lose its phosphate through 5'-nucleotidase. The nucleoside, adenosine, is then deaminated and hydrolyzed to form hypoxanthine via adenosine deaminase and nucleosidase respectively. Hypoxanthine is then oxidized to form xanthine and then uric acid through the action of xanthine oxidase. The other purine nucleoside, guanosine, is cleaved to form guanine. Guanine is then deaminated via guanine deaminase to form xanthine which is then converted to uric acid. Oxygen is the final electron acceptor in the degradation of both purines. Uric acid is then excreted from the body in different forms depending on the animal.

Free purine and pyrimidine bases that are released into the cell are typically transported intercellularly across membranes and salvaged to create more nucleotides via nucleotide salvage. For example, adenine + PRPP --> AMP + PPi. This reaction requires the enzyme adenine phosphoribosyltransferase. Free guanine is salvaged in the same way except it requires hypoxanthine-guanine phosphoribosyltransferase.

Defects in purine catabolism can result in a variety of diseases including gout, which stems from an accumulation of uric acid crystals in various joints, and adenosine deaminase deficiency, which causes immunodeficiency.

Interconversion of Nucleotides

Once the nucleotides are synthesized they can exchange phosphates among one another in order to create mono-, di-, and tri-phosphate molecules. The conversion of a nucleoside-diphosphate (NDP) to a nucleoside-triphosphate (NTP) is catalyzed by nucleoside diphosphate kinase, which uses ATP as the phosphate donor. Similarly, nucleoside-monophosphate kinase carries out the phosphorylation of nucleside-monophosphates. Adenylate kinase is a specific nucleoside-monophosphate kinase that functions only on adenosine-monophosphate.

References

- International Human Genome Sequencing Consortium (2001). "Initial sequencing and analysis of the human genome" (PDF). Nature. 409 (6822): 860–921. Bibcode:2001Natur.409..860L. doi:10.1038/35057062. PMI D 11237011
- "Definitions of Genetic Testing". Definitions of Genetic Testing (Jorge Sequeiros and Bárbara Guimarães). EuroGentest Network of Excellence Project. 2008-09-11. Archived from the original on February 4, 2009. Retrieved 2008-08-10
- Samarsky, DA; Fournier MJ; Singer RH; Bertrand E (1998). "The snoRNA box C/D motif directs nucleolar targeting and also couples snoRNA synthesis and localization". The EMBO Journal. 17 (13): 3747–3757. doi:10.1093/emboj/17.13.3747. PMC 1170710. PMID 9649444
- Brock, Thomas D.; Madigan, Michael T. (2009). Brock biology of microorganisms. Pearson / Benjamin Cummings. ISBN 0-321-53615-0
- Venter, JC; et al. (2001). "The sequence of the human genome" (PDF). Science. 291 (5507): 13041351. Bibcode:2001Sci...291.1304V. doi:10.1126/science.1058040. PMID 11181995
- Richard Owczarzy (2008-08-28). "DNA melting temperature - How to calculate it?". High-throughput DNA biophysics. owczarzy.net. Retrieved 2008-10-02
- Bailey, CJ (2009). "Orotic aciduria and uridine monophosphate synthase: a reappraisal". Journal of Inherited Metabolic Disease. 32: S227–33. doi:10.1007/s10545-009-1176-y. PMID 19562503
- Budowle B, van Daal A (April 2009). "Extracting evidence from forensic DNA analyses: future molecular biology directions". BioTechniques. 46 (5): 339–40, 342–50. doi:10.2144/000113136. PMID 19480629
- Owczarzy, Richard (2008-08-28). "DNA melting temperature - How to calculate it?". High-throughput DNA biophysics. owczarzy.net. Retrieved 2008-10-02
- Roe DR, Chaka AM (2009). "Structural basis of pathway-dependent force profiles in stretched DNA". Journal of Physical Chemistry B. 113: 15364–15371. doi:10.1021/jp906749j. PMID 19845321
- Krieger M, Scott MP, Matsudaira PT, Lodish HF, Darnell JE, Lawrence Z, Kaiser C, Berk A (2004). "Section 4.1: Structure of Nucleic Acids". Molecular cell biology. New York: W.H. Freeman and CO. ISBN 0-7167-4366-3
- Elson D (1965). "Metabolism of nucleic acids (macromolecular DNA and RNA)". Annu. Rev. Biochem. 34: 449–86. doi:10.1146/annurev.bi.34.070165.002313. PMID 14321176
- "The Analysis of DNA or RNA using Its Wavelengths: 230 nm, 260 nm, 280 nm". Bioteachnology.com. 2010-01-13. Archived from the original on 2012-09-06. Retrieved 2010-03-12
- Sankoff D (1985). "Simultaneous solution of the RNA folding, alignment and protosequence problems". SIAM Journal on Applied Mathematics. 45: 810–825. doi:10.1137/0145048
- Anthony-Cahill SJ, Mathews CK, van Holde KE, Appling DR (2012). Biochemistry (4th Edition). Englewood Cliffs, N.J: Prentice Hall. ISBN 0-13-800464-1
- National Institute of Health (September 28, 2007). "Discovering DNA: Friedrich Miescher and the early years of nucleic acid research". Hum. Genet. nih.gov. 122(6): 565–81. doi:10.1007/s00439-007-0433-0. PMID 17901982
- Nussinov R, Pieczník G, Grigg JR and Kleitman DJ (1978) Algorithms for loop matchings. SIAM Journal on Applied Mathematics
- Alberts B, Johnson A, Lewis J, Raff M, Roberts K, Wlater P (2002). Molecular Biology of the Cell (4th ed.). New York NY: Garland Science. ISBN 0-8153-3218-1
- Verma S, Eckstein F (1998). "Modified oligonucleotides: synthesis and strategy for users". Annu. Rev. Biochem. 67: 99–134. doi:10.1146/annurev.biochem.67.1.99. PMID 9759484

Permissions

All chapters in this book are published with permission under the Creative Commons Attribution Share Alike License or equivalent. Every chapter published in this book has been scrutinized by our experts. Their significance has been extensively debated. The topics covered herein carry significant information for a comprehensive understanding. They may even be implemented as practical applications or may be referred to as a beginning point for further studies.

We would like to thank the editorial team for lending their expertise to make the book truly unique. They have played a crucial role in the development of this book. Without their invaluable contributions this book wouldn't have been possible. They have made vital efforts to compile up to date information on the varied aspects of this subject to make this book a valuable addition to the collection of many professionals and students.

This book was conceptualized with the vision of imparting up-to-date and integrated information in this field. To ensure the same, a matchless editorial board was set up. Every individual on the board went through rigorous rounds of assessment to prove their worth. After which they invested a large part of their time researching and compiling the most relevant data for our readers.

The editorial board has been involved in producing this book since its inception. They have spent rigorous hours researching and exploring the diverse topics which have resulted in the successful publishing of this book. They have passed on their knowledge of decades through this book. To expedite this challenging task, the publisher supported the team at every step. A small team of assistant editors was also appointed to further simplify the editing procedure and attain best results for the readers.

Apart from the editorial board, the designing team has also invested a significant amount of their time in understanding the subject and creating the most relevant covers. They scrutinized every image to scout for the most suitable representation of the subject and create an appropriate cover for the book.

The publishing team has been an ardent support to the editorial, designing and production team. Their endless efforts to recruit the best for this project, has resulted in the accomplishment of this book. They are a veteran in the field of academics and their pool of knowledge is as vast as their experience in printing. Their expertise and guidance has proved useful at every step. Their uncompromising quality standards have made this book an exceptional effort. Their encouragement from time to time has been an inspiration for everyone.

The publisher and the editorial board hope that this book will prove to be a valuable piece of knowledge for students, practitioners and scholars across the globe.

Index